PLANT GENOME: BIODIVERSITY AND EVOLUTION

Volume 1, Part D

Phanerogams (Gymnosperm) and (Angiosperm-Monocotyledons)

Plant Genome

Biodiversity and Evolution

Series Editors
A.K. Sharma and Archana Sharma
Department of Botany
University of Calcutta, Kolkata
India

Advisory Board

J. Dolezel, Laboratory of Molecular Cytogenetics and Cytometry, Institute of Experimental Botany, Sokolovska, Czech Republic

K. Fukui, Department of Biotechnology, Graduate School of Engineering, Osaka University, Osaka, Japan

R.N. Jones, Institute of Biological Sciences, Aberystwyth, Ceredigion, Scotland, UK

G.S. Khush, 416 Cabrillo Avenue, Davis California 95616, USA

Ingo Schubert, Institüt für Pflanzengenetik and Kulturpflanzenforschung, Gatersleben, Germany

Canio G. Vosa, Scienze Botanische, Pisa, Italy

Volume 1, Part A : **Phanerogams**
1, Part B : **Phanerogams**
1, Part C : **Phanerogams (Angiosperm-Dicotyledons)**
1, Part D : **Phanerogams (Gymnosperm) and (Angiosperm-Monocotyledons)**
2, Part A : **Lower Groups**
2, Part B : **Lower Groups**

Plant Genome

Biodiversity and Evolution

Volume 1, Part D

Phanerogams (Gymnosperm) and (Angiosperm-Monocotyledons)

Editors

A.K. SHARMA and ARCHANA SHARMA

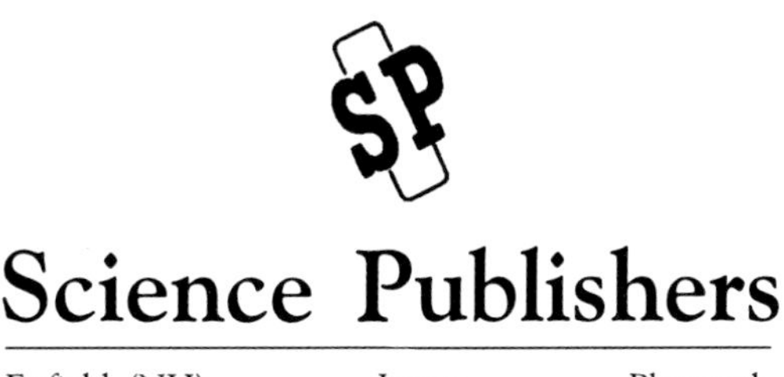

Science Publishers

Enfield (NH) Jersey Plymouth

CIP data will be provided on request

SCIENCE PUBLISHERS
An Imprint of Edenbridge Ltd., British Isles.
Post Office Box 699
Enfield, New Hampshire 03748
United States of America

Website: *http://www.scipub.net*

sales@scipub.net (marketing department)
editor@scipub.net (editorial department)
info@scipub.net (for all other enquiries)

ISBN 1-57808-420-2 [10 digits]
978-1-57808-420-3 [13 digits]

© 2006, Copyright Reserved

All rights reserved. No part of this publication may be reproduced, stored in a retrieval system, or transmitted in any form or by any means, electronic, mechanical, photocopying, recording or otherwise, without the prior written permission.

This book is sold subject to the condition that it shall not, by way of trade or otherwise, be lent, re-sold, hired out, otherwise circulated without the publisher's prior consent in any form of binding or cover other than that in which it is published and without a similar condition including this condition being imposed on the subsequent purchaser.

Published by Science Publishers, Enfield, NH, USA
An Imprint of Edenbridge Ltd.
Printed in India.

Preface to the Series "Plant Genome"

The term *genome*, the basic gene complement of an individual, is almost synonymous with the chromosome complement of both nucleus and organelles. Refinements in cellular, genetic and molecular methods in recent years have opened up unexplored avenues in genome research. The modern tools of gene and genome analyses, coupled with analysis of finer segments of gene sequences in chromosomes utilizing molecular hybridization, are now applied on a wider scale in different groups of plants, ranging from algae to angiosperms. This synergistic approach has made the study of biodiversity highly fascinating, permitting a deep insight into the molecular basis of genetic diversity. Simultaneous to the enrichment of fundamentals in systematics and phylogeny, the plant system, because of its inherent flexibility, has permitted genetic engineering and horizontal transfer of genes with immense importance in agriculture, horticulture and medicine.

Despite the fact that the data on plant genomics with its impact on the assessment of biodiversity and evolution show a logarithmic increase, a comprehensive series on the aspect covering all groups of plant kingdom is sadly lacking. In view of this lacuna, the present series on Plant Genomics: Biodiversity and Evolution has been planned. It aims to cover, in successive volumes, *comprehensive reviews, concepts and discussions on the results of genome analysis and their impact on systematics, taxonomy, phylogeny and evolution of all plant groups*. We have not gone out of our way to seek original articles, but in course of reviews and discussions, research articles, if any, are welcome.

A.K. Sharma
Archana Sharma
Series Editors

Preface to this Volume

The coverage of this volume is rather wide ranging from Gymnosperms with the two important groups the Cycads and the Pines, to monocotyledons represented by genera having special significance in terms of phylogeny, agriculture, horticulture and commercial importance. The dicotyledonous genera have not been included as the preceding volume I C dealt exclusively with dicotyledons.

In gymnosperms, the DNA sequence data have aided in ascertaining the uniqueness of Cycads as distinct from *Ginkgo*, *Gnetum* and Pines. The phylogenetic relationship within Cycadophytes has been studied from chloroplastid sequences and the monophyletic nature of *Cycas*, basic to Encephalarteae and Zamieae has been clearly indicated. On the other hand, in *Pinus* a correlation between genomic diversity and microhabitat characteristics has been recorded suggesting that climatic selection and stress have played a prominent role in adaptation of genetic variants.

In the monocotyledons, RAPD, ISSR and SSR have been employed in establishing diversity in *Phoenix dactylifera*, the oil palm, simultaneously bringing out evidence of a common genetic basis characterizing ecotypes of this species. The significant finding in date palm on the other hand, is the single gene control of shell thickness and the origin of all products of commercial value through hybridization of dura mother and pisifera pollen.

In *Allium*, along with other approaches, for analyzing genome complexity, the use of molecular markers in determining trends of diversification, hybridity, genetic map and genome analysis has been recorded. Detailed chromosome and *in situ* hybridization as reported has been presented.

Of the two reviews on Orchidaceae, one deals exclusively with phylogeny and evolution in *Cymbidium* as ascertained through nrITS and RAPD data. Molecular data shows partial congruence with current taxonomy and ITS and plastid matK could delineate only a few subgenera. The other review deals with genome evolution and population biology in the family as a whole. The results of study with an array of unclear and organellar markers indicate distinctive features of terrestrial, epiphytic and lithophytic orchids.

Of the grasses, cytological and molecular characteristics along with morphology have been utilized in determining phylogeny, affinities and status in *Bromus*. Moreover allozyme pattern and cpDNA and rDNA have aided in highlighting controversial issues in this genus.

In the subgroup Loliineae, cytogenetic, morphoanatomical and morphological data in *Festuca* in particular have been taken to suggest that evolution of broad leaf, large genomes with low heterochromatin have led to fine leaved taxa with small genome and high heterochromatin. Simultaneously, polyploidy and hybridization have led to the evolution of Annual from Perennial habit. Discrepancies in phylogeny reconstruction between morphological and molecular data have been recorded though karyotype evolution in subtribe Loliineae which is concordant with molecular data.

Finally, in *Avena*, the oats, cytological, genetic and molecular data specially including two sat DNA sequences As12O a and Am 1 have been utilized to identify the genomes of a new tetraploid species. The putative progenitors of the hexaploid AACCDD have also been delineated indicating steps in their evolution. This volume on plants of economic importance and of phylogenetic value would be an asset to anyone interested in genomics and evolution of plants of commercial and agricultural importance.

July 28, 2006

A.K. Sharma
A. Sharma

Contents

List of Contributors

Araceli Fominaya
Department of Cell Biology and Genetics, University of Alcalá, Campus Universitario, ES-28871-Alcalá de Henares, Madrid, Spain. E-mail : araceli.fominaya@uah.es;

Cai-Yun Sun
South China Institute of Botany, The Chinese Academy of Sciences, Guangzhou 510650, P.R. China

Cheng-Ye Liang
South China Institute of Botany, The Chinese Academy of Sciences, Guangzhou 510650, P.R. China

Don MacDonald
Cambridge University, Department of Genetics, Downing Site, Downing Street, Cambridge CB2 2EH, UK

Elizabeth Ann Veasey
Escola Superior de Agricultura "Luiz de Queiroz", Departamento de Genética, Av. Pádua Dias, 11, Caixa Postal 83, CEP: 13400-970, Piracicaba, SP, Brasil. E-mail: eaveasey@esalq.usp.br; gcxolive@esalq.usp.br

Ellen B. Peffley
Texas Tech University, Department of Plant & Soil Science, PO Box 42122 (post) Corner of 15th and Boston, Ag Science 101 (UPS), Lubbock, Texas 79409, USA. Email: ellen.peffley@ttu.edu

Esther Ferrer

Department of Cell Biology and Genetics, University of Alcalá, Campus Universitario, ES-28871-Alcalá de Henares, Madrid, Spain.

Farah Hafeez

Cambridge University, Department of Genetics, Downing Site, Downing Street, Cambridge CB2 2EH, UK.

Frederic Dumortier

DAMI, OPRS, New Britain Palm Oil Ltd., P.O. Box 165, Kimbe, West New Britain Province, Papua New Guinea.

Gabriel Schiler

Agricultural Research Organization, The Volcani Center, Department of Agronomy and Natural Resources, Forestry Section, Bet Dagan, Israel.

Giancarlo Conde Xavier Oliveira

Escola Superior de Agricultura "Luiz de Queiroz", Departamento de Genética, Av. Pádua Dias, 11, Caixa Postal 83, CEP: 13400-970, Piracicaba, SP, Brasil.

Kuai-Fei Xia

South China Institute of Botany, The Chinese Academy of Sciences, Guangzhou 510650, P.R. China.

Leonid Korol

Agricultural Research Organization, The Volcani Center, Department of Agronomy and Natural Resources, Forestry Section, Bet Dagan, Israel E-mail: vckorol@agri.gov.il.

M. Luisa Irigoyen

Department of Cell Biology and Genetics, University of Alcalá, Campus Universitario, ES-28871-Alcalá de Henares, Madrid, Spain.

M.C. García-Herran

Institut de Botanique, Laboratoire de botanique évolutive, Université de Neuchâtel. 11, rue Emile-Argand, 2000 Neuchâtel, Switzerland. E-mail: karmegherran@hotmail.com.

Marrakchi Mohamed

Laboratoire de génétique moléculaire, immunologie & biotechnologie, Faculté des Sciences de Tunis, Campus Universitaire, 2092 El Manar Tunis, Tunisie.

Michael Wink

Institute of Pharmacy and Molecular Biotechnology, Heidelberg University, INF 364, 69120 Heidelberg, Germany. E-mail: wink@uni-hd.de.

Ming-Yong Zhang

South China Institute of Botany, The Chinese Academy of Sciences, Guangzhou 510650, P.R. China E-mail: zhangmy2005@yahoo.com.cn.

Norbert Billote

CIRAD (CIRAD-CP) TA 80/03 Avenue Agropolis 34398 Montpellier Cedex, France.

Ould Mohamed Salem Ali

Département de Biologie, Faculté des Sciences et Techniques, Université de Nouakchott, B.P. 5026, Mauritanie.

Pilar Catalán

University of Zaragoza, High Polytechnic School of Huesca, Department of Agriculture (Botany), Ctra. Cuarte km 1, E-22071 Huesca (Spain), Email: pcatalan @unizar.es.

Rhouma Abdelmajid

IPGRI, Centre de Recherches Phoénicicoles, 2260 Degache, Tunisie.

Rhouma Soumaya

Laboratoire de génétique moléculaire, immunologie & biotechnologie, Faculté des Sciences de Tunis, Campus Universitaire, 2092 El Manar Tunis, Tunisie.

Sakka Hela

Laboratoire de génétique moléculaire, immunologie & biotechnologie, Faculté des Sciences de Tunis, Campus Universitaire, 2092 El Manar Tunis, Tunisie.

Sean Mayes

Nottingham University, Division of Biosciences, Sutton Boninghton Campus, Loughborough, Leicester LE12 5RD, UK.

Tatjana Oja

Institute of Botany and Ecology, University of Tartu, 40 Lai Str., 51005, Tartu, Estonia. E-mail: tatjana.oja@ut.ee.

Trifi Mokhtar

Laboratoire de génétique moléculaire, immunologie & biotechnologie, Faculté des Sciences de Tunis, Campus Universitaire, 2092 El Manar Tunis, Tunisie.

Xiu-Lin Ye

South China Institute of Botany, The Chinese Academy of Sciences, Guangzhou 510650, P.R. China.

Yolanda Loarce

Department of Cell Biology and Genetics, University of Alcalá, Campus Universitario, ES-28871-Alcalá de Henares, Madrid, Spain.

Zehdi-Azouzi Salwa

Laboratoire de génétique moléculaire, immunologie & biotechnologie, Faculté des Sciences de Tunis, Campus Universitaire, 2092 El Manar Tunis, Tunisie

Zuzana Price

Cambridge University, Department of Genetics, Downing Site, Downing Street, Cambridge CB2 2EH, UK, E-mail z.price@gen.cam.ac.uk,

CHAPTER

1

Evolution and Phylogeny of Cycads

MICHAEL WINK

Institute of Pharmacy and Molecular Biotechnology, Heidelberg University, Heidelberg, Germany

ABSTRACT

Cycads represent a monophyletic plant group that is traditionally grouped in 3 families and 11 genera with a total of 250 species. DNA sequence data suggest that extant cycads evolved during the last 60 million years and they represent a unique group besides *Ginkgo*, Gnetophyta and Pinophyta. Sequences from *cpDNA* (chloroplast DNA) are used to determine the phylogenetic relationship within the Cycadophyta. The genus *Cycas* is monophyletic and clusters basal to the other cycad genera. *Encephalartos, Macrozamia* and *Lepidozamia* form a well recognized clade, that is commonly included in the tribe Encephalarteae. *Zamia, Chigua* and *Microcycas* also form a well-supported clade, recognized in the tribe Zamieae. The position of *Ceratozamia, Dioon* and the almost monotypic genera *Stangeria* and *Bowenia* cannot be resolved unambiguously with the present data sets. No evidence could be found for a family Stangeriaceae that includes both *Bowenia* and *Stangeria.*

A molecular clock approach is employed to estimate the age of extant cycads. The Cycadaceae/Zamiaceae split may have occurred about 50 million years ago (mya). The African-Australian disjunction of *Encephalartos, Lepidozamia,* and *Macrozamia* is much younger and can be explained by long-distance dispersal across the ocean in the Miocene rather than by continental drift. The extant species of *Encephalartos* evolved about 6.5 mya and spread over Southern and Central Africa in the late Pleistocene and Pliocene (5-1.6 mya). The colonization of Madagascar by *Cycas thouarsii,* far from the distribution areas of the other *Cycas* species, seems to be very recent and might probably even be due to

Address for correspondence: Institute of Pharmacy and Molecular Biotechnology, Heidelberg University, INF 364, 69120 Heidelberg, Germany. E-mail: wink@uni-hd.de.

human intervention. Also, the colonization of several islands in the Pacific and Indian oceans by other members of the *Cycas* subgroup Rumphiae is apparently correlated with the presence of the spongy endocarp that allows seeds of these species to be transported via the sea.

Key Words: Cycads, molecular phylogeny, gymnosperms, molecular clock, long-distance dispersal, Gondwana

Abbreviations: ML= Maximum likelihood, MP= Maximum parsimony, NJ= Neighbour-joining.

INTRODUCTION

Cycads, also known as palm ferns, represent a small group of slow-growing woody perennials with slow recruitment and population turnover rates that are united by several unique characters [12, 22]. As they have large divided leaves, the cycads resemble tree ferns or palms to some degree. Cycads generally have pinnate leaves that are spirally arranged in crowns on the stem apex and pubescent when young. Another characteristic trait is the pachycaul stem consisting of storage tissue rich in starch. Cycads are dioecious and reproduce by seeds that are produced on open carpophylls or seed-bearing leaves. Except for Cycadaceae, sporophylls are arranged into cones. Male plants carry several sporangia on the under- or abaxial surface of the sporophylls. Pollen is released from slits in the sporangia. Male gametophytes have several flagellae and are motile. Female sporophylls produce large seeds with a two-layered testa; the outer layer is often coloured and fleshy, which serves to attract animals. The animals such as birds, rodents small marsupials and fruit-eating bats eat the seeds, thereby help in seed dispersal. Another unique trait is coralloid and contractile roots that occur in addition to normal roots. The coralloid roots harbour nitrogen fixing symbiotic Cyanobacteria, which allow cycads to live on soils that are poor in nutrients—ranging from dense tropical forests to semideserts in tropical or subtropical climates with summer rainfall.

Cycads have naked ovules, a character shared with other "gymnosperms" such as the *Welwitschia* group (Gnetophyta), the conifers (Pinophyta) and *Ginkgo* (Ginkgophyta). All "gymnosperms" represent ancient (arising in the Permian era) seed plants, of which many are now extinct, except for the four major extant groups. Molecular evidence for the relationships within these basal groups is elaborated in this chapter.

Presently more than 250 species are recognized in the order Cycadales in the class Cycadophyta, grouped in 11 genera, which make them a small group within the flowering plants with more than 300,000 species [7, 8, 23]. Figure 1 shows an overview of the classification of cycads into 3 families, 4 subfamilies, and 11 genera [23]. A concise history of cycad taxonomy is given by Hill et al. [7].

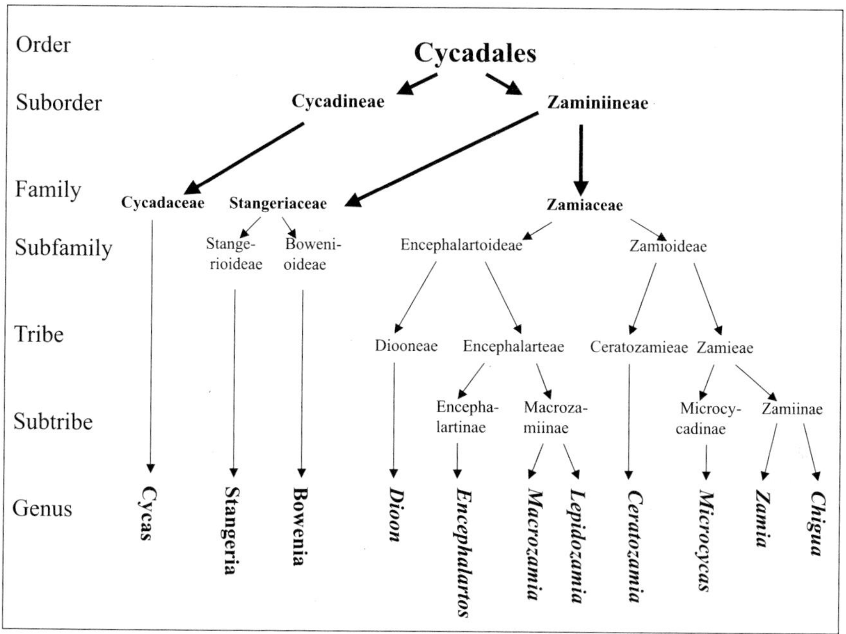

Fig. 1. *Classification of cycads.*

The cycad origins go back to the Early Permian era, about 280 mya; it has been suggested that they evolved as a sister-group to all other seed plants. Their ancestors could have been the more ancient (i.e. Palaeozoic) seed ferns. The cycads had their maximum expansion during the Triassic to Tertiary era. In these times, cycads were widely distributed [12], the Jurassic period, is often termed as the 'Age of Cycads'. Cycads are often called living fossils because their overall morphology has changed negligibly from their ancestors in Mesozoic times. However, the three extant families of today show similarities only to fossils from Tertiary about 50-60 mya, whereas more than 19 extinct cycad genera are only known as fossils of the Palaeozoic and Mesozoic era [12]. As described in this review, cycads did not stop their phylogenetic development 200 mya ago but underwent phylogenetic speciation as any other plant order. The consequence is that species living today may, in fact, be quite young [24, 25]. Fossil cycads have been detected in Mesozoic deposits of every continent and every latitude, ranging from Siberia to the Antarctic. Cycads are now only found sparsely distributed in the tropics and subtropics (Fig. 2) of the formerly united, but later separated supercontinents Laurasia and Gondwana. They are now reduced in numbers and distribution. It would be plausible to explain the extant distribution in terms of plate tectonics, but as discussed later in this review, also long-distance dispersal via the sea was another mechanism for the present distribution of cycads.

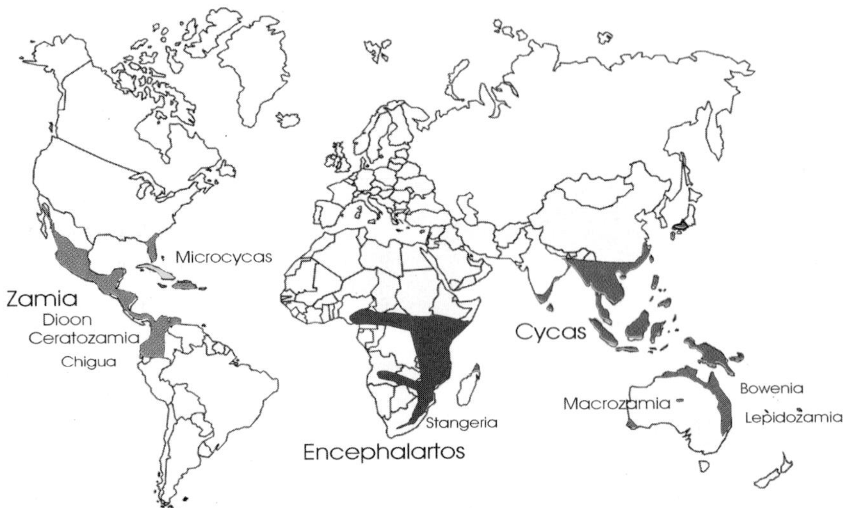

Fig. 2. *Present day distribution of cycads.*

SECONDARY METABOLITES IN CYCADS

Cycads produce a series of unique secondary metabolites, as well as dimeric flavones and the nitrogen containing methylazoglycosides cycasin, and macrozamin with effective repellent and toxic properties [2, 20]. There is some direct and additional circumstantial evidence that the cycasins are the central protective means of cycads against herbivores. The effective, toxic component of these glycosides is their aglycon, methylazomethanol (MAM). MAM is a methylating agent that can covalently modify DNA. It is, therefore, mutagenic and potentially carcinogenic. The azoglycoside levels in cycads range from 0.01 to 5% of fr. wt [11, 20].

Another cycad toxin is a non-protein amino acid ***β-N-methylamino-L-alanine*** (BMAA) [2], which in higher concentrations was found to be neurotoxic for mammals and chickens, and also suspected to be the cause of neurological syndromes in man. BMAA occurs, like cycasin, in nearly all cycad genera analysed so far, yet with a wide range of genus-specific, quantitative differences. BMAA (fr. wt.) contents range below those of the azoglycosides: leaves and seeds 0.0002-0.17%: *Zamia* cone tissue 0.001%; cone storage cells (ideoblasts) 0.04% [20]. The ubiquitous occurrence of cycasin and BMAA in all extant cycad genera strongly suggests that these defence compounds evolved early in the history of these plants [12, 20].

Cycads are largely avoided by herbivores, yet some insects are adapted to cycads and so are some vertebrates which eat cycad leaves and fruits. These insects are (besides some aphids, thrips and scale insects), leaf-eating larvae of Lepidoptera and Coleoptera of several families [20] and play a role in the pollination of cycads [20]. The cycads have been generally thought to be wind pollinated. However, several recent studies in different regions indicate that cycads are mostly insect pollinated, often by weevils that are closely dependent on the cycads. This contrasts with both *Ginkgo* and the conifers (the other primitive seed plants), all of which are wind pollinated [14]. Chemistry of the pollinator-attractants in cycads is markedly different from that of other flowering plants, suggesting that insect pollination has evolved independently in the two groups. Species of several beetle taxa were observed in a variety of interactions with the cycads, reaching from plain herbivory to symbiotic, mutualistic pollination. Curculionid beetles of several families represent a majority of the associations with cycads. The currently best understood cases of such mutualism are species-specific relations between two New World cycads and two snout-weevils in Florida. The first pair comprises the endemic cycad *Zamia integrifolia (=Z. pumila)* and its symbiotically colonizing weevil *Rhopalotria slossonae* (Belidae) and the second pair consists of the originally Mexican *Z. furfuracea* and its also introduced partner *R. mollis*. Interestingly, another beetle, *Pharaxonothus zamiae* (Languriidae) is an additional pollinator/ parasite of the Floridan *Zamia integrifolia* ([12, 20].

MOLECULAR PHYLOGENY OF CYCADS

Phylogenetic Position of Cycadophyta

A number of research groups have used DNA sequences of cpDNA, mtDNA (mitochondrial DNA), nuclear ITS (nuclear ribosomal DNA internal transcribed spacer) or 16S and 18S rDNA to elucidate the evolutionary position of cycads [1, 3, 15, 21, 24, 25]. A summary of the results obtained from different marker genes is shown in Fig. 3. It is apparent that the phylogenetic position of cycads strongly depends on the marker genes that have been analyzed.

Ribosomal genes place cycads at the base of the tree leading to seed plants. Since rDNA-genes were the first ones to be analyzed, the view that cycads are basal to the rest of the seed plants derives from the early days of molecular systematics. Subsequent studies on chloroplast genes changed the overall topology and placed the Gnetophyta at the basal position, followed by a bifurcation that split angiosperms and the rest of the "gymnosperms". Cycads are basal to the residual gymnosperms and clusters as a sister to *Ginkgo* and conifers (Fig. 3). Phytochrome genes also revealed a similar topology [19]. Mitochondrial genes that have been somewhat neglected in plant phylogeny, reveal angiosperms as a basal clade followed by cycads that is basal to *Ginkgo*, conifers and Gnetophyta. Soltis et al. [21] have combined the three data set to form one combined data set. Here, similar to the situation in the mtDNA data set, the angiosperms are basal, followed by a monophyletic gymnosperm cluster that is divided into a cycad/*Ginkgo* and Gnetophyta/conifer clade.

Since Cycads share the plesiomorphic character "naked ovules" with other "gymnosperms" such as the *Welwitschia* group (Gnetophyta), the conifers (Pinophyta) and *Ginkgo* (Ginkgophyta), the results of the mtDNA and combined data set have the advantage that they concur with the traditional perspective that gymnosperms and angiosperms are monophyletic assemblages.

Phylogenetic Structure within the Cycadophyta

DNA sequences of marker genes can also be used to elucidate the phylogenetic relationships within the monophyletic Cycad ensemble. The following discussions are mainly based on results of my own laboratory [24, 25, 26, 27] and those of [7].

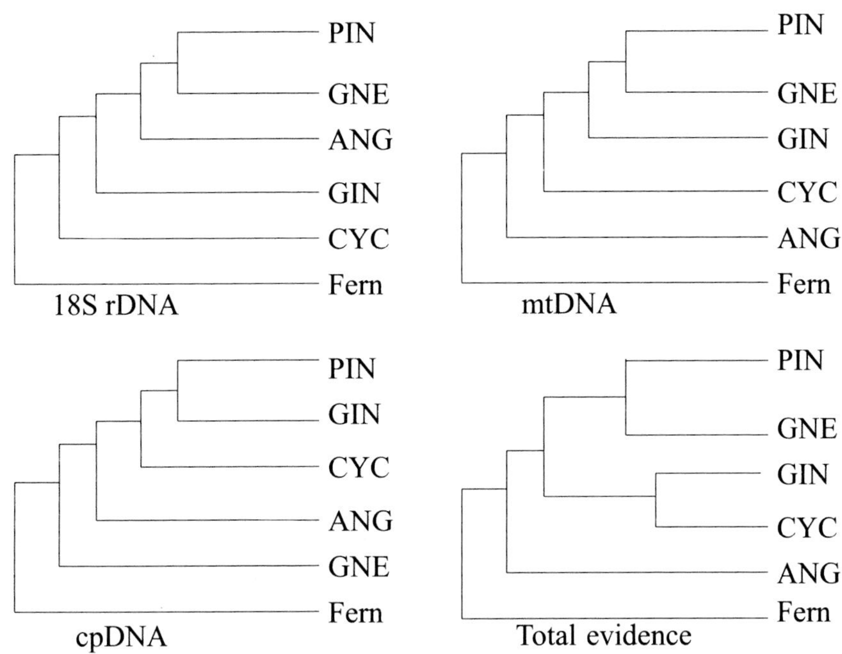

Fig. 3. *Phylogenetic position of Cycadales among gymnosperms and angiosperms based on nucleotide sequences of different sets of marker genes.*

GNE= Gnetophyta, ANG= Angiospermae, CYC= Cycadophyta, GIN = Ginkgophyta, PIN= Pinophyta

For the total evidence trees, the data from 18SrDNA, mtDNA and cpDNA were combined (from Soltis et al. [21].

Figures 4A, B and C show a phylogenetic reconstruction based on *rbc*L (RuBisCo large subunit gene) sequences of all cycad genera, *Ginkgo* and Gnetales and selected genera from conifers, in comparison to some mono- and dicots. These main groups cluster as expected and are known from other studies [1, 3, 15, 21, 24] as monophyletic clades. The Gnetophyta are always at the base of the clade leading to higher plants (bootstrap support 100%). The next bifurcation divides branches leading to angiosperms (with mono- and dicots) and residual gymnosperms (bootstrap support 98%). Within the residual gymnosperms, conifers (Pinophyta) cluster as a sister to Ginkgophyta and Cycadophyta, which cluster as a terminal sister pair (bootstrap support 99-100%). These relationships are always recovered independent from the method of phylogeny reconstruction (i.e. MP, NJ, ML). Almost identical topology

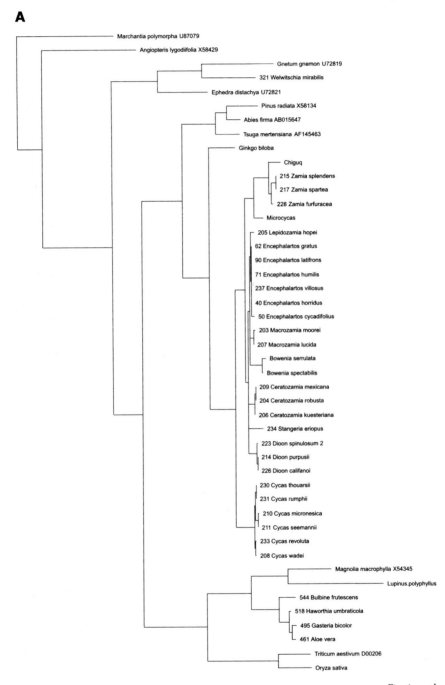
A
Marchantia polymorpha U87079
Angiopteris lygodiifolia X58429
Gnetum gnemon U72819
321 Welwitschia mirabilis
Ephedra distachya U72821
Pinus radiata X58134
Abies firma AB015647
Tsuga mertensiana AF145463
Ginkgo biloba
Chiguq
215 Zamia splendens
217 Zamia spartea
228 Zamia furfuracea
Microcycas
205 Lepidozamia hopei
62 Encephalartos gratus
90 Encephalartos latifrons
71 Encephalartos humilis
237 Encephalartos villosus
40 Encephalartos horridus
50 Encephalartos cycadifolius
203 Macrozamia moorei
207 Macrozamia lucida
Bowenia serrulata
Bowenia spectabilis
209 Ceratozamia mexicana
204 Ceratozamia robusta
206 Ceratozamia kuesteriana
234 Stangeria eriopus
223 Dioon spinulosum 2
214 Dioon purpusii
226 Dioon califanoi
230 Cycas thouarsii
231 Cycas rumphii
210 Cycas micronesica
211 Cycas seemannii
233 Cycas revoluta
208 Cycas wadei
Magnolia macrophylla X54345
Lupinus.polyphyllus
544 Bulbine frutescens
518 Haworthia umbraticola
495 Gasteria bicolor
461 Aloe vera
Triticum aestivum D00206
Oryza sativa

Fig. 4 contd.

B
Strict 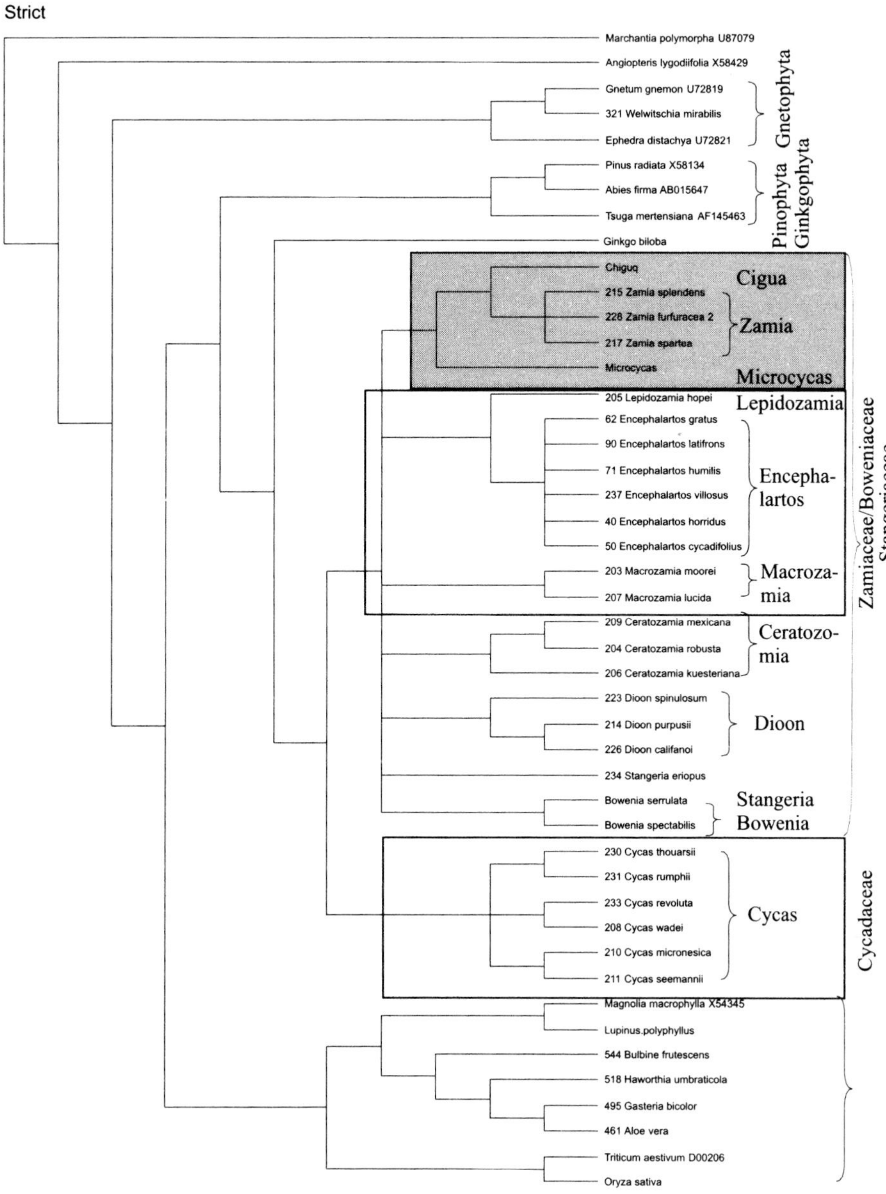
Marchantia polymorpha U87079
Angiopteris lygodiifolia X58429
Gnetum gnemon U72819
321 Welwitschia mirabilis
Ephedra distachya U72821
Gnetophyta
Pinus radiata X58134
Abies firma AB015647
Tsuga mertensiana AF145463
Pinophyta
Ginkgo biloba
Ginkgophyta
Chiguq
Cigua
215 Zamia splendens
228 Zamia furfuracea 2
217 Zamia spartea
Zamia
Microcycas
Microcycas
205 Lepidozamia hopei
Lepidozamia
62 Encephalartos gratus
90 Encephalartos latifrons
71 Encephalartos humilis
237 Encephalartos villosus
40 Encephalartos horridus
50 Encephalartos cycadifolius
Encepha-
lartos
203 Macrozamia moorei
207 Macrozamia lucida
Macroza-
mia
209 Ceratozamia mexicana
204 Ceratozamia robusta
206 Ceratozamia kuesteriana
Ceratozo-
mia
223 Dioon spinulosum
214 Dioon purpusii
226 Dioon califanoi
Dioon
234 Stangeria eriopus
Bowenia serrulata
Bowenia spectabilis
Stangeria
Bowenia
Zamiaceae/Boweniaceae
Stangeriaceae
230 Cycas thouarsii
231 Cycas rumphii
233 Cycas revoluta
208 Cycas wadei
210 Cycas micronesica
211 Cycas seemannii
Cycas
Cycadaceae
Magnolia macrophylla X54345
Lupinus.polyphyllus
544 Bulbine frutescens
518 Haworthia umbraticola
495 Gasteria bicolor
461 Aloe vera
Triticum aestivum D00206
Oryza sativa

Fig. 4 contd.

C

Bootstrap

Marchantia polymorpha U87079
Angiopteris lygodiifolia X58429
Gnetum gnemon U72819
321 Welwitschia mirabilis
Ephedra distachya U72821
Gnetophyta
Pinus radiata X58134
Tsuga mertensiana AF145463
Abies firma AB015647
Pinophyta
Ginkgo biloba
Chiguq — Cigua
215 Zamia splendens
217 Zamia spartea
228 Zamia furfuracea
Zamia
Microcycas — Microcycas
205 Lepidozamia hopei — Lepidozamia
62 Encephalartos gratus
90 Encephalartos latifrons
50 Encephalartos cycadifolius
71 Encephalartos humilis
237 Encephalartos villosus
40 Encephalartos horridus
Encephalartos
203 Macrozamia moorei
207 Macrozamia lucida
Macrozamia
209 Ceratozamia mexicana
204 Ceratozamia robusta
206 Ceratozamia kuesteriana
Ceratozamia
223 Dioon spinulosum
214 Dioon purpusii
226 Dioon califanoi
Dioon
234 Stangeria eriopus — Stangeria
Bowenia serrulata
Bowenia spectabilis
Bowenia
230 Cycas thouarsii
231 Cycas rumphii
210 Cycas micronesica
211 Cycas seemannii
233 Cycas revoluta
208 Cycas wadei
Cycas
Cycadaceae
Magnolia macrophylla X54345
Lupinus polyphyllus
Triticum aestivum D00206
Oryza sativa
544 Bulbine frutescens
518 Haworthia umbraticola
495 Gasteria bicolor
461 Aloe vera
Angiosperms

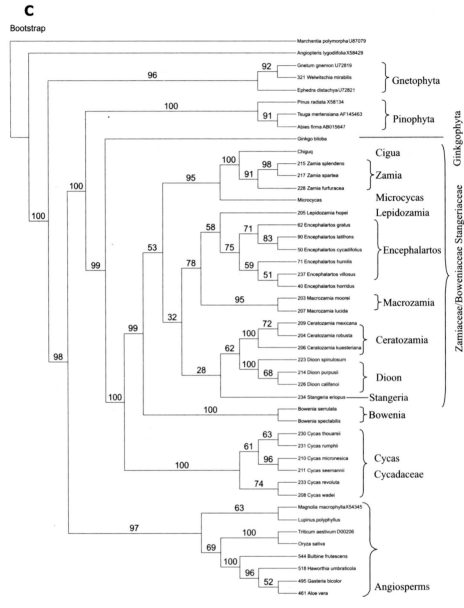

Fig. 4. *Phylogenetic relationships within cycads based on sequences of the rbcL gene. A liver moss (*Marchantia*) and a fern (*Angiopteris*) were selected as far distant outgroups. Tree length: 167 steps; CI=0.554, RI= 0.739; RC= 0.409, HI= 0.446; of 1,428 characters 408 are parsimony informative.*

A. One of 270 most parsimonious trees; B: MP strict consensus; C: NJ with Kimura 2 parameter as distance algorithm; bootstrap values (1,000 replications) are shown above branches.

had been recovered by Soltis et al. [21] and especially Rai et al. [15], who had employed sequences of 17 chloroplast genes and associated non-coding regions.

The phylogenetic trees (Figs. 4A, B and C) also illustrate the relationships within the Cycadales. Members of particular cycad genera cluster as monophyletic clades. Among them, *Cycas* forms the most basal group (100% bootstrap support), justifying its treatment as a separate family Cycadaceae. This is in agreement with previous morphological studies [9, 12, 23] suggesting that the *rbc*L phylogram reflects the true phylogeny, even if the full picture of species phylogeny cannot be concluded solely from a single gene. However, *Cycas* had a basal position also in an analysis involving sequences from cpDNA, ITS and 26S rDNA [7].

More complicated is the situation with the two other cycad families, the large Zamiaceae and the very small Stangeriaceae (with only three species).

Both Cycadaceae and Zamiaceae form well-supported clades. *Stangeria* clusters within Zamiaceae, which questions its systematic status as a monotypic family. *Encephalartos, Macrozamia* and *Lepidozamia* form a well recognized clade (Fig. 4C) that is commonly included in the tribe Encephalarteae. This finding is corroborated by an analysis involving sequences from cpDNA, ITS and 26S rDNA [7]. *Zamia, Chigua* and *Microcycas* also form a well-supported clade, recognized in the tribe Zamieae. This finding is also supported by an analysis involving sequences from cpDNA, ITS and 26S rDNA [7].

The position of *Ceratozamia, Dioon* and the almost monotypic genera *Stangeria* and *Bowenia* cannot be resolved unambiguously with the present data set. However, an analysis involving sequences from cpDNA, ITS and 26S rDNA [7] implied that *Ceratozamia* might be a sister to the tribe Zamieae. *Dioon* clusters at the base of a clade including the tribe Encephalarteae and the genus *Bowenia.* These phylogenetic relationships are, however, only weakly supported by bootstrap values. According to [7], *Stangeria* takes a basal position of the clade comprising the Zamiaceae and *Bowenia*. No support was found for a family Stangeriaceae (sensu Stevenson [23]) that includes the genera *Bowenia* and *Stangeria*. In conclusion, the molecular data are in conflict with all previously proposed classifications of the Cycadophyta at one or more levels. More work is clearly needed to understand the relationships of *Stangeria*,

Bowenia and *Dioon* in order to establish a natural classification of Cycadophyta.

Phylogenetic Relationships within the Genus *Encephalartos*

Encephalartos, which is the largest genus of the Cycadales comprising 55 species [6, 12], is geographically restricted to Africa (south of the Sahara). A significant number of species occur in the tropical regions of central and east Africa, but more than half in South Africa. They occupy a wide range of climatic regimes and habitats, and show a high degree of endemism [4]. Research on relationships between these taxa has been done at a morphological and biochemical level [26, 27, 28, 29], but nucleotide sequencing has been applied only to a few taxa [24, 25, 26, 27].

*rbc*L sequences (Fig. 4) clearly indicate that the genus *Encephalartos* forms a monophyletic clade with *Lepidozamia* and *Macrozamia* [7, 24]. Nuclear internal transcribed spacer regions (ITS 1 and 2), have been used to resolve the phylogenetic history of the genus *Encephalartos*; the chloroplast encoded *rbc*L gene was of little use since only few nucleotide positions differ between species (Table 2) [25].

ITS 1 and 2 resolve *Encephalartos* in three distinct clusters (indicated as ITS-groups 1, 2, 3 and their subgroups, respectively) (Fig. 5) that agree well with the intragenic structure based on morphological and geographical characters (Fig. 6).

MP and ML trees of ITS 1 and 2 are almost congruent in topology, indicating that the phylogenies are well-supported by the sequence data (Fig. 5). The taxa of ITS-group 1 form the most basal group (within-group maximal p-distance: 0.8%) and are also separated from the rest of the taxa by morphological as well as ecological data. The monophyly of group 1 is statistically well-supported by 89% bootstrap value. The branch leading to the sister-group of ITS-group 1 (composed of clusters 2 and 3) is even better supported (bootstrap value of 91%). Taxa of clusters 2 and 3 show a maximal p-distance of 3.7% to cluster 1 and are resolved by 63 & 95% bootstrap, respectively. ITS-cluster 2 consists (maximal 1.7% p-distance) of 6 subgroups and further 5 ungrouped species, and is widely distributed from the Eastern Cape (*E. caffer*) to Uganda (*E. septentrionalis*), Kenya (*E. tegulaneus*) and Nigeria (*E. barteri allochrous*). Subgroups 2.0, 2.1 and 2.2 are well-supported by 70, 91 and

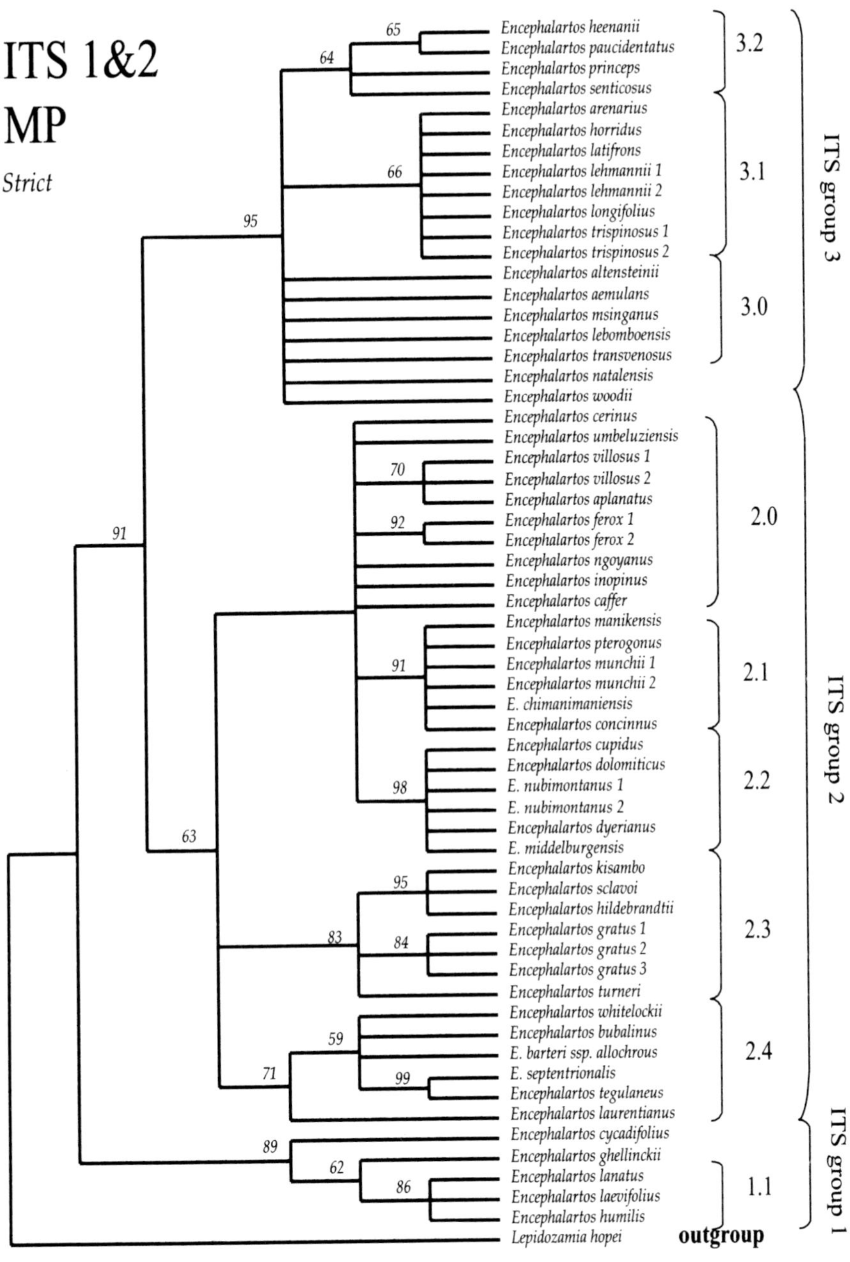
ITS 1&2
MP
Strict
Encephalartos heenanii
Encephalartos paucidentatus
Encephalartos princeps
Encephalartos senticosus
Encephalartos arenarius
Encephalartos horridus
Encephalartos latifrons
Encephalartos lehmannii 1
Encephalartos lehmannii 2
Encephalartos longifolius
Encephalartos trispinosus 1
Encephalartos trispinosus 2
Encephalartos altensteinii
Encephalartos aemulans
Encephalartos msinganus
Encephalartos lebomboensis
Encephalartos transvenosus
Encephalartos natalensis
Encephalartos woodii
Encephalartos cerinus
Encephalartos umbeluziensis
Encephalartos villosus 1
Encephalartos villosus 2
Encephalartos aplanatus
Encephalartos ferox 1
Encephalartos ferox 2
Encephalartos ngoyanus
Encephalartos inopinus
Encephalartos caffer
Encephalartos manikensis
Encephalartos pterogonus
Encephalartos munchii 1
Encephalartos munchii 2
E. chimanimaniensis
Encephalartos concinnus
Encephalartos cupidus
Encephalartos dolomiticus
E. nubimontanus 1
E. nubimontanus 2
Encephalartos dyerianus
E. middelburgensis
Encephalartos kisambo
Encephalartos sclavoi
Encephalartos hildebrandtii
Encephalartos gratus 1
Encephalartos gratus 2
Encephalartos gratus 3
Encephalartos turneri
Encephalartos whitelockii
Encephalartos bubalinus
E. barteri ssp. allochrous
E. septentrionalis
Encephalartos tegulaneus
Encephalartos laurentianus
Encephalartos cycadifolius
Encephalartos ghellinckii
Encephalartos lanatus
Encephalartos laevifolius
Encephalartos humilis
Lepidozamia hopei
outgroup
65
64
66
95
91
70
92
91
98
63
95
83
84
59
71
99
89
62
86
3.2
3.1
3.0
2.0
2.1
2.2
2.3
2.4
1.1
ITS group 3
ITS group 2
ITS group 1

Fig. 5A

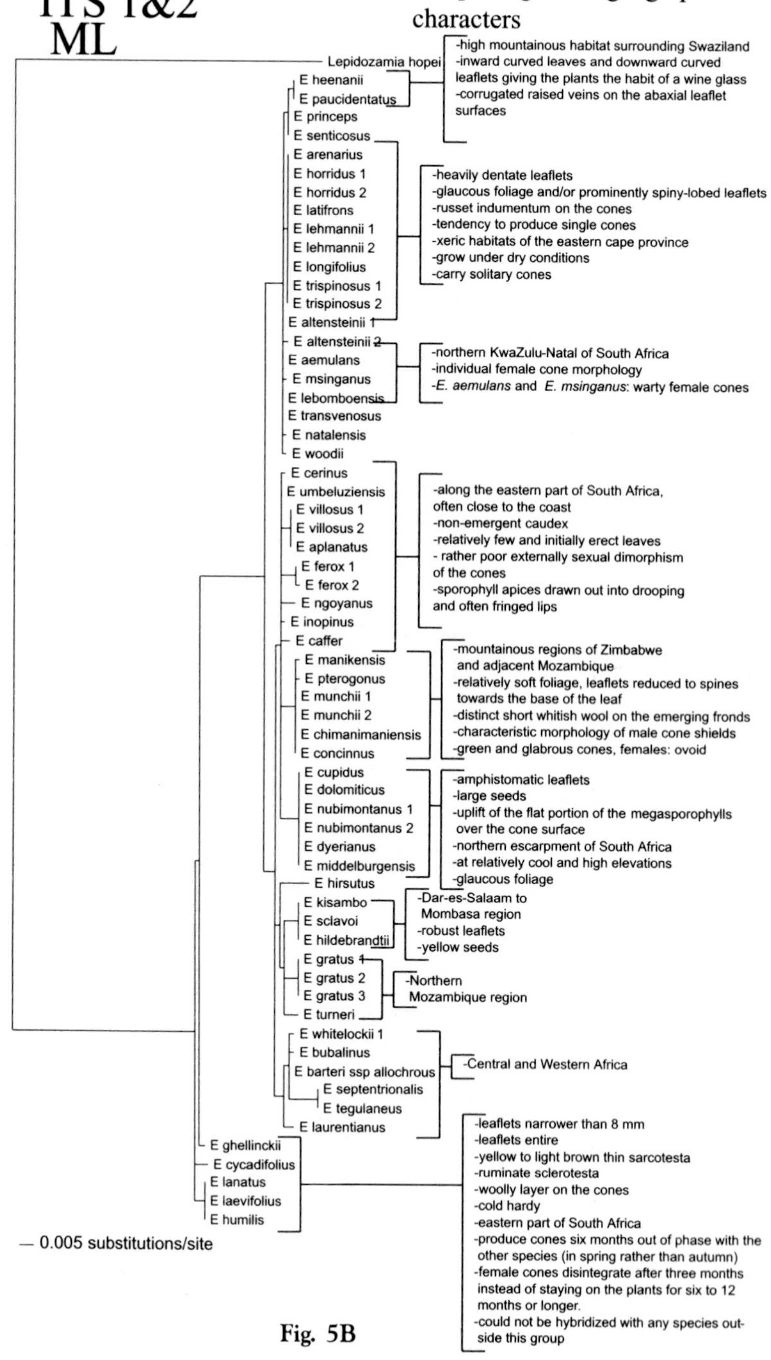
ITS 1&2
ML
Morphological & geographical characters
Lepidozamia hopei
E heenanii
E paucidentatus
E princeps
E senticosus
E arenarius
E horridus 1
E horridus 2
E latifrons
E lehmannii 1
E lehmannii 2
E longifolius
E trispinosus 1
E trispinosus 2
E altensteinii 1
E altensteinii 2
E aemulans
E msinganus
E lebomboensis
E transvenosus
E natalensis
E woodii
E cerinus
E umbeluziensis
E villosus 1
E villosus 2
E aplanatus
E ferox 1
E ferox 2
E ngoyanus
E inopinus
E caffer
E manikensis
E pterogonus
E munchii 1
E munchii 2
E chimanimaniensis
E concinnus
E cupidus
E dolomiticus
E nubimontanus 1
E nubimontanus 2
E dyerianus
E middelburgensis
E hirsutus
E kisambo
E sclavoi
E hildebrandtii
E gratus 1
E gratus 2
E gratus 3
E turneri
E whitelockii 1
E bubalinus
E barteri ssp allochrous
E septentrionalis
E tegulaneus
E laurentianus
E ghellinckii
E cycadifolius
E lanatus
E laevifolius
E humilis
0.005 substitutions/site
-high mountainous habitat surrounding Swaziland
-inward curved leaves and downward curved leaflets giving the plants the habit of a wine glass
-corrugated raised veins on the abaxial leaflet surfaces
-heavily dentate leaflets
-glaucous foliage and/or prominently spiny-lobed leaflets
-russet indumentum on the cones
-tendency to produce single cones
-xeric habitats of the eastern cape province
-grow under dry conditions
-carry solitary cones
-northern KwaZulu-Natal of South Africa
-individual female cone morphology
-E. aemulans and E. msinganus: warty female cones
-along the eastern part of South Africa, often close to the coast
-non-emergent caudex
-relatively few and initially erect leaves
- rather poor externally sexual dimorphism of the cones
-sporophyll apices drawn out into drooping and often fringed lips
-mountainous regions of Zimbabwe and adjacent Mozambique
-relatively soft foliage, leaflets reduced to spines towards the base of the leaf
-distinct short whitish wool on the emerging fronds
-characteristic morphology of male cone shields
-green and glabrous cones, females: ovoid
-amphistomatic leaflets
-large seeds
-uplift of the flat portion of the megasporophylls over the cone surface
-northern escarpment of South Africa
-at relatively cool and high elevations
-glaucous foliage
-Dar-es-Salaam to Mombasa region
-robust leaflets
-yellow seeds
-Northern Mozambique region
-Central and Western Africa
-leaflets narrower than 8 mm
-leaflets entire
-yellow to light brown thin sarcotesta
-ruminate sclerotesta
-woolly layer on the cones
-cold hardy
-eastern part of South Africa
-produce cones six months out of phase with the other species (in spring rather than autumn)
-female cones disintegrate after three months instead of staying on the plants for six to 12 months or longer.
-could not be hybridized with any species outside this group

Fig. 5B

98% bootstrap value, and subgroup 2.3 by 83% bootstrap support. Taxa of subgroup 2.3 are distributed from Mozambique (*E. gratus*) to Kenya (*E. kisambo*), and are internally further structured: *E. sclavoi, E. hildebrandtii*, and *E. kisambo* (95% bootstrap support) within subgroup 2.3 grow in the very restricted region of Dar-es-Salaam to Mombasa, and are characterized by leathery leaflets and yellow seeds. Subgroup 2.4 (71% bootstrap support) is the most diverse among all the subgroups and distributed along the equator in Africa from the west (*E. barteri allochrous*) to the east (*E. whitelockii* in Uganda, *E. tegulaneus* in Kenya, and *E. bubalinus* in Tanzania). Consequently, ITS-group 2 is distributed over a wider distribution range than any other group or subgroup. All the taxa of ITS-group 3 (95% bootstrap support, maximal p-distance 0.5%) occur in the area from Willowmore in the south (*E. lehmannii* and *E. longifolius*) to Piggs Peak in Swaziland in the north (*E. paucidentatus*); taxa of subgroup 3.1 usually grow in dry habitats of the Eastern Cape and share glaucous leaflets as a common character. Detailed morphological features and biogeographical affinities are given in Table 1.

A genomic fingerprinting technique was employed using ISSR-PCR (inter simple sequence repeat-polymerase chain reaction) to study the taxonomy of *Encephalartos* [25]. ISSR was able to support only two of the relationships found by ITS 1&2. The group ITS 1.1, consisting of *E. lanatus*, *E. laevifolius* and *E. humilis*, is supported by ISSR-band 4 and in part by ISSR-band 7. Interestingly, the taxa *E. cycadifolius* and *E. ghellinckii*, that clearly belong to ITS-group 1 by ITS 1&2 and morphology, show neither band 4 nor band 7 fragments and are clustered differently by ISSR. A second ISSR group that represents the ITS 1&2

Fig. 5. *Molecular phylogeny of Encephalartos based on nucleotide sequences of ITS I and II [from 25]*

A. Strict consensus parsimony cladogram of ITS 1&2 constructed by MP. Parameters: sequence addition: closest; branch swapping option: tree bisection-reconnection. Bootstrap values (1,000 replications) above 50% are presented. 1,152 most parsimonious trees of the length 214 were obtained, of which the strict consensus cladogram is shown here (CI 0.916; RI 0.944; RC 0.865; HI 0.084).

B. Reconstruction by ML. Two trees -ln L = 1798.26007 were obtained. Results are presented as phylogram with branch lengths representing the genetic distance under the GTR+G+I algorithm. User-specified substitution rate matrix: AC=0.642793 AG=4.437940 AT=1.687732, CG=1.496376, CT=3.590283, GT=1.000000; Assumed proportion of invariable sites = 0.341177; shape parameter (alpha)= 401.87.

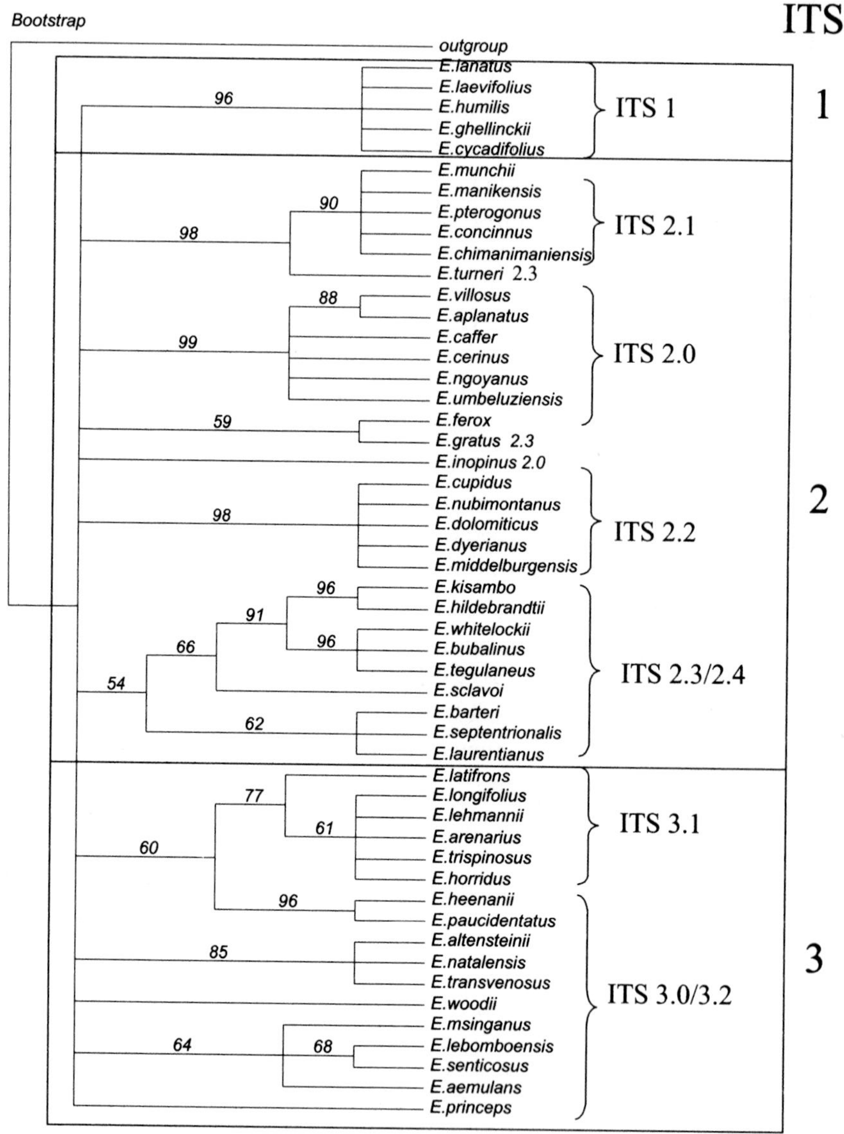

Fig. 6. *Most parsimonious tree of* Encephalartos*—morphological and geographical traits [after 25]: from a matrix of 27 morphological and biogeographical characters. Bootstrap values (1,000 replications) above 50% are presented.*

Table 1. Morphological and geographical characters of the cycad genera (from [7, 8, 22]), N = numbers of species

Family	*Genus*	*N*	*Occurrence*	*Characteristic traits*
Cycadaceae	*Cycas* L.	90	Australian (26 species); Indo-Chinese region (30 species); Malaysian region; Japan; SE Asia; Micronesia; Polynesia; Madagascar; E. Africa	**Megasporophylls**: not in cones, in a terminal rosette; **ovules**: 2 or many; **seeds**: yellow, orange or brown fleshy outer sarcotesta; **microsporophyll:** numerous microsporangia assembled in determinate male cones; **stem:** aerial or subterranean, pachycaul & cylindrical; **leaves**: pinnate or rarely bipinnate; **leaflets**: midrib, lacking secondary veins; not articulated.
Zamiaceae	*Dioon* Lindl.	11	Mexico, Honduras & Nicaragua	**Megasporophylls**: in spirally assembled stalked female cones; ovules: 2 (rarely 3), stalked; **microsporophylls**: spirally assembled in stalked male cones; **seeds**: white or cream fleshy outer sarcotesta; **stem**: aerial or subterranean, pachycaul & cylindrical; **leaves**: pinnate, spirally arranged; lower leaflets often reduced to spines; **leaflets**: simple, often with spiny margins & numerous bifurcating parallel veins and without distinct midrib; broad-based, not articulated; **basal offsets**: common.

Table 1 contd.

Table 1 contd.

	Encephalartos Lehm.	62	Africa	**Megasporophylls**: in spirally assembled stalked female cones; ovules: 2 (rarely 3), sessile; **microsporophylls:** spirally assembled in determinate, stalked or sessile male cones; **seeds**: with a red, yellow, orange or brown fleshy outer sarcotesta; **stem:** often large aerial or subterranean, pachycaul & cylindrical, with usually many leaves and persistent leaf bases; **leaves**: pinnate, spirally arranged, lower leaflets often as spines; **leaflets**: simple, frequently with spiny, dentate or lobed margins, with numerous bifurcating parallel veins and no distinct midrib, leaflets not articulated; **basal offsets**: common.
	Chigua D.W.Stev.	2	Colombia	**Megasporophylls**: in spirally assembled stalked female cones; ovules: 2 (rarely 3), sessile; **microsporophylls**: spirally assembled into stalked male cones; **seeds:** with a red or pink fleshy outer sarcotesta; **stem**: subterranean, pachycaul & globose; with few leaves; **leaves**: pinnate, spirally arranged, lower leaflets not reduced to spines; **leaflets:** articulated, simple, with dentate margins and numerous bifurcating parallel veins, and a distinct midrib.

Table 1 contd.

Table 1 contd.

	Ceratozamia Brongn.	16	Mexico, Guatemala & Belize	**Megasporophylls**: in spirally assembled stalked female cones; ovules: 2 (rarely 3), sessile; **microsporophylls**: spirally assembled in stalked male cones; **seeds:** with a cream or white fleshy outer sarcotesta; **stem:** aerial or subterranean, pachycaul, cylindrical or globose; with few to many leaves; **leaves**: pinnate, spirally arranged, lower leaflets not reduced to spines; **leaflets:** articulated, simple, entire, with numerous bifurcating parallel veins and no distinct midrib; leaf bases mostly not persistent; new leaves emerging singly or in flushes; **basal offsets**: in some cases.
	Lepidozamia Regel	2	E Australia	**Megasporophylls:** in spirally assembled stalked female cones**;** **ovules:** 2 (rarely 3), sessile; **microsporophylls**: spirally assembled in determinate, sessile male cones; **seeds**: with a red outer sarcotesta; **stem**: erect, aerial, & cylindrical, with many leaves; new leaves emerging in flushes; **leaves**: pinnate, spirally arranged, lower leaflets not reduced to spines; **leaflets**: simple, with numerous parallel veins and no distinct midrib.

Table 1 contd.

Table 1 contd.

	Macrozamia Miq.	38	E Australia (34 species), Central Australia (1 species), SW Australia (3 species)	**Megasporophylls**: in spirally assembled stalked female cones; ovules: 2 (rarely 2), sessile; **microsporophylls**: spirally assembled in determinate, stalked male cones; **seeds**: with a red or less commonly yellow, orange or brown fleshy outer sarcotesta; **stem**: aerial or subterranean, pachycaul & cylindrical, with few to many leaves and persistent leaf bases; **leaves:** pinnate, spirally arranged, lower leaflets often reduced to spines; **leaflets:** simple or dichotomously divided, with numerous parallel veins and no distinct midrib.
	Zamia L.	50	South, Central & North America	**Megasporophylls**: in spirally assembled stalked female cones; ovules: 2 (rarely 3), sessile; **microsporophylls:** spirally assembled in stalked or sessile male cones; **seeds**: with a red, orange, yellow or rarely white fleshy outer sarcotesta; **stem:** aerial or subterranean, pachycaul, cylindrical or globose, with few to many leaves; **leaves**: pinnate, spirally arranged, lower leaflets: not reduced to spines; dichotomous branching common in geophytic species; **basal offsets:** common.

Table 1 contd.

Table 1 contd.

	Microcycas (Miq.) A. DC	1	Cuba	**Megasporophylls:** in spirally assembled stalked female cones; ovules: 2 (rarely 3), sessile; **microsporophylls:** spirally assembled in sessile male cones; **seeds:** with a red, fleshy outer sarcotesta; **stem:** aerial, pachycaul, & cylindrical, with many leaves; **leaves:** pinnate, spirally arranged, lower leaflets not reduced to spines; petioles lacking prickles; **leaflets:** simple, entire, with numerous bifurcating parallel veins and no distinct midrib, leaflets articulated, dichotomous branching uncommon, **basal offsets:** rare.
Stangeriaceae	*Stangeria* T. Moore, Hook	1	S. Africa	**Megasporophylls:** in spirally assembled stalked ovoid female cones; ovules: 2 (rarely 3), sessile; **microsporophylls:** spirally assembled in stalked, ovoid male cones; **seeds:** with a dark red sarcotesta; **stem:** naked and often dichotomously branched & subterranean; new leaves single; **leaves**: pinnate; **leaflets:** with a terminal leaflet, lower leaflets not reduced to spines; petioles lacking spines or prickles; flat, penniveined, with a large midrib and numerous sub-parallel, bifurcating lateral veins.

Table 1 contd.

Table 1 contd.

Stangeriaceae or Zamiaceae	*Bowenia*	2	Australia (E. Queensland)	**Megasporophylls**: cones stalked but not elevated above soil surface, ovoid to globose; **ovules:** 2 (rarely 3); **microsporophylls**: spirally assembled into stalked and ovoid male cones; **seeds**: with a white aging to purple sarcotesta; **stem**: naked & subterranean with 1 to many short, slender, determinate leaf- and cone-bearing branches; **leaves**: bipinnate, and often a true terminal leaflet, lower leaflets not reduced to spines; petioles lacking spines or prickles; **leaflets:** flat, lacking a midrib, with numerous sub-parallel longitudinal veins.

Table 2. Relationships among the 31 *rbc*L sequences of *Encephalartos*. The three parsimony informative positions were used to characterize chloroplast DNA types.

cp-DNA type	*rbcL position* 573	*rbcL position* 673	*rbcL position* 688	*Taxon/specimen*
1	A	A	G	*E. cupidus, E. cycadifolius, E. dyerianus, E. ghellinckii, E. gratus, E. hildebrandtii E. humilis, E. inopinus, E. kisambo, E. laevifolius, E. lanatus, E. latifrons, E. lehmannii, E. longifolius, E. manikensis, E. munchii, E. princeps, E. villosus, E. whitelockii*
1	A	A	A	*E. aplanatus, E. cerinus, E. ferox, E. heenanii, E. ngoyanus, E. senticosus, E. umbeluziensis*
2	G	C	G	*E. natalensis, E. woodii*

relationships is composed of *E. aplanatus* and *E. villosus* 1, and defined by ISSR band 9. All other relationships found by ISSR support neither ITS 1&2 nor *rbc*L topologies [25].

The differences of the phylogenies of nuclear DNA, cpDNA and genomic fingerprinting (ISSR) suggest that hybridization and introgression played an important role in the phylogenetic history of *Encephalartos*. The differences in topology may be a consequence of the diverse behaviour of the three molecular markers to hybridization between species of *Encephalartos* and subsequent backcrossing. Even today, when the distribution of the species is strongly disjunct, several cases of natural hybrids were found [4, 28], and in phases of global cooling all the occurrences of *Encephalartos* may have been contracted and expanded repeatedly from eastern South Africa. The occurrence of hybrids from the three species *E. altensteinii*, *E. arenarius* and *E. lehmanii* [28] suggest that hybrids in *Encephalartos*, at least between several of the species, are fertile. The absence of hybridization barriers, together with the small genetic distances of ITS 1&2 (maximally 3.7% p-distance), *rbc*L (maximally 0.3% p-distance) and allozyme markers [26, 27], suggest that the formation of the extant lineages of *Encephalartos* occurred recently rather than in Cretaceous times when the first fossils were assigned to this genus [13].

ITS 1&2 reveal that the genus *Encephalartos* displays its maximum genetic diversity (3.7% ITS 1&2 p-distance) in the mountainous regions of eastern South Africa, where the most basal ITS-group 1 as well as most taxa of ITS-groups 2 and 3, are located. This region, extending from the catchment area of the Limpopo to Port Elizabeth, is well-known as the Pleistocene refugium and diversification centre in several angiosperm groups [30]. Taking into account, the genetic diversification of *Encephalartos* in this region and the evidence from other plant genera, this area might also have been the centre of origin of the extant lineages of *Encephalartos* and refuge area of the ancestors in the Pliocene/Pleistocene global cooling period. Occurrence of all three ITS-groups in this region, together with the small genetic distances between the tropical African species and those from eastern South Africa, suggest that the spatial disjunction over large regions of Africa may have been caused by recent long-distance dispersal (LDD) rather than by shrinkage of a former wide distribution area into island-like occurrences and subsequent differentiation [10].

MOLECULAR CLOCK: AGE ESTIMATE OF CYCAD SPECIATION

Since cycads represent an old plant family with a former Gondwana distribution, it is tempting to explain the vicariance of extant cycad genera on different continents by plate tectonics as nucleotide and amino acid sequences change with time [5, 17], the molecular clock approach offers a chance to discuss the evolutionary scenario of cycads.

For dating the divergence times, the approach of Savard et al. [18] was applied, using non-synonymous substitutions to calibrate the *rbc*L clock. Pairwise non-synonymous p-distances and divergence times were taken from Savard et al. [18] for calibration (taxon pairs: non-synonymous distance – divergence time): Liverworts-vascular plant split: 5.67% – 440 mya; fern-seed plant split: 4.67% - 395 mya; monocot-dicot split: 3.31% - 200 mya; divergence within Pinaceae: 0.71% - 140 mya.

For *rbc*L, rates were found to be homogeneous between gymnosperms, perennial angiosperms, fern and liverwort [18].

Using non-synonymous substitutions and four landmark events [18], a linear regression was calculated and used to date the divergence times within and between *Encephalartos, Lepidozamia, Macrozamia* and *Cycas* (Fig. 4) [24, 25]. Since cycads are gymnosperms with a long generation time, compared to that of the pine family (the latter factor is supposed to influence the substitution rate), the non-synonymous substitution rate variation between cycads and Pinaceae was calculated using the formula of Gaut et al. [5]. The distances within *Encephalartos* are unexpectedly small in *rbc*L (Table 3), as well as in the ITS region and between *Encephalartos*, *Lepidozamia* and *Macrozamia* (Table 3). The maximum

Table 3. Pairwise non-synonymous p-distances and calculated divergence times between Cycads and other vascular plants based on the *rbc*L gene. * Divergence times taken from Savard et al. [18], used to calibrate the molecular clock (from [24]).

Taxon-pair	*Genetic distances of non-synonymous bases (%)*	*Maximal divergence time (mya)*
Liverworts-vascular plants*	4.44	440*
Ferns-seed plants*	3.40	395*
Monocots-Dicots*	3.81	200*
Diversification of Pinaceae*	1.21	140*
Cycas-other Cycads	1.48	132
Cycads except *Cycas*	1.05	94
Within Cycas	0.40	36
Lepidozamia-Macrozamia-Encephalartos	0.22	20
Within Encephalartos	0.11	9.8

non-synonymous distances and divergence times found were (average and standard deviations in brackets): Cycadaceae–Zamiaceae/Stangeriaceae-split: 1.19% (0.65 ± 0.28) – maximally 92.5 mya (50.2 ± 21.7); within Zamiaceae/Stangeriaceae: 0.95% (0.45 ± 0.26) – maximally 73.9 mya (35.2 ± 20.5); within *Cycas*: 0.12% (0.08 ± 0.05) – maximally 9.33 mya (6.2 ± 4.4); *Lepidozamia-Encephalartos*-split: 0.25% (0.16 ± 0.06) – maximally 19.4 mya (12.7 ± 4.8); within *Encephalartos*: 0.13% (0.08 ± 0.06) – maximally 10.1 mya (6.5 ± 4.6).

To confirm that the *rbc*L molecular clock is not irregularly slow in Cycads, the Relative-Rate Test choosing *Magnolia* as reference taxon was applied. Since the Relative-Rate Test could be sensitive to taxonomic sampling [16], four GenBank sequences of the Pinaceae were chosen to check for rate variation within the lineages leading to the four cycads *Encephalartos septentrionalis, Zamia spartea, Stangeria eriopus* and *Cycas revoluta*. The Relative-Rate Test revealed that the *rbc*L evolutionary rate is approximately 0.64 times slower in Cycads than in Pinaceae, which would not change the dimensions of divergence times and the main conclusion of our results. Due to the minor reduction of the substitution rate compared to the Pinaceae, it seems more likely to us that the extant species of cycads are the result of recent, rather than old, radiation.

Our molecular clock estimates agree with the fossil record showing that extant cycads evolved in the Tertiary era [7].

The African-Australian disjunction with low *rbc*L-divergence between the species of *Encephalartos*, *Lepidozamia* and *Macrozamia* (Figs. 4 & 5, Table 3) could be explained by long-distance dispersal in the Miocene, rather than by vicariance and continental drift. Another argument against a possible Gondwana origin is the observation that the Zamiaceae are absent from other Gondwanan crustal fragments, such as India, New Zealand, New Caledonia and South America.

The colonization of Madagascar by *Cycas thouarsii*, far from the distribution areas of the other *Cycas* species, seems to be even younger. The *rbc*L sequences of *Cycas thouarsii* and *Cycas rumphii* are identical, suggesting a recent dispersal from Indonesia to Madagascar. Since this cycad has a spongy endocarp, a transoceanic distribution appears plausible. Dispersal via the sea is also relevant for other members of *Cycas* subsection Rumphiae, since *Cycas* seeds remain viable even after long time in seawater. It is, therefore, noteworthy that the members of the subgroup Rumphiae occur both on the mainland of Southeast Asia and are widely distributed on islands of Indian and western Pacific oceans. An alternative idea for the settlement of Madagascar has been communicated to me by an ethnologist. Anthropologists suggest that the natives who colonized Madagascar came from Southeast Asia. In this region, Cycads are used for food. Therefore, it could be possible that the early settlers brought C. *thouarsii* directly to Madagascar as a potential food plant.

Acknowledgement

The author would like to thank Dr. J. Treutlein (Heidelberg University), Prof. Dr. Herman van der Bank, Dr. Michelle van der Bank (University of Johannesburg) and Dr. P. Vorster (Stellenbosch University) for fruitful collaboration during the last seven years. Plant material was supplied by several botanical gardens (Heidelberg, Nelspruit, Montgomery Foundation and other gardens); their support is gratefully acknowledged.

References

[1] Bowe LM, Coat G, Depamphilis CW. Phylogeny of seed plants based on all three genomic compartments: extant gymnosperms are monophyletic and gnetales' closest relatives are conifers. PNAS USA: 2000; 97: 4092-4097.

[2] Brenner ED, Stevenson DW, Twigg RW. Cycads. Evolutionary innovations and the role of plant-derived neurotoxins. Trends in Plant Science 2003; 8: 446-452.

[3] Chaw SM, Parkinson CL, Cheng Y, et al. Seed plant phylogeny inferred from all three plant genomes: monophyly of extant gymnosperms and origin of Gnetales from conifers. PNAS USA: 2000; 97: 4086-4091.

[4] Dyer RA. The Cycads of Southern Africa. Bothalia: 1965; 8: 405-514.

[5] Gaut BS, Clark LG, Wendel JF, Muse SV. Comparisons of the molecular evolutionary process at *rbc*L and *ndh*F in the Grass Family (Poaceae). Mol Biol Evol 1997; 14: 769-777.

[6] Grobbelaar N. Phylogeny of *Encephalartos* species. *Encephalartos*. Cycad Society South Afri 2001; 66: 17-18.

[7] Hill KD, Chase MW, Stevenson DW, et al. The families and genera of cycads: a molecular phylogenetic analysis of Cycadophyta based on nuclear and plastid DNA sequences. Int. J. Plant Sci 2003; 164: 933-948.

[8] Hill KD, Stevenson DW. World List of Cycads. http://plantnet.rbgsyd.gov.au/ 2005

[9] Johnson LAS. The families of cycads and the Zamiaceae of Australia. Proc Linn Soc New South Wales 1959; 84: 64-117.

[10] Melville R. *Encephalartos* in Central Africa. Bulletin Kew 1957; 12: 237-257.

[11] Moretti A, Sabato S, Siniscalco GG. Taxonomic significance of methylazoxymethanol glycosides in the cycads. Phytochemistry 1983; 22: 115-117.

[12] Norstog KJ, Nicholls TJ. The Biology of the Cycads. Ithaca, USA Cornell University Press, 1998.

[13] Pant DD. The fossil history and phylogeny of the Cycadales. Geophytology; 1987; 17: 125-162.

[14] Pellmyr O, Tang W, Groth I, et al. Cycad cone and angiosperm floral volatiles: inferences for the evolution of insect pollination. Biochem Syst Ecol 1991; 19: 623-627.

[15] Rai HS, O'Brien HE, Reeves PA, et al. Inference of higher-order relationships in the cycads from a large chloroplast data set. Mol Phyl Evol 2003; 29: 350-359.

[16] Robinson M, Gouy M, Gautier C, Mouchiroud D. Sensitivity of the Relative-Rate Test to taxonomic sampling. Mol Biol Evol 1998: 15; 1091-1098.

[17] Sarich VM, Wilson AC. Generation time and genome evolution in primates. Science 1973; 179: 1144-1147.

[18] Savard L, Li P, Strauss SH, et al. Chloroplast and nuclear gene sequences indicate Late Pennsylvanian Time for the last common ancestor of extant seed plants. PNAS USA 1994; 91: 5163-5167.

[19] Schmidt M, Schneider-Poetsch HAW. The evolution of gymnosperms redrawn by phytochrome genes: The Gnetatae appear at the base of the gymnosperms. J Mol Evol 2002; 54: 715-724.

[20] Schneider D, Wink M, Sporer F, Lounibos P. Cycads: Their evolution, toxins, herbivores and insect pollinators. Naturwissenschaften 2002; 89: 281-294.

[21] Soltis DE, Soltis PS, Zanis MJ. Phylogeny of seed plants based on evidence from eight genes. Amer J Bot 2003; 89: 1670-1681.

[22] Stevenson DW. Morphology and Systematics of the Cycadales. In: Memoirs of the New York Botanical Garden. 1990; 57: 8-55.

[23] Stevenson DW. A formal classification of the extant cycads. New York Botanical Garden Press, New York, USA Brittonia 1992; 44: 220-223.

[24] Treutlein J, Wink M. Molecular phylogeny of cycads inferred from *rbc*L sequences. Naturwissenschaften 2002; 89: 221-225.

[25] Treutlein J, Vorster P, Wink M. Molecular relationships in *Encephalartos* (Zamiaceae, Cycadales) based on nucleotide sequences of nuclear ITS 1&2, *rbc*L, and genomic ISSR fingerprinting. Plant Biol 2005; 7: 79-90.

[26] van der Bank H, Wink M, Vorster P, et al. Allozyme and DNA sequence comparisons of nine species of *Encephalartos* (Zamiaceae). Bioch Syst Ecol 2001; 29: 241-266.

[27] Van der Bank H, Vorster P, van der Bank M. Phylogenetic relationships, based on allozyme data, between six cycad genera indigenous to South Africa. S Afr J Bot 1998; 64: 182-188.

[28] Vorster P. Hybridisation in *Encephalartos*. Excelsa 1986; 12: 101-106.

[29] Vorster P. Taxonomy of *Encephalartos* (Zamiaceae): Taxonomically useful external characteristics. In: Stevenson DW, Norstog KJ eds. Proc Second Intl Confce on Cycad Biol. Palm & Cycad Societies of Australia, Milton, Queensland: 1995: 294-299.

[30] White F, Moll EJ. The Indian Ocean costal belt. In: Werger MJA ed. Biogeography and Ecology of Southern Africa. The Hague: 1978: 561-598.

CHAPTER

2

Molecular Variability and Diversity of Mediterranean Pines: *Pinus halepensis* Mill. and *Pinus brutia* Ten.

LEONID KOROL and GABRIEL SCHILER
Agricultural Research Organization, The Volcani Center, Department of Agronomy and Natural Resources, Forestry Section, Bet Dagan, Israel

ABSTRACT

Knowledge of the life history and ecological characteristics of woody plant species helps in predicting the level and distribution of genetic diversity within and among populations. The plant communities, which grow in a particular place, are influenced by their phytogeographical position, climatic factors, soil and human activities. The acquired changes are reflected in the genomic structure of forest trees during evolution. Genetic diversity in nature is the result of evolutionary processes, and investigation of molecular polymorphism is a quick way of understanding genetic changes in forest populations and the evolutionary development. Trees, which survived in new margins of environmental conditions, assist in identifying the forest boundaries and the evolutionary processes. Thus, in response to environmental changes, some of tree forest species may acquire new traits. These traits may allow many tree species to survive predicted global climatic changes while preserving much of their genetic diversity, advancing their evolutional path in the new habitat.

Key Words: Genetic diversity, *P. halepensis*, *P. brutia*, pine evolution, ecological adaptation, population genetics

Address for correspondence: Leonid Korol—Agricultural Research Organization, The Volcani Center, Department of Agronomy and Natural Resources, Forestry Section, Bet Dagan, P. O. Box 6, Dagan 50250, Israel. E-mail: vckorol@agri.gov.il.

INTRODUCTION

Knowledge of the life history and ecological characteristics of woody plant species assists in prediction of the level and distribution of genetic diversity within and among populations. Generalizations developed from such analyses can be used to develop sampling strategies for the preservation of genetic diversity.

Lately, evolutionary history of plants has been studied by investigating mitochondrial DNA (mtDNA) and chloroplast DNA (cpDNA). The main advantage of these genomes is that they allow separate research for maternal and paternal lines, as mtDNA is transferred to the offspring usually only from the mother. This analysis helps in the revelation of the evolutionary past of plants. Undoubtedly, the mtDNA and cpDNA represent only a small part of the plant genome, and these parts differently reflect migration and selection events. However, minor details might be of great importance for revealing the complete picture. Accordingly, research of these genomes, together with the nuclear genome, may reconstruct the objective picture of the evolutionary paths for plants.

The first gymnosperms arose in the Middle Devonian (~365 million years ago). Palaeobotanical discoveries have shown that ancestors of *Pinaceae* family evolved by the mid-Jurassic period (~160 million years ago). More than half the species in the *Pinaceae* are included in the genus *Pinus* (100 species) [73] representing 20% of all gymnosperms. The genus *Pinus* is divided into two main subgenera, *Pinus* (hard pines) and *Strobus* (soft pines), diversified by the end of the Cretaceous (66 mya) [54, 76]. Several sections (e.g. *Strobus* and *Pinus*) and further subsections (e.g. *Sylvestres*, *Attenuatae* and *Strobi*) have evolved since the diversification of these two subgenera [44, 73].

The impact of the Eocene had the effect of dissecting the genus and concentrating pines into widely disconnected regions. During the Pleistocene (1.7 - 0.01 mya), pine populations shifted first south, then north. The climatic fluctuation during the Pleistocene may have played an important role in the speciation and preservation of distinctive genotypes [76]. For the last 10,000 years after the last glacial period, the current distribution of pines has been shaped. Pines are the most widespread tree genus in the world and are of ecological and economic significance. The natural distribution of pines encompasses the area from North America and Eurasia south to subtropical and tropical regions of Central America and Asia [54, 73].

GENOME ORGANIZATION OF CONIFER

Pines form an important group of evergreen plant species, which are extensively distributed worldwide. This group is characterized by high genetic variability in comparison to other higher plants [31]. Although conifers are the largest and most diverse group of living gymnosperms, the relationship between families and species is not clear [91]. An important characteristic of the conifer genome is its large size compared with other plant species. Among eukaryotic organisms genome size varies several hundredfold, thus among angiosperms the DNA content variation is from 2C = 0.1 to 254 pg. [4]. The variation of the genome size between 20 *Pinus* species collected in North America, Europe, and Eurasia was 1.73-fold [67] and only in 18 North American *Pinus* species alone the variation was 1.5 fold [94]. In spite of the large size of the DNA content, the evolution of the conifer chromosome appears to be quite conservative. The number of chromosomes among conifers is not highly varied and is normally equal to 11 or 12. The number of genes, which cover genomic traits, their spatial organization and degree of variability among different trees, populations, provenances and species, remains largely unknown.

The size of the pine genome (20,000-30,000 million nucleotide base pairs [bp], for example, is 6 to 8 times larger than the human genome (3,400 million bp), and 150 to 200 times larger than the genome of model plant species, *Arabidopsis thaliana* (125 million bp). Even a relatively small physical size of the *Populus* genome (500 million bp), which is 40 times smaller than the best-studied conifer, *Pinus taeda*, and, therefore, can be a good forest tree model species, is still about 4 times as large as that of *Arabidopsis* (although similar to rice and 6 times smaller than maize. In the nucleus of the conifers, DNA content is also the largest among higher plants. In pine chromosome, the DNA content is approximately 3.7 times larger than in maize, 5.2 times than in lettuce, 15 times than in tomato, and 110 times than in *Arabidopsis* [60]. Although the size of the conifer genome in terms of DNA content appears very large compared with the size of other plant genomes, in terms of map units it is not much greater than in other crops. Neale and Williams [60] estimated the pine genome to be approximately 2,500 cantimorgans (cM), which is only two times larger than tomato, corn or lettuce. Since the main limitation of the pine genome measurement with allozyme technique is a small number of markers, it is possible to elevate the resolution in genome evolution research by applying modern technologies. Studies of conifers based on allozyme markers have mapped only about 10% of the genome (226.4 cM) [13]. With the help of random amplified polymorphic DNA (RAPD) for mapping of *Pinus elliottii,* it was shown that 64-75% of the total genome size is approximately 2,160 cM [63]. In *Pinus taeda* 191 RAPDs were mapped to 12 linkage groups with a distance of 1,687 cM, and in *Pinus silvestris* 282 RAPDs formed 14 linkage groups with a total distance 2,638 cM [104]. For *Pinus brutia* the total map covers a distance of 662.8 cM [34]. Sewell et al. [86] estimated the genome size of loblolly pine (*Pinus taeda*) as 1,227 cM., whereas a constructed genetic map of loblolly pine from all available genotypic data for comparative analyses among pine genomes (i.e. 12 linkage groups) consisting of 155 RFLPs, 75 ESTPs, and 5 isozyme loci is similar to the conservative estimation from Sewell et al. [86] - 1165 cM [8].

The chloroplast genome structure has been studied in a variety of plants. In many species it generally consists of homogeneous circular double-stranded DNA molecules with a size of 120-160 kb. An outstanding feature of chloroplast DNAs (cpDNAs) found in most plants is the presence of a large inverted repeat (IR) of 6-76 kb [71]. The complete nucleotide sequence of the black pine (*Pinus thunbergii*) chloroplast genome is 119, 707 bp [96]. The chloroplast genome is highly conservative and has a much lower mutation rate than that of plants with nuclear genomes. However, the level of polymorphism depends on the taxa or species. Some regions in the chloroplast genome are highly conserved and display an absence of polymorphism. The other regions are more variable. In Angiosperms a small variation between populations was discovered, and no intra-populational variations were found, possibly due to the maternal mode of inheritance [72]. In view of the importance of conifers among particular forest trees, their evolutionary distance and divergence may be determined from chloroplast-specific polymorphic assay.

In all genomes only a very small amount of DNA apparently has coding functions and most of the pine genome is made of repetitive sequences. Changes in the genome size possibly involve a variation in the amount of repeated DNA sequences [6, 22, 24, 97].

There are two hypotheses, which may explain why DNA is larger in conifers than in other plants. One hypothesis is that conifers may have a higher content of repetitive DNA than other plants [19, 43]. The repetitive DNA fraction in conifers is approximately 75%. The other 25% of DNA is equivalent to DNA of plants with much smaller genomes.

An alternative hypothesis is that there must also be an extra amount of the single-copy DNA [43]. This hypothesis is supported by the fact that rRNA genes in conifers are more repetitive than in other plants. Flavell et al. [23] showed in *Triticum dicoccoides*, that rDNA diversities are correlated with climatic variables in Israel. Significant correlations between the *Pinus* genome size and climatic factors (temperature and precipitation) were observed [66, 94, 95]. According to these correlations, it was suggested that the large genome size and its variation in *Pinus* was a response to the habitat conditions of these species. Thus, the model for genome evolution presumes that the plant genome is flexible and has repeatedly undergone amplification and deletion of DNA sequences over evolutionary time. However, the functions of repetative DNA and its role in genome organization are not clear.

MOLECULAR POLYMORPHISM

Genetic diversity may be estimated by using measurable traits, but they are unable to provide information about which particular genes or how many of them are involved in adaptation. Another, and generally complementary, approach for estimating adaptive genetic diversity is by measuring genetic variation using molecular genetic markers. Numerous molecular markers were used in forest trees to understand their genetic structure and evolution, as well as to find changes for evaluation of large numbers of adaptive genes and genetic variations. Pines are diploid organisms with a constant number of haploid chromosomes equal to 11 or 12. That may be the reason why small karyotype differentiation was shown within the genus by cytological studies [70, 78]. Accordingly, the use of genetic markers for the detection of adaptive traits - which are very complex and are controlled by many genes, each with relatively small effect – is acceptable and may be very successful.

The first widely used method of polymorphism detection in pines was allozymes. In most studies allozyme markers have been used as a basic tool for providing genetic information [38]. Of approximately 50 species studied with these markers, at least one estimate of expected heterozygosity was obtained. These markers have provided most of the data on the genetic structure of plants in general, and trees in particular. Unfortunately, the sensitivity of allozymes is about one-fourth of the total number of bases sampled. The benefit is that these markers are codominant. Studies of biochemical markers have shown that forest trees, particularly conifers, are among the most variable organisms known. Therefore, a great interest of ecologists and foresters is associated with studying native variability by molecular markers and understanding mechanisms of genetic differentiation and evolution of genomes.

In angiosperms, the mitochondrial and chloroplast genomes are generally inherited maternally. In conifers the situation is different. According to the genus and the species, the organelle genome can be inherited maternally, paternally or both ways. The plant mitochondrial genome is highly variable in structure but shows a very slow rate of nucleotide substitution, despite its large size – due to the presence of introns, intergenic sequences, duplicate sequences and sequences of plastid and nuclear origin in higher plant mtDNA. mtDNA is as conservative as the chloroplast genome, and as a result only a few mtDNA markers are available and this slows down the evolutionary studies in conifers.

The molecular biological advances are focused on DNA markers, such as random amplified polymorphic DNA (RAPD) [34, 63, 104] and simple sequence repeat (SSR) markers [91, 92]. Molecular markers, such as restriction fragment length polymorphisms (RFLP), are more powerful in revealing genetic polymorphism; however they are technically more demanding and expensive.

RAPD markers are a type of genetic markers which are based on the polymerase chain reaction (PCR) [102]. The markers are inherited in a Mendelian manner and can be successfully generated for any species without prior DNA sequence information. The ability of RAPD primers to produce multiple bands using a single primer means that a relatively small number of primers can be used to generate a very large number of fragments. These fragments are usually generated from different regions of the genome, hence they appear as multiple loci. RAPD markers are especially well suited for indicating genotypic distinctions at the clonal level, since an increasing proportion of the total genome can be sampled by the use of additional primers until genotypic differences are discovered.

The single stranded repeats (SSRs) are DNA sequences that are repeated in a tandem number of times. The larger repeats (up to 5 Mb) are called satellites. Intermediate repeats (the repeated units that are bigger than 10 bp and form blocks of 0.5 to 30 kb) are called minisatellites. Microsatellites have repeated units of 1-8 bp and form structures that have 20 to 100 bp. Microsatellites are co-dominantly inheritable, which allow for the discrimination of homo- and heterozygotic states in diploid organisms during population analysis [45]. The nuclear and chloroplast microsatellites were identified for population analysis in *Pinus contorta* [32], *Pinus radiata* [21, 87], *Pinus strobus* [20], *Pinus silvestris* [41, 88], *Pinus brutia* and *Pinus halepensis* [9, 35].

Lately, the measurement of genetic diversity using single nucleotide polymorphisms - SNPs - has been acquiring much popularity. According to the accepted definition [7], SNP is a single-nucleotide position in genomic DNA which has various variants of sequences (alleles) in some populations, while the rare allele's frequency is at least 1%. Sometimes SNPs with frequency of rare allele higher than 20% are defined as "abundant SNPs". However, quite often, all the small changes in the genome sequences, which were detected during SNP screening, are placed in same databases. Theoretically, the presence of two-, three- and four-allelic polymorphic forms is possible. However, in practice, presence of even three-allele SNPs is extremely rare in genome (less than 0.1% of the whole individuals SNP). Four different types of two-allele SNPs can be found. Practical interest in SNPs had increased during realization of projects, in which the full nucleotide sequences of some organisms were researched. A huge quantity of SNPs in the human genome (about 3-10 million of SNP) allows for a selection of about 100,000 SNP markers [48]. Thus, in every known or assumed gene, there are at least two appropriate markers.

The above methods of molecular polymorphism investigation are not exclusive, because the design and development of new methods in this field are continuing.

DIFFERENCES IN *P.HALEPENSIS* AND *P.BRUTIA* AS EXAMPLES IN THE EVOLUTIONARY HISTORY OF MEDITERRANEAN PINES

Until 1952, *P. halepensis* and *P. brutia* were thought to be two varieties of a single species (i.e. *P. halepensis* var. *halepensis,* and *P. halepensis* var. *brutia*). By analyzing morphological parameters in seedlings of *P. halepensis*, *P. brutia*, *P. eldarica*, *P. stankeweczii* and *P. pithyusa*, Debazae et al. [18] concluded that *P. halepensis* and *P. brutia* are separate species, with three relict pine taxa, namely *eldarica, pithyusa* and *stankeweczii,* being subspecies of *P. brutia*. There are significant differences between *P. brutia* and *P. halepensis* trees in their morphological characteristics, i.e. bark color, needle length, width and cross-section structure, cone structure and 1,000-seed weight [84, 93). By using data obtained from the physical and chemical analysis of the resin turpentine as genetic markers, Mirov et al. [53] concluded that *P. halepensis* and *P. brutia* are two distinct species. According to Conkle et al. [14] allozymes indicate a highly significant divergence between *P. brutia* Ten. and *P. halepensis* Mill.

A number of differences were also revealed in phenological trials (i.e. periodical phenomena of flowering and vegetative growth) of *P. halepensis* and related species, *P. brutia* and *P. eldarica* [100]. In the phenological stages, which were observed during an annual cycle, changes in the development of reproductive and vegetative organs were found. The phenological phenomena, being subject to natural selection pressures such as the timing of frosts and droughts and insect behavior, reflect an adaptive response to environmental heterogeneity [10]. A different approach to distinguishing between these two species was based on karyotype analysis. According to Saylor [79] and Kammacher [33] significant differences occur mainly in the 11th and 12th chromosomes.

The relationships between *P. halepensis* and other Mediterranean pines are still not evident, the origin and the migration pathways and the past areas of distribution are unclear. On the basis of palaeobotanical data, Nahal [58] concluded that in the Tertiary *P. halepensis* was growing in what is now known as the Baltic Sea area. Panetsos [68] postulated the existence of two centers of distribution of *P. halepensis* and *P. brutia*: the former in northwestern Europe, and the latter in Eastern Europe. Changes in the climate forced a southward shift of these two species and, according to their physiological-ecological characteristics, *P. halepensis* established itself mainly in the western Mediterranean, and *P. brutia* in the eastern Mediterranean region. The eastern group is represented by the Israeli composite sample, and probably includes the native Jordanian and Lebanese-Syrian provenances. There are distinct additional alleles, which are widespread in Israel and rarely or even not detected in other *P. halepensis* or *P. brutia* populations [80]. Pines of the *P. brutia* group are polymorphic and are presumed to have more variation than *P. halepensis*. The allozyme similarities of *Pinus eldarica* (Medw.) to the easternmost *P. brutia* subspecie *brutia* (Nahal) population and its reduced diversity provide evidence of its derivation from subspecie *brutia* [14].

The evolutionary history of *P. halepensis*, reconstructed from allozyme evidence, indicated that the center of origin included the regions bordering the Black Sea and easternmost Anatolia, with eastward extensions into the lands between the Black and Caspian Seas. Early populations may have been more widespread, larger in size, and more closely adjacent. *P. brutia* is a characteristic species of the eastern Mediterranean, whereas *P. halepensis* generally occupies the western and middle Mediterranean - except for local occurrence in the southern parts of the eastern basin. *Brutia,* the modern subspecies around the Aegean and Mediterranean Seas (western population), is a widespread subspecies that maintains significant levels of allozyme variation throughout its geographic distribution. The eastern population of *brutia* subspecies is now geographically isolated from the main distribution. Several of the eastern populations resemble *stankeweczii* (Suk.), *pithyusa* (Steven), and *eldarica* (Medw.) subspecies in possessing rare alleles and in having allele frequencies, which distinguish them from western populations of *brutia* subspecies. The morphological differentiation among subspecies *brutia* was sufficient to assign species status to them, but enzyme allele frequencies of these subspecies closely resemble those of subspecies *brutia*. Analysis of chloroplast microsatellites in natural populations of *P. halepensis* and *P. brutia* identified "species-specific" haplotypes in the two species [9]. At species level, two species are clearly divided into two main clusters with further subdivision, showing genetic divergence among populations of *P. halepensis* and *P. brutia*.

The research of sequence divergence of chloroplast rbcL. MatK trnV intron and rpl120-rps18 spacer regions among 32 *Pinus* species and members of six other *Pinacea* genera showed that within the Mediterranean pine clade, *P. halepensis* and *P. brutia* formed a highly supported group of Wang *et al.* [98]. A clear resemblance in their seed protein profiles [85] and allozyme pattern [14], and their ability to hybridize in nature [69] - indicate a close relationship between the two. *Pinus brutia* is even described as a variety of *P. halepensis* by some authors [84, 100]. Allozyme [14] and morphology [25] studies have suggested that *P. halepensis* is derived from a *P. brutia*-like ancestor and that *P. brutia* has retained greater ancestral variation, showing affinities not only to *P. halepesis*, but also to other Mediterranean pines, e.g. *P. pinaster* and *P. canariensis* [25]. *Pinus pinaster*, *P. pinea* and *P. canariensis* formed one group, albeit with weak (<50%) bootstrap support. Many authors consider *Pinus pinea* as an enigmatic and isolated species [36, 53]. Traditionally, *P. pinea* is placed in a monotypic subsection *Pinea* [50]. However, Wang et al. [98] did not observe such a distinct separation of *P. pinea* from other Mediterranean pines. Klaus [36] noted that *P. pinea*, *P. pinaster* and *P. canariensis* share many cone and vegetative characters. Frankis [25] combined *P. pinaster*, *P. canariensis*, *P. halepensis* and *P. brutia* into one subsection, *Pinaster*, but both authors still placed *P. pinea* in a separate subsection. The results [98] lend additional support to the grouping of these species into one subsection, *Pinaster*, as suggested by Frankis [25], but indicate that *P. pinea* may also belong to this subsection.

GENETIC ALTERATIONS AS ADAPTATION TO ENVIRONMENTAL CHANGES

Genetic diversity is the basis for the ability of organisms to adapt to changes in their environment through natural selection. During their evolutionary and ecological histories, forest tree species have experienced numerous environmental changes. Shifts in climate could take place for a thousand years (glacial epochs), or last for the lifetime of an individual– a couple of decades. Environmental changes could be smooth, or it could be rapid climatic shifts taking place for only a few years. For the past few years much attention has been focused on the research of genetical diversity of tree species due to the environmental changes related to human activity [2, 16]. Simultaneously, both the degree and the rate of climatic changes are of relation to the future of tree species. Firstly, the rate of environmental changes could be too rapid for trees with their long generation time to adapt to it [17]. Secondly, there are many concerns that the extension of environmental changes could be much bigger than the genetic adaptation abilities of the plants [16]. Thirdly - and finally - even if trees could have proper genetic diversity for an adequate response to the rate and strength of climatic changes, it is possible that they might not be capable of spreading out into newly available habitats quickly enough to match the rate of environmental change [12].

Aleppo pine (*Pinus halepensis* Mill) is due to one of the most widespread forest tree species in the Mediterranean its optimum adaptation to the microclimates of the area, i.e. its high tolerance to drought and adverse ecological conditions [59, 99]. This species has some features of its reproductive cycle and the production of serotinous cones, which are important for natural distribution and regeneration following fire passage [46, 77]. Its area of growth stretches from Morocco (longitude about 9°W) to Jordan (longitude 36°E) and widens from France (latitude about 45°N) to Israel (31°30'N) [15, 84]. Within the Mediterranean Basin alone, natural forests and forest plantation of *P. halepensis* (Mill) cover over 3.5 millions ha. [29, 58, 90]. However, the main area of *P. halepensis* distribution is southern Europe and North Africa.

Natural populations of *Pinus halepensis* have high survival capacity and occupy mountain sites as well as low-altitude sites, reaching elevations up to 2,100 m [68]. Individual populations grow in deserts with a minimal amount of annual rainfall. However, natural populations showed a high degree of similarity based on results of researching the characteristics of seeds, seedlings, needles and cones [51, 52]. Trees are subjected to more variation due to drought than any other plants, and the effect is often more devastating [28]. Tree diameter growth [42], height growth [106], and the number of shoot flushes produced in a growing season [105] can be substantially reduced by drought. Drought stress is a common cause of seedling mortality in both naturally regenerated and planted stands of *Pinus* species. Williston [103] noted that over a 16-year period, 57% of the first year mortality in pine plantations was due to drought stress. In established stands, drought stress accounts for 80% of the variation in annual ring width of conifers in humid temperate climates [107], and up to 90% in semi-arid regions [26]. The ability to survive drought stress has been shown to be variable among geographic seed sources of *Pinus* [101]. Various authors have suggested adaptation to local environmental conditions as an explanation for differences in allele frequencies in forest tree species covering large geographic distances. In other words, genetic differentiation in species is suggested to be a basis of micro geographic adaptation. The survival and adaptive possibility of *Pinus halepensis* in Israel supposedly led to formation of new provenances adapted to local environmental conditions [31, 39, 40, 65]. Unfortunately, the short distance between separate provenances of *Pinus halepensis* may lead to an appearance of mixed stands due to "introversive hybridization". The creation of intermediate morphologic forms is an indirect evidence of natural hybridization. Also, an indirect evidence of natural hybridization in populations of *P. halepensis* and *P. brutia* is based on morphological traits [68], allozymes [37] and microsatellites [9].

Within the Mediterranean Basin *Pinus halepensis* grows under various ecological conditions, which are the result of topography, climate and human activity. These conditions have always played a significant role in moving the geographic borders of this species. Therefore, there is a need to be very cautious in the interpretation of genetic diversity data in the geographical context.

ENVIRONMENTAL EFFECTS ON ADAPTIVE GENETIC VARIATION IN *PINUS HALEPENSIS*

Genetic diversity in nature is the result of evolutionary processes, and it is apparent within species at different levels in both the enzymes and the DNA [64, 65]. Nevo, in his "Evolution Canyon" model, showed correlations between environmental conditions and diversity of plant species. This parallelism between diversity and microhabitats suggests that genetic and physiological diversity should be represented as arising from a complex of adaptive factors, related to environmental heterogeneity. Natural climatic selection and stress appear to be the major differentiating factors. Apparently, the proportion of heterozygotes in a population is related to its adaptive effectiveness; accordingly, stressful conditions may change the adaptive capability of plant populations [31, 49]. Furnier and Adams [27] found correlation between allele frequencies and adaptation to ultramafic soils, and Guries and Ledig [30] established correlation between allele frequencies and climatic variables, such as winter temperature. In the Swiss sub-alpine stands of *Picea abies* and *Fagus sylvatica*, Müller-Starck [57] found relatively large intra-populational and average inter-populational genetic variation in comparison with reference populations in Europe. According to Müller-Stark [56, 57] and Bergmann and Ruetz [5] levels of genetic diversity in stressed plantations of forest tree species are similar or higher than those in favorable natural populations. Similarly, the numbers of heterozygous phenotypes present in natural populations were found to be considerably lower than those in the offspring stressed populations in Israel [39, 40]. Climatic natural selection through water deficit and existence of significant selective factors in populations of *P. halepensis* in Spain were found when phenotypic traits and molecular markers were compared [1]. Heterozygous individuals are believed to be more stable than their homozygous counterparts because of some inherently superior biochemical efficiency possessed by heterozygotes [47].

Allozymes and RAPD markers were used to analyze the genetic variability among natural *Pinus halepensis* (Mill.) populations, and between their offspring populations, and to relate the intra-specific variability to the ecological differences among the sites. Both methods showed an increase of heterozygosity in populations growing in new environments. Most of the trees in provenances that survived under the stressful ecological conditions (an average of 44% of the number planted) were heterozygous (Fig. 1). Also, the fixation indices were negative at most of the enzyme loci, i.e. they indicated an excess of heterozygotes in

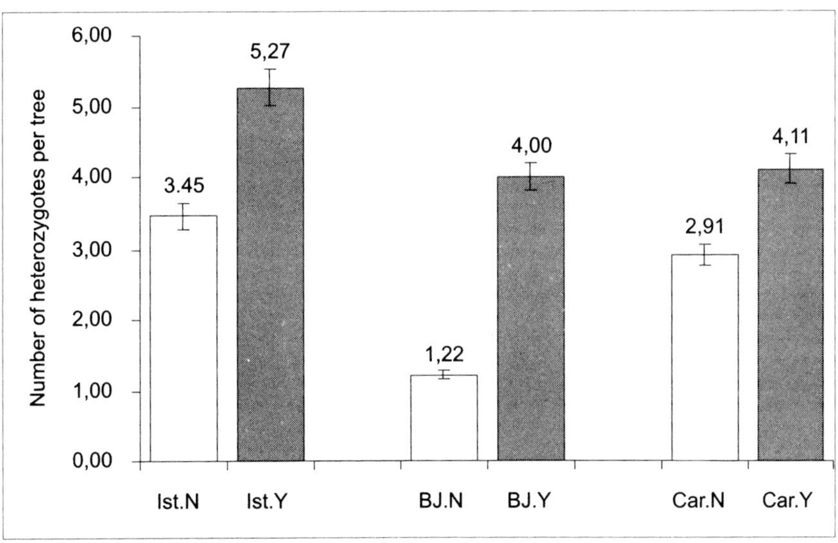

Fig. 1. *Alteration of heterozygote number in main provenances of P. halepensis.*
N-number of heterozygote in natural populations
Y-number of heterozygote in offspring populations in the stressful site
Abbreviations and population names
BJ - Bet J'ann –provenance (Israel), Car - Carmel provenance (Israel), Ist – provenance Istiaia (Greece)

the F1 offspring populations. These results gave rise to the hypothesis that as a population adapts to new climatic conditions, an increase of genetic variability may be expected and the number of heterozygotes would correlate with environmental features. We consider that the correlation between gene diversity and survival rates, with negative correlation coefficients, suggests a model of heterozygosity in which heterozygous individuals are developmentally more stable than their homozygous counterparts [47, 89]. All these indicate that at least some of the inferred changes may be reflected in selective responses to environmental conditions that are also reflected in the low survival rate 15 years after planting.

Further natural selection should promote divergence within provenances at a micro-site; therefore, theoretically it seems reasonable to suppose that micro-site ecological-genetic differences can promote a tendency towards site-specific differentiation for stress-resistance traits. It is obvious that population size, history, and past and present genetic flow values are important in the determination of genetic heterogeneity and structure within and between plant populations. Nevertheless, adaptive selection is receiving increasing attention: non-selective processes, which cause loss of allele heterozygosity, probably influence all components of population adaptation.

Diversity Between Circum-Mediterranean

By analyzing levels of within-population genetic diversity in natural populations of *P. halepensis*, published by Korol et al. [40], we have compared our results with the results of the distribution of genetic variability in the same populations measured by nine cpSSR [9] Fig. 2. In all probability, the correlation ($r = 0.804$, $P = 0.046$) between data obtained by different methods could represent an indirect confirmation of our supposition on the evolutionary history of this species in the Mediterranean Basin.

The analysis of the population diversity in Mediterranean *Pinus halepensis* revealed two main geographic aggregations that were very similar to the groups published in the earlier study by Schiller et al. [81] and Bucci et al. [9]. We have compared different types of the dendrograms constructed by methods of the nearest neighbor and UPGMA, on the basis of Nei's genetic distances [61, 62], Cavalli-Sforza's chord measure [11] and Reynolds, Weir, and Cockerham's [75] genetic

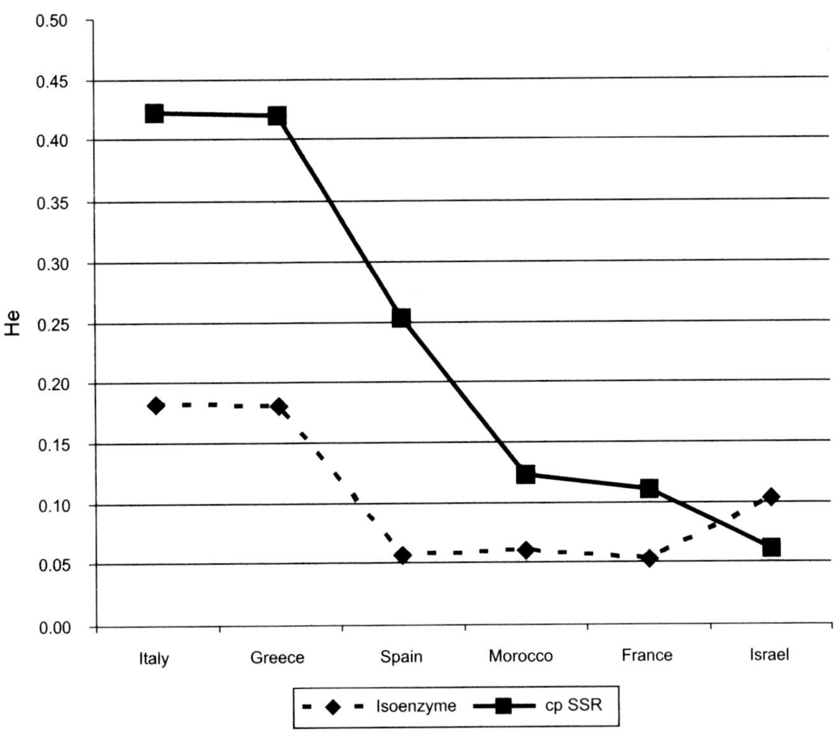

Fig. 2. *Levels of within-population genetic diversity (He) in natural populations of* Pinus halepensis *measured by two types of molecular markers.*

distances, and have found that as a rule, populations of the same region are grouped together, regardless of the method of genetic distance measure used.

In Fig. 3 a dendrogram of *Pinus halepensis* is shown, which is created on the basis of Cavalli-Sforza's chord distance. We chose Cavalli-Sforza's chord distances, whose main advantage compared to other genetic distances is that its estimations do not depend on variations of population sizes during the time pace, and are very sensitive even to weak genetic drift in population. This is a very important advantage, for populations always vary in size.

The results of the analysis showed that all natural Aleppo pine populations in eastern Mediterranean, i.e. Turkey, Lebanon, Syria, Israel and Jordan, belong to the same group.

The eastern Mediterranean group differs from the western Mediterranean group mainly by the frequencies in the Cat2 and the Aap

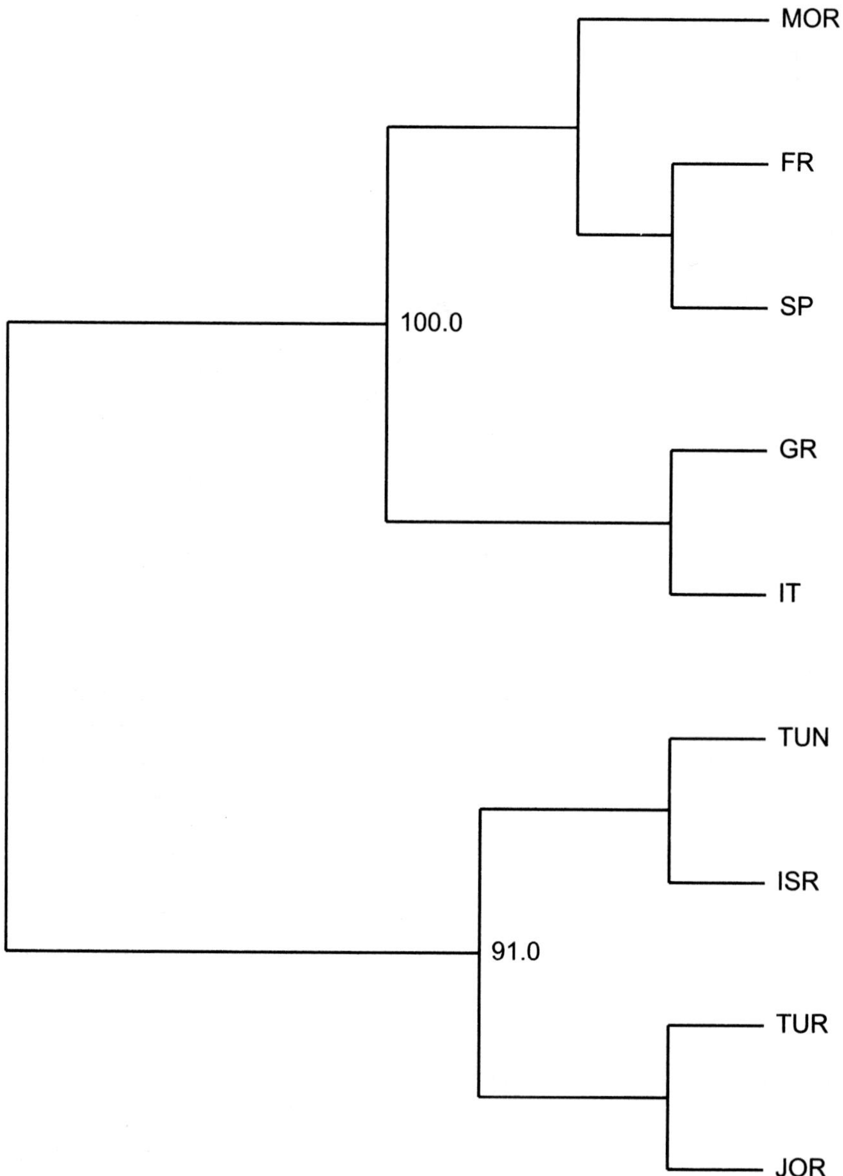

Fig. 3. *UPGMA dendrogram of Pinus halepensis populations from different regions of Mediterranean Basin, created on the basis of Cavalli-Sforza's chord distance. Numbers on branches leading to the main clusters are bootstrap values in percentage (of 100 replicates).*

Abbreviations and population names

MOR – Morocco, FR – France, SP – Spain, GR – Greece, IT – Italy, TUN – Tunisia, ISR – Israel, TUR – Turkey, JOR – Jordan

loci. This divergence was also supported by the results of research by chloroplast microsatellites [9]. Similarity of allele frequencies in AAP and Cat2 loci points to a connection between Tunisian populations and eastern Mediterranean group, which strengthens former conclusions about the migration path of the Aleppo pine into the African continent. Native populations of *P. halepensis* from Tunisia have differences in genetic variability from the other populations forming a particular cluster in the dendrogram. A split among the Tunisian populations was also evident from the results presented in previous studies by Schiller and Grunwald [82] and Baradat et al. [3] using resin monoterpene composition as a genetic marker.

Western-Mediterranean group contains two subgroups: stands from Morocco were very similar to the southern French and Spanish stands on the one hand, and to the group comprising Greek and Italian stands of *P. halepensis* on the other hand. According to Schiller and Mendel [83], the Albanian, Greek and Italian peninsula Aleppo pine populations were considered as the eastern-European subgroup of the western-Mediterranean group, including an introgression from *Pinus brutia* [81]. Partially, evidence may be found in the work of Bucci et al. [9], where a possibility of hybridization between halepensis-complex pine species was proved by using microsatellites, although the authors themselves were supporting the hypothesis of Nahal [58], that genetic similarity of Greek and southern Italian populations is based on the migration of a limited number of individuals (founder effect) and regulation of population size by fire (bottlenecks).

Genetic similarity between Greek and Italian populations, having the highest genetic diversity in the whole range of this species may be represented as an importation of individuals by humans [74]. Schiller and Mendel [83] suggested that Aleppo pine in the Balkan Peninsula is "a direct descendant of Tertiary Aleppo pine in central Europe, which migrated into the Balkan Peninsula due to climate changes, keeping its relative high heterozygosity", which might be more believable. Such a hypothesis is supported by the findings of Morgante et al. [55], who categorically rejected the idea of introgression because of their interpretation of chloroplast microsatellites analysis results.

CONCLUSION

Tree species have passed through large-scale global environmental changes many times during their evolutionary history. It is most likely that although these changes occurred very quickly, many of the tree populations have survived. There are at least two main differences between the conditions today and during these historical events. Firstly, we currently experience quick global climatic changes during a period when many landscapes are very fragmented. As a result, secure sites for colonization may be so dispersed, that species might be unable to extend their dispersion in response to global ecological changes.

Secondly, however major areas of agricultural grounds were colonized by trees, following the abandonment of agriculture, in a very fast tempo. Also, it is likely that the ranges of commercially valuable species will be extended artificially into previously unoccupied areas. The establishment of plantations may expedite the colonization of adjacent habitats by the cultivated species and may also provide a suitably large target for non-commercial tree species to become established.

Moreover, as a response to environmental changes, some of tree forest species may acquire new traits. These traits may allow many tree species to survive predicted global climatic changes while preserving much of their genetic diversity and advancing their evolutional path in the new habitat.

Acknowledgement

The authors express deep gratitude to Dr A. Weinstein for his helpful suggestions on the manuscript, and to Dr Darina Korol for assistance in preparation of the manuscript.

[1] Alia R, Gomez A, Agundez AD, et al. Levels of genetic differentiation in *Pinus halepensis* Mill. In Spain using quantitative traits, isozymes, RAPDs and cp-microsatellites. In: Uller-Stark G, Schubert T, eds. Genetic Response of Forest Systems to Changing Environmental Conditions. Boston/London: Kluwer Acad Pub Doedrecht, 2002; 70: 151-160.

[2] Alig RJ, Adams DM, McCarl BA. Projecting impacts of global climate change on the US forest and agricultural sectors and carbon budgets. For Ecol Manage 2002; 169: 3-14.

[3] Baradat Ph, Lambardi M, Michelozzi M. Terpene composition in four Italian provenances of Aleppo pine (*Pinus halepensis* Mill.). Journal of Genetics and Breeding 1989; 43: 195-200.

[4] Bennett HJ, Smith JB. Nuclear DNA amounts in angiosperms. London: Phil Trans R Soc 1976; 274: 227-232.

[5] Bergmann F, Ruetz W. Isozyme genetic variation and heterozygosity in random tree samples and selected orchard clones of the same Norway spruce populations. For Ecol Manage 1991; 46: 39-47.

[6] Britten RJ, Kohne DE. Repeated sequences in DNA. Science 1968; 161: 529-543.

[7] Brookes A.J. The essence of SNP. Gene 1999; 234: 177-186.

[8] Brown GR, Kadel III EE, Bassoni DL, et al. Anchored reference loci in loblolly pine *Pinus taeda* L. for integrating pine genomics. Genetics 2001; 159: 799-809.

[9] Bucci G, Anzidei M, Madachiele A, Vandramin GG. Detection of haplotypic variation and natural hybridization in halepensis-complex pine species using chloroplast simple sequence repeat (SSR) markers. Mol Ecol 1998; 7: 1633-1643.

[10] Campbell RK. Genecology of Douglas-fir in watershed in the Oregon Cascades. Ecology 1979; 60: 1036-1050.

[11] Cavalli-Sforza LL, Edwards AWF. Phylogenetic analysis: models and estimation procedures. Evolution 1967; 32: 550-570.

[12] Clark JS. Why trees migrate so fast: confronting theory with dispersal biology and the paleorecord. Am Nat 1998; 152: 204-224.

[13] Conkle MT. Isozyme variation and linkage in six conifer species. In: Conkle MT, ed. Proc Symp Isozyme of North Am. Forest Trees and Insects. Berkeley/California: USDA Forest Service Gen Tech Rep PSW 1981; 48: 89-96.

[14] Conkle MT, Schiller G, Grunwald C. Electrophoretic analysis of diversity and phylogeny of *Pinus brutia* and closely related taxa. Syst Bot 1988; 13: 411-424.

[15] Critchfield WB, Little EL. Geographic Distribution of the Pines in the World. USDA Forest Service, Miscellaneous Publication No. 991, 1996.

[16] Davis MB, Zabinski C. Changes in geographical range resulting from greenhouse warming: effects on biodiversity in forests. In: Peters RL, Lovejoy TE, eds. Global Warming and Biological Diversity. Yale University Press: 1992. 297-308.

[17] Davis MB, Shaw RG. Range shifts and adaptive responses to Quaternary climate change. Science 2001; 292: 673-679.

[18] Debazac EF, Tomassone E. Contribution a une etude comparee des Pins Mediteraneens de la section Halepensis. Annals Science Forestier Nancy 1965; 22: 213-256.

[19] Dhillon SS. DNA in tree species. In: Bonga JM, Durza DY, eds. Cell and tissue culture in forestry. Martinus Nijhoff Publish 1987; 1: 298-333.

[20] Echt CS, May-Marquardt P, Hseih M, Zahorchak R. Characterization of microsatellite markers in eastern white pine. Genome 1996; 39: 1102-1108.

[21] Fisher PJ, Gardner RC, Richardson TE. Single locus microsatellites isolated using 5' anchored PCR. Nucl Acids Res 1996; 24: 4369-4372.

[22] Flavell RB, Bennett MD, Smith JB, Smith DB. Genome size and proportion of repeated nucleotide sequence DNA in plants. Bioch. Genet 1974; 12: 257-269.

[23] Flavell RB, O'Dell M, Sharp P, et al. Variation in the intergenic spacer of ribosomal DNA of wild wheat, *Triticum dicoccoides*, in Israel. Mol Biol Evol 1986; 3 (6): 547-558.

[24] Flavell RB. Repetitive DNA and chromosome evolution in plants. London: Phil Trans R Soc B 1986; 312: 227-242.

[25] Frankis MP. Morphology and affinities of *Pinus brutia*. In Intl Symp on *Pinus brutia* Ten. Ankara: Ministry of Forestry: 1993; 1-18.

[26] Fritts HC, Smith DG, Stokes MA. The biological model for the paleoclimatic interpretation of Mesa Verde tree-ring series. Amer Antiq 1965; 31: 101-121.

[27] Furnier GR, Adams WT. Geographic patterns of allozyme variation in Jeffery pine. Amer J Bot 1986; 73: 1009-1015.

[28] Gaertner EE. Water relations of forest trees. In: Rutter AJ, ed. The Water Relations of Plants. Oxford, London, Edinburgh, Boston and Melbourne. Brit Ecol Soc Symp Blackwell, 1964, 366-378.

[29] Gomez A, Aravanopulos FA, Bueno MA, Alia R. Linkage of Random Amplified Polymorphic DNA Markers in *Pinus halepensis* Mill. Silvae Genetica 2002; 51: 196-201.

[30] Guries RP, Ledig FT. Genetic structure of populations and differentiation in forest trees. In: Conkle MT, ed. Pro Symp Isozymes of North Am. Forest Trees and Forest Insects. USDA Forest Service Gen Tech Rep PSW 48 1981; 48: 42-48.

[31] Hamrick JL, Godt MJW. Allozyme diversity in plant species. In: Brown AHD, Clegg MT, Kahler AL, Weir BS, eds. Plant population genetics, breeding and genetic resources. Sunderland/Massachusetts: Sinauer Associates Inc. 1989: 43-63.

[32] Hicks M, Adams D, O'Keepe S, et al. The development of RAPD and microsatellite markers in lodgepole pine (*Pinus contoria* var. latifolia). Genome 1998; 41: 797-805.

[33] Kammacher P, Zygomala AM. Demonstration of divergence in chromosomal structures between related genomes of *Pinus*. Comptes Rendus de Academie des Sciences. Paris: 1989; 308: 27-30.

[34] Kaya Z, Neale DB. Utility of random amplified polymorphic DNA (RAPD) markers for linkage mapping in Turkish Red pine (*Pinus brutia* Ten.). Silvae Genetica 1995; 44: 110-116.

[35] Keys RN, Autino A, Edwards KJ, et al. Characterization of nuclear microsatellites in *Pinus halepensis* Mill and their inheritance in *Pinus halepensis* and *Pinus brutia* Ten. Mol Ecol 2000; 9: 2157-2159.

[36] Klaus W. Mediterranean pines and their history. Pl Sys Evol 1989; 162: 133-163.

[37] Korol L, Madmony A, Riov Y, Schiller G. *Pinus halepensis* X *Pinus brutia* subsp. brutia hybrids? Identification using morphological and biochemical traits. Silvae Genetica 1995; 44: 186-190.

[38] Korol L, Gil S, Climent MJ, et al. Canary Islands pine (*Pinus canariensis* CHR.SM.EX.DC.) 2. Gene flow among native populations. Forest Genetics 1999; 6: 277-282.

[39] Korol L, Shklar G, Schiller G. Site influences on the genetic variation and structure of *Pinus halepensis* Mill. Provenances. Forest Genet 2001; 8: 295-305.

[40] Korol L, Shklar G, Schiller G. Genetic variation within *Pinus halepensis* Mill. provenances growing in different microenvironments in Israel. Israel J Plant Sci 2002; 50: 135-143.

[41] Kostia S, Varvio S, Vakkari P, Pulkkimen P. Microsatellite sequences in a conifer, *Pinus sylvestris*. Genome 1995; 38: 1244-1248.

[42] Kozlowski TT. Water relations and growth of trees. J For 1958; 56: 498-502.

[43] Kribbel HB. DNA sequence components of the *Pinus strobus* nuclear genome. Can J For Res 1985; 15: 1-4.

[44] Krupkin AB, Liston A, Strauss SH. Phylogenetic analysis of hard pines: Pinus subgenus Pinus, Pinaceae from chloroplast DNA restriction site analysis. Amer J Bot 1996; 83: 489-498.

[45] Lefort F, Echt C, Streiff R, Vendramin GG. Microsatellite sequences: a new generation of molecular markers for forest genetics. Forest Genet 1999; 6: 15-20.

[46] Leone V, Borghetti M, Saracino A. Ecology of post-fire recovery in *Pinus halepensis* in southern Italy. In: Trabaud L, ed. Life and environment on the Mediterranean. Southampton: Advances in Ecological Sciences Series vol 3 WIT press, 2000: 129-154.

[47] Lerner IM. Genetic homeostasis. New York: Wiley, 1954.

[48] Li J, Butler JM, Tan J, et al. Single nucleotide polymorphism determination using primer extension and time off mass spectrometry. Electrophoresis 1999; 20: 1258-1265.

[49] Linhart YB, Mitton JB, Sturgeon KB, Davis ML. Genetic variation in space and time in a population of ponderosa pine. Heredity 1981; 46: 407-426.

[50] Little EL, Critchfield WB. Subdivisions of the genus Pinus (pines). Washington DC: USDA Forest Service, 1969; 1144.

[51] Melzack RN, Crunwald C, Schiller G. Morphological variation on Aleppo pine (*Pinus halepensis* Mill.) in Israel. J Bot 1981; 30: 199-205.

[52] Melzack RN, Schiller G, Crunwald C. Seed size, germination and seedling growth in *Pinus halepensis* Mill. and their relation to provenances in Israel. a Leafl Dept For Agric Res Organ, Ilanot 72 1982.

[53] Mirov NT, Zavarin E, Snajberk K. Chemical composition of the turpentines of some eastern Mediterranean pines in relation to their classification. Phytochemistry 1966; 5: 97-102.

[54] Mirov NT. The genus *Pinus*. New York: Ronald Press 1967.

[55] Morgante M, Felice N, Vedramin GG. Analysis of hypervariable chloroplast microsatellites in *Pinus halepensis* reveals a dramatic genetic bottleneck. In: Karp A, Isaac GP, Ingram DS, eds. Molecular Tools for Screening Biodiversity. London: Chapman and Hall, 1998; 407-412.

[56] Mueller-Starck G. Genetic differentiation among samples from provenances of *Pinus sylvestris*. Silvae Genet 1987; 39: 40-42.

[57] Mueller-Starck G. Genetic variation under extreme environmental conditions. In: Baradat Ph, Adams WT, Müller-Starck G, eds. Population Genetics and Genetic Conservation of Forest Trees. Amsterdam: SPB Acad Pub 1995: 201-210.

[58] Nahal I. Le pin D'Alep (*Pinus halepensis* Mill.): Etude taxonomique, phytogeographique ecologique et sylvicol. Annales de l'Ecole National des Eaux et Forets, Nancy 1962; 19: 479-687.

[59] Nahal I. The Mediterranean climate from a biological viewpoint. In: Casrti FD, ed. Ecosystems of the Wold 1981; 3: 63-86.

[60] Neale DB, Williams JGK. Restriction fragment length polymorphism mapping in conifers and applications to forest genetics and tree improvement. Can J For Res 1991; 21: 545-553.

[61] Nei M. Genetic distance between populations. American Naturalist 1972; 106: 283-292.

[62] Nei M. Molecular Evolutionary Genetics. New York: Columbia University Press, 1987.

[63] Nelson CD, Nance WL, Doudrick RL. A partial genetic linkage map of slash pine (*Pinus elliottii* Engelm. Var. *elliottii*) based on random amplified polymorphic DNAs. Theor Appl Genet 1993; 87: 145-151.

[64] Nevo E. Evolution in action across phylogeny caused by microclimatic stress at "Evolution Canyon". Theor Pop Biol 1997; 52: 231-243.

[65] Nevo E. Molecular evolution and ecological stress at global, regional and local scales: The Israeli perspective. J Exper Zool 1998; 282: 95-119.

[66] Newton RJ, Wakamiya I, Price HJ. Genome size and environmental stress in gymnosperm crops. In: Pessarakli M, ed. Handbook of Crop Stress. Marcell Dekker Inc., 1993; 321-345.

[67] Ohri D, Khoshoo TN. Genome size in gymnosperms. Pl Sys Evol 1986; 153: 119-132.

[68] Panetsos CP. Monograph of *Pinus halepensis* (Mill) and *P. brutia* (Ten). Annales Forestales, Zagreb 1981; 9: 39-77.

[69] Panetsos K, Scaltsoyiannes A, Aravanopoulos FA, et al. Identification of *Pinus brutia* Ten., *P halepensis* Mill. and their putative hybrids. Silvae Genetica 1997; 46: 253-257.

[70] Pederick LA. Chromosome relationships between *Pinus* species. Silvae Genet 1970; 19: 171-180.

[71] Plamer J. In: Bogorad L, Vasil IK, eds. The Molecular Biology of Plastids. San Diego: Academic, 1991: 5-53.

[72] Powell W, Morgante M, MnDevitt R, et al. Polymorphic simple sequence repeat regions in chloroplast genomes: Applications to the population genetics of pines. Proc Natl Acad Sci USA 1995; 92: 7759-7763.

[73] Price RA, Liston A, Strauss SH. Phylogeny and systematics of Pinus. In: Richardson DM, ed. Ecology and Biogeography of *Pinus*. Cambridge: Cambridge University Press, 1998: 49-68.

[74] Puglisi S, Lovreglio R, Attolico M, Leone V. Allozyme variation within and between nine Italian populations of Aleppo Pine (*Pinus halepensis* Mill). Forest Genetics 2002; 9: 87-102.

[75] Reynolds JB, Weir BS, Cockerham CC. Estimation of the coancestry coefficient: basis for a short-term genetic distance. Genetics 1983; 105: 767-779.

[76] Richardson DM, Rundel PW. Ecology and biogeography of *Pinus:* and introduction. In: Richardson DM, ed. Ecology and Biogeography of *Pinus*. Cambridge: Cambridge University Press, 1998: 3-46.

[77] Saracino A, Pacella R, Leone V, Borgheti M. Seed Dispersal and changing seed characteristics in a *Pinus halepensis* Mill. Forest after fire. Plant Ecology 1997; 130: 13-19.

[78] Sax K, Sax HJ. Chromosome number and morphology in conifers. J Arnold Arbor 1933; 14: 356-375.

[79] Saylor LG. Karyotype analysis of Pinus–group Lariciones. Silvae Genetica 1964; 13: 165-170.

[80] Schiller G, Conkle MT, Grunwald C. Local differentiation among Mediterranean populations of Aleppo pine in their isoenzymes. Silvae Genetica 1986; 35: 11-19.

[81] Schiller G, Grunwald C. Xylem resin composition of *Pinus halepensis* Mill. in Israel. Isr J Bot 1986; 35: 23-33.

[82] Schiller G, Grunwald C. Resin monoterpene in range-wide provenance trials of *Pinus halepensis* Mill in Israel. Silvae Genetica 1987; 36: 109-114.

[83] Schiller G, Mendel Z. Is the overlap of ranges of Allepo pine and *Brutia* pine in the east Mediterranean natural or due to human activity. In: Baradat Ph, Adams WT, Muller-Starck, eds. Population Genetics and Genetic Conservation of Forest Trees. Academic Publishing; 1995: 159-163.

[84] Schiller G. Inter-and intra-specific genetic diversity of *Pinus halepensis* Mill. and P. brutia Ten. In: Ne'eman G, Trabaud L. eds. Ecology, Biogeography and Management of *Pinus halepensis* and *P. brutia* Forest Ecosystems in the Mediterranean Basin. Leiden: Backhuys Publishers, 2000: 13-35.

[85] Schirone BG, Piovesan R, Bellarosa, Pelosi C. A taxonomic analysis of seed proteins in *Pinus* spp. *Pinaceae*. Plant Systematics and Evolution 1991; 178: 43-53.

[86] Sewell MM, Sherman BK, Neale DB. A consensus map for loblolly pine *Pinus taeda* L. I. Construction and integration of individual linkage maps from two outbreed three-generation pedigree. Genetics 1999; 151: 321-330.

[87] Smith DN, Devey ME. Occurrence and inheritance of microsatellites in *Pinus radiata*. Genome 1994; 37: 977-983.

[88] Soranzo N, Provan J, Powell W. Characterization of microsatellite loci in *Pinus sylvestris*. Molecular Ecology 1998; 7: 260-1261.

[89] Soule ME. Heterozygosity and developmental stability-another look. Evolution 1979; 33: 396-401.

[90] Teisseire H, Fady B, Pichot C. Allozyme variation in five French populations of Aleppo pine (*Pinus halepensis* Mill.). Forest Genetics 1995; 2: 225-236.

[91] Tsumura T, Yoshimura K, Tomaru N, Ohba K. Molecular phylogeny of conifers using RFLP analysis of PCR-amplified specific chloroplast genes. Theor Appl Genet 1995; 91: 1222-1236.

[92] Vendramin GG, Lelli L, Rossi P, Morgante M. A set of primers for the amplification of 20 chloroplast microsatellites in *Pinaceae*. Mol Eco 1996; 5: 595-598.

[93] Vidakovic M. Conifers: morphology and variation. Zagreb: Graficki Zavod Hrvatske, 1991.

[94] Wakamiya I, Newton RJ, Johnston JS, Price HJ. Genome size and environmental factors in the genus *Pinus*. Amer J Bot 1993; 80: 1235-1241.

[95] Wakamiya I, Price HJ, Messina MG, Newton RJ. Pine genome size diversity and water relations. Physiol Plant 1996; 96: 13-20.

[96] Wakasugi T, Tsudzuki J, Ito S, et al. Loss of all *ndh* genes as determined by sequencing the entire chloroplast genome of the pine *Pinus thunbergii*. Proc Natl Acad Sci USA 1994; 91: 9704-9798.

[97] Walbot V. Genome organization in plants. In: Rubenstein I, Phillips RL, Green RL, Gengenbach BG, eds. Molecular Biology of Plants. New York: Academic Press, 1979: 31-72.

[98] Wang X, Tsumura Y, Yoshimaru H, et al. Phylogenetic relationships of Eurasian pines (*Pinus*, *Pinaceae*) based on chloroplast RBCL, matK, rpl20-rps18 spacer, and trnV intron sequences. Amer Bot 1999; 12: 1742-1753.

[99] Weinstein A. The species of *Pinus* and *Eucalyptus* used for afforestation in Israel. In: Hartman H ed. Afforestation in Israel. Allgemeine Forst Aeitschrift. 1989; 24-26: 627-628.

[100] Weinstein A. Geographic variation and phenology of *Pinus halepensis*, *P. brutia*, and *P. eldarica* in Israel. For Ecol Manage 1989; 27: 99-108.

[101] Wells O. Results of the Southwide pine seed source study through 1968-9. Proc 10th So For Tree Improv Conf 1969: 117-129.

[102] Williams JGK, Kubelic AR, Livak KJ, et al. DNA polymorphisms amplified by arbitrary primers are useful as genetic markers. Nucl Acid Res 1990; 18: 6531-6535.

[103] Williston HL. The question of adequate stocking. Tree Planters Notes 1972; 23: 1-2.

[104] Yazdani R, Yeh FC, Rimsha J. Genomic mapping of *Pinus sylvestris* (L.) using random amplified polymorphic DNA markers. Forest Genetics 1995; 2: 109-116.

[105] Zahner R. Terminal growth and wood formation by juvenile loblolly pine under two soil moisture regimes. For Sci 1962; 8: 345-352.

[106] Zahner R, Stage AR. A procedure for calculating daily moisture stress and its utility in regressions of tree growth on weather. Ecology 1966; 47: 64-74.

[107] Zahner R, Donnelly JR. Refining correlations of rainfall and radial growth in young red pine. Ecology 1967; 48: 525-530.

CHAPTER

3

Survey of Genetic Diversity and Phylogenic Relationships in Tunisian Date-palms (*Phoenix dactylifera* L.) by Molecular Methods

TRIFI MOKHTAR[1*], *ZEHDI-AZOUZI SALWA*[1], *OULD MOHAMED SALEM ALI*[2], *RHOUMA SOUMAYA*[1], *SAKKA HELA*[1], *RHOUMA ABDELMAJID*[3] *and MARRAKCHI MOHAMED*[1]

[1]Laboratoire de Génétique Moléculaire, Immunologie & Biotechnologie, Faculté des Sciences de Tunis, Tunis, Tunisie

[2]Departement de Biologie, Faculté des Sciences et Techniques, Université de Nouakchott, B.P. 5026, Mauritanie, Africa

[3]International Plant Genetic Resources Institute, Centre de Recherches Phoénicicoles, 2260 Degache, Tunisie

ABSTRACT

This study portrays the investigation of the genetic diversity and the phylogenic relationships in Tunisian date-palms *Phoenix dactylifera* with the help of molecular markers (RAPDs, ISSRs and SSRs). The data provide evidence of a high genetic diversity in the local germplasm and prove the reliability of the designed methods as attractive approaches to examine the relationships in this crop. Compared to the RAPD, the ISSR procedure seems to be more useful in assessing the molecular polymorphism of this crop. As a result, the genetic diversity is typically continuous and the ecotypes studied are clustered independently from their geographic origin and from the sex of trees. Thus, it

Address for correspondence: *Laboratoire de Génétique Moléculaire, Immunologie & Biotechnologie, Faculté des Sciences de Tunis, Campus Universitaire 2092 El Manar Tunis, Tunisie. E-mail: mokhtar.t@fst.rnu.tn; trifimokhtar@yahoo.fr.

is assumed that a common genetic basis characterizes the ecotypes of the date palm despite their morphological distinctiveness. Moreover, the data are in agreement with the unique Mesopotamian origin of this crop.

Key Words: *Phoenix dactylifera,* molecular markers, polymorphism, phylogeny

INTRODUCTION

The date-palm (*Phoenix dactylifera* L.) (2n = 36), a dioecious perennial monocotyledon, is the most important fruit crop growing in arid and semi-arid areas of North Africa and several tropical countries. This is one of the oldest fruit crops known for its Mesopotamian origin in the fertile crescent [46, 83] and constitutes the major factor of oases' environmental and economic stability. The date-palm is cultivated either for food or for many other commercial purposes, and it maintains appropriate habitat for animals, menkind, subjacent vegetables, and other fruit crops (pepper, saffron, melon, fig, apricot, etc.) and constitutes the main financial resources of oasians. About 10% of Tunisians are dependent on its culture [21, 34]. Utilization of the palm consisted of locally adapted ecotypes; female trees, artificially pollinated and selected according to agronomic criteria, mainly the productivity and the date quality. These cultivars or varieties are clonally reproduced via offshoots [45, 50]. Recent investigations have proved that Tunisian date-palm germplasm is characterized by a large diversity since more than 250 cultivars have been identified [59, 60]. Unfortunately, this phytogenetic patrimony is seriously menaced either by a severe genetic erosion due to the predominance of the elite cultivar "Deglet Nour", or by plagues locally called the "brittle leaves", a disease of unknown origin, and the "bayoud disease", a fusariosis due to the fungus *Fusarium oxysporum* f. sp. *albedinis* [20, 80]. Hence, it has been imperative to elaborate a strategy aimed at the evaluation of the genetic diversity and the preservation of this important germplasm. For this purpose, many studies have been designed and have described the use of either morphological traits or isozyme makers to identify the Tunisian date-palm varieties [12, 51, 52, 58, 59]. However, these studies are less rewarding due to the limitation in the number of morphological markers that are highly sensitive to the environmental variations and despite their usefulness, the generation of isozyme markers is time consuming, their numbers are limited and their expression is often restricted to specific development states or tissues.

Therefore, to overcome these difficulties, DNA based techniques have been developed and have proved effective to assess genetic polymorphisms. It should be stressed that little is known about the date-palm's genetics and genome organization [16, 17]. These authors have proved that data based on the Restriction Fragment Length Polymorphisms (RFLPs) are suitable in the characterization of date-palm genotypes. Thus, a relatively important genetic variability among cultivars has been evidenced and the reliability of several parameters as efficient criteria to discriminate the cultivars studied has been also suggested. Thus, we became interested in the development of other molecular methods in order to characterize date-palm genotypes and to have a deeper insight of the genetic diversity in the Tunisian germplasm.

A large panel of available DNA based methods has been designed to examine the genetic organization in higher plants. Among these techniques, the Random Amplified Polymorphic DNA (RAPD) [82], the Inter Simple Sequence Repeats (ISSR) [32] and the Simple Sequence Repeats (SSR) methods are the most appropriate and widely used either in higher plants [4, 13, 75, 78] or animals [26, 27, 41, 57]. In fact, the RAPD and ISSR, a based DNA Polymerase Chain Reaction (PCR) technologies [65] are of several benefits over the other techniques since: firstly, little amounts of either plant material or DNA are required to perform DNA variation analyses; secondly, these methods are known either as informative about the molecular diversity or suitable in the discrimination of closely related genotype variants and the establishment of their DNA fingerprinting [2, 23, 33, 35, 85]; and thirdly, they have enabled the detection of polymorphisms without any previous knowledge of any crop DNA sequence. Moreover, RAPD and ISSR provide a nearly unlimited potential of markers to reveal differences at the molecular level. In addition, microsatellites or SSR consist of variable numbers of tandemly repeated units and represent a class of repetitive DNA commonly found in eukaryotic genomes [77]. These markers are characterized by a great abundance [15, 62], a high variability [26, 70] and a large distribution throughout different genomes [40, 63, 76]. Microsatellites are typically multi-allelic loci since more than five alleles per locus are commonly observed either in plant or in animal populations [36, 41, 73]. In addition, automated PCR procedures, which enable high-throughput data collection and good analytical resolution at a low cost, have been developed for microsatellites [37, 42]. Hence, taking into account advantages of these methods, we have designed their use in date-palms to assess genetic distances and to survey evolutionary genetic diversity.

MATERIALS AND METHODS

Plant Material

A set of Tunisian accessions collected from three main oases, namely: Tozeur, Gabès and Kébili were used in the study. These consisted of cultivars and male trees. Varieties were chosen for their good fruit quality and are the most common genotypes cultivated in the main plantations located in the South of Tunisia (Fig. 1). Other fruit characteristics and morphological traits for most varieties were reported by Rhouma [59, 60]. Among these varieties, the two that were recently introduced in Tozeur groves ("Zehdi" from Iraq and "Ghars Mettig" from Algeria) are also included in the study.

The plant material consists of young leaves from adult trees randomly sampled from the mentioned oases.

DNA Preparation

Total cellular DNA was isolated and purified from frozen young leaves according to the procedure of Dellaporta et al. [18]. After purification, DNA concentrations were determined using a GeneQuant spectrometer (Amersham-Pharmacia Biotech, France) and its quality was checked on analytic agarose minigel electrophoresis according to Sambrook et al. [68].

Primers and RAPD Assays

Universal random primers purchased from Operon Technologies (Alameda, USA) were tested to generate RAPD banding profiles from total cellular DNA used as templates. For PCR amplification, DNA templates (25 ng) from each sample were tested in a total volume of 25 μl containing 50 pM of primer, 200 mM of each dNTP (DNA polymerization mix, Pharmacia), 2.5 μl of enzyme buffer and 1.5 unit of Taq DNA polymerase (QBIOgene, France). The reaction mix was overlaid with a drop of mineral oil to avoid evaporation during the cycling. PCR reactions were carried out with a Crocodile III thermal cycler (QBIOgène, France) programmed to execute: 1 cycle at 94°C for

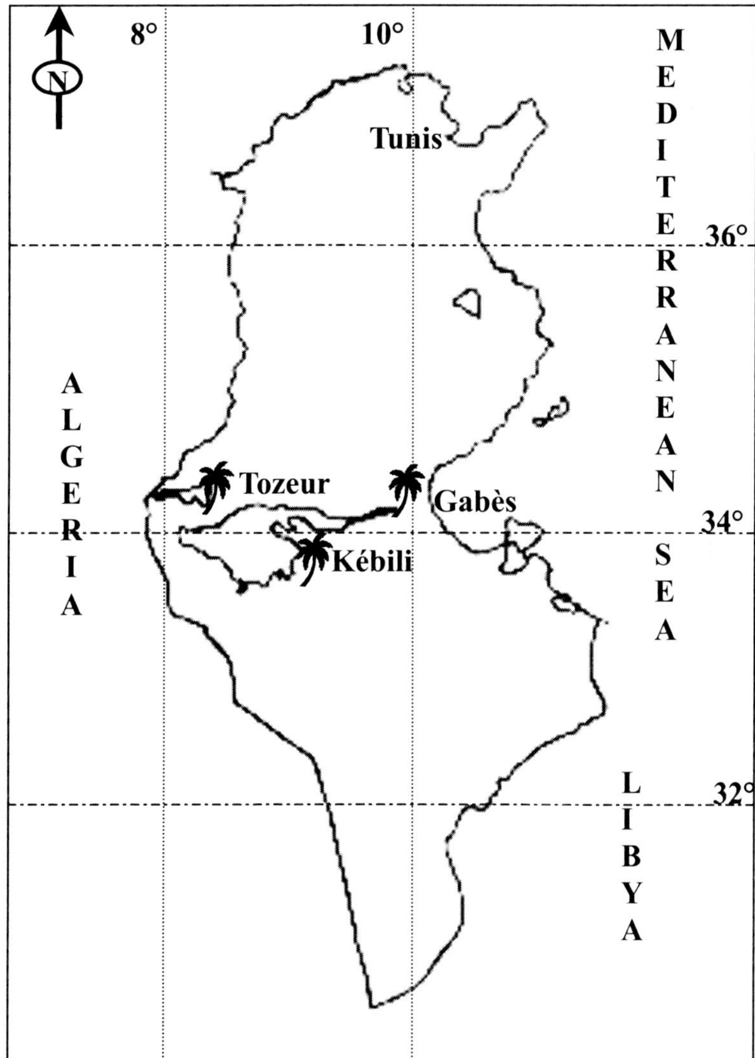

Fig. 1. *Tunisia map showing the geographic origin of the analyzed date-palm ecotypes. Scale 1/100 000*

5 minutes, followed by 25 cycles, each consisting of a denaturation step at 94°C for 30 seconds; an annealing step at 52°C for 1 minute and an elongation step at 72°C for 1 minute. A 5 minutes final extension was performed at the end of the last PCR cycle.

Primers and ISSR Assays

A total of 12 primers, shown in Table 2, were tested to amplify DNA banding patterns using the varieties' total cellular DNA.

PCR reaction mixture (25 μl) is composed of: 20-30 ng of total cellular DNA (1 μl), 60 pg of primer (1 μl), 2.5 μl of Taq DNA polymerase reaction buffer, 1.5 unit of Taq DNA polymerase (QBIOgène, France) and 200 μM of each dNTP (DNA polymerization mix, Amersham-Pharmacia, France). Each reaction mixture was overlaid with 25 μl of mineral oil to avoid evaporation during PCR cycling. Amplifications were performed in a Crocodile III thermal cycler (QBIOgène, France). The apparatus was programmed to execute the following conditions: a denaturation step of 5 minutes at 94°C, followed by 35 cycles each composed of 30 seconds at 94°C, 90 seconds at the primer's specific melting temperature (Tm) (Table 2), and 90 seconds at 72°C. A final extension step of 5 minutes was run at the end of the last PCR cycle.

To reduce the possibility of cross contamination and variation either in the RAPD or ISSR amplification batches, master mixes of the reaction constituents were always used and a negative control, reaction mix without any DNA was also included.

After amplification, the reaction mixture was electrophoresed on a 1.5% agarose gel, stained with ethidium bromide and photographed under UV light [68].

SSR Genotyping

A total of 16 date-palms specific primer pairs developed by Billotte et al. [10] were tested in the present study. SSRs were screened on Li-Cor IR2 automated DNA sequencer (Li-Cor, Lincoln, NE, USA) by loading 0.2 μl of PCR product diluted 10 × in loading buffer, on 6.5% polyacrylamide gel. Automatic genotyping and allele size scoring were performed by the SAGA-GTTM software (Li-Cor, Lincoln, NE, USA).

Data Analysis

For each primer, the total number of bands and the polymorphic ones were calculated. Data were then transformed into a binary matrix where the presence of the generated band is scored as 1 and its absence is notified by 0. A genetic distance matrix was estimated from the data matrix by using the Genedist (version 3.572c) programme based on the formula of Nei and Li [48]. The genetic distance matrix was then analyzed with the Neighbour program to produce a dendrogram using the Unweighted Pair Group Method with the Arithmetic Averaging Algorithm (UPGMA) cluster analysis [74], or using the Neighbour-Joining (NJ) algorithm [24]. Bootstrap values were computed over 100 replications using MSA 3.10 [19]. The TreeView program was used to draw the between and among accessions' dendrogram. Appropriate programs in PHYLIP (Phylogeny Inference Package, version 3.5c) and TreeView (Win32, version1.5.2) were used to carry out all these analyses [25,53]. The subsequent ISSR and RAPD's similarity matrices obtained with ISSR and RAPD were compared according to the Pearson's and Spearman's correlation coefficients using the Statistical Analysis System Version 6.1.2 software [69].

In addition, the sampling was subdivided into three geographical groups of cultivars (female), and a group composed of the male individuals. For each group of date-palm genotypes, the genetic diversity was estimated by the determination of the allelic diversity (total number of alleles, and allele frequencies per group), the observed and expected heterozygosity (H_{obs} and H_{exp}) [49]. The total genetic diversity (H_t), the mean genetic diversity within population (H_s) and the genetic differentiation among populations (G_{st}) were also calculated using the program Genetix 4.04 [5]. The allelic richness, corresponding to the evaluated number of alleles independently from the sample size, was assessed using FSTAT 2.9.1 [31]. A Wilcoxon-Mann-Whitney test was applied to differentiate the allelic richness scored values using Statistica 6.0.

Data were also computed using GENEPOP 3.1 [56] and FSTAT 2.9.1 programs to test pairwise linkage equilibrium (LE) at all loci over the four groups to calculate Fis and pairwise estimates of Fst according to the formula of Weir and Cockerham [81].

In order to compute ordinations and hierarchical classifications, the Populations 1.2.28 Software of Langella [39] was used to calculate the shared allele distance: DAS [14]. This genetic distance is known to be appropriate for recently diverged populations [29].

Establishment of the Cultivars' Identification Key

Ecotypes identification was performed as follows: genotypes were hierarchically organized according to the greater number of alleles per locus. Accessions were then classified and those of similar fingerprints were grouped.

RESULTS

A preliminary experiment was performed starting from total cellular DNA to generate RAPD and ISSR patterns. Due to logistical constraints, only five primers that generate banding profiles over independent PCR runs were chosen to analyze variation levels in each variety. These primers were tested in presence of DNAs used as templates and extracted from different trees of two varieties. As shown in Fig. 2, the RAPD generated fragments are uniformly observed between the different palm trees of a given variety. In this case, the OPA-10 primer was used to generate the patterns from diverse trees of Deglet Nour variety. This result suggests that no variation is scored at the intra-variety level. Hence, it is assumed that each date-palm tree sampled from any variety is characterized by a consensus profile presenting substantial monomorphic DNA bands and corresponding to the variety pattern. This finding is strongly supported since date-palms are clonally propagated throughout by offshoots naturally produced by the parent tree

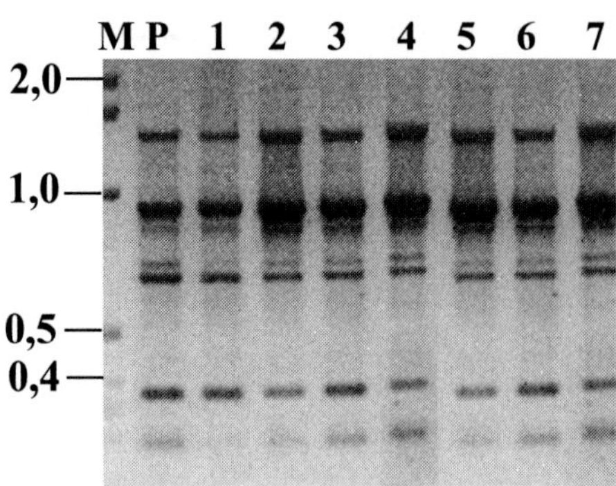

Fig. 2. *RAPD banding patterns generated from diverse trees of Deglet Nour variety using the OPH-07 primer. Lanes 1 to 7: samples, P: mother plant, M: molecular standard size (1 Kb Ladder, GIBCO-BRL, France). DNA bands are in bp.*

[45, 50]. Moreover, multilocus SSR genotypes concur with this assumption since two accessions of the Deglet Nour variety originated from Kebili, and Tozeur oases have exhibited similar fingerprints [84].

Cultivars' Identification Key

For each date-palm accession, the detected genotypes for mPdCIR78, mPdCIR85 and mPdCIR25 microsatellites loci are scored. A total of 25 alleles are identified in these loci: 10 alleles labelled (a1 to a10) for the mPdCIR78 locus, 8 alleles denoted (b1 to b8) for the mPdCIR85, and 7 alleles (c1 to c7) for the mPdCIR25 locus. Taking into consideration the identified alleles, we have established an ecotypes' identification key (Fig. 3). This precise diagram confirmed our assumption about the Deglet Nour accessions and those of Arichti and Rochdi since they presented identical genotypes. In fact, these cultivars sampled from different oases exhibited identical fingerprints taking into account the microsatellite loci examined. This assumption is in agreement with the clonally propagation mode in date-palms [45, 50]. Consequently, the constructed identification key helped to unambiguously discriminate 47 over 47 accessions studied (100%). This result confirms the ability of microsatellites to fingerprint genotypes. Varietal identification keys have been previously reported in date-palms based on isozymes [11, 52] and organellar DNA haplotypes [66]. These authors have reported 69, 93 and 71% resolving power values, respectively. Therefore, our results produced neutral molecular markers powerfully suitable in cultivars' identification. In fact, a maximum of 55,440 theoretically different multilocus genotypes could be distinguished taking into account only the three mentioned loci (i.e. mPdCIR78, mPdCIR85 and mPdCIR25).

Genetic Diversity Analysis as Inferred by RAPDs

Starting from total cellular DNAs extracted from the accessions studied, a large number of reproducible and polymorphic bands were obtained and scored as different RAPD markers. Depending on the DNA origin primer combination, 9 to 15 reproducible bands were evidenced in the 0.2-2.5 kb size range. A maximum of 15 bands were generated using the OPA-15 and OPM-05 primers, while only 10 fragments were produced with the OPA-01 and OPM-15 primers.

A genetic distance matrix based on the evidenced RAPDs expressed the large genetic diversity within the varieties studied since the scored

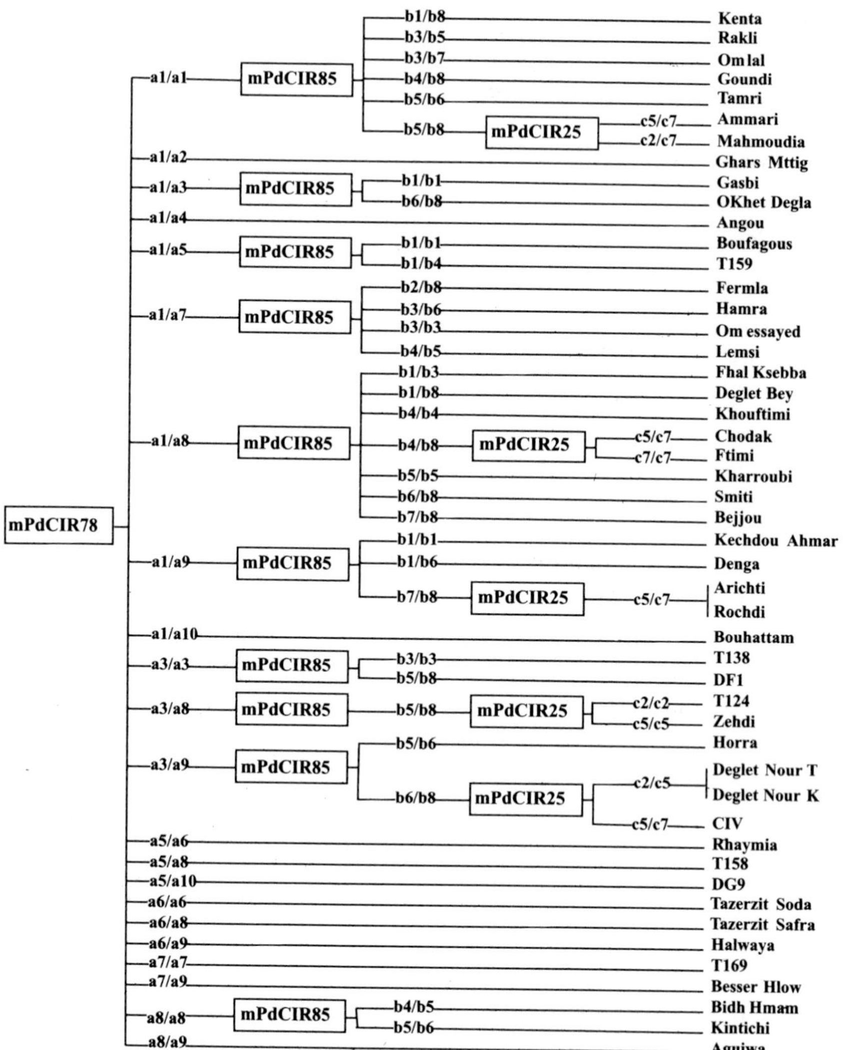

Fig. 3. *Diagram illustrating the varietal identification key of 48 Tunisian date-palm ecotypes based on multilocus genotypes.*

average genetic distance ranged from 0.15 to 0.86. Thus, it is assumed that the RAPD procedure is an efficient method for the detection of the date-palm polymorphism at the DNA level. The smallest distance value of 0.15 was scored either between Boufagous and Deglet Hassen, Cheknet Etterzi and Khad Khadem, or Cheknet Etterzi and Chekent

Ebbay Hamed varieties. These appear to be the most similar varieties and are closely clustered. However, Deglet Nour, Chekent Ebbay Hamed and Hlaoui varieties that have presented the greatest genetic distance of 0.86 are characterized by the maximum divergence at the DNA level. All the remaining varieties displayed intermediate levels of similarity.

Phylogenic Relationships as Revealed by RAPD Data

In order to draw the precise relationships between the genotypes tested, the genetic distance matrix was computed using the Neighbour UPGMA algorithm. The distinctiveness of the clusters identified in the derived phenogram exhibits five main cluster groups. The first one, labelled **a**, is composed of a single variety (i.e. Bouaffar). The Besser Hlou and Tazerzit Soda varieties are clustered in the second group, labelled **b**. All the remaining genotypes are clustered in the other groups, labelled **c**, **d** and **e**, that present auxiliary ramifications (Fig. 4).

Genetic Diversity as Inferred by ISSRs

Among the primers screened for their ability to generate consistently banding patterns and to assess polymorphisms in the tested varieties, only seven have revealed polymorphic and unambiguously scorable bands (Table 1). These ranged from 7 to 11 with a mean of 9.57 polymorphic DNA bands per primer in a ranging size from 200-2,500 bp. A minimum of 7 and a maximum of 11 DNA fragments were obtained using $(CT)_{10}A$ and $(AG)_{10}C$, respectively. All the remaining primers generated an intermediate number of polymorphic DNA bands suggesting that the ISSR procedure constitutes an alternative approach suitable for the date-palm DNA diversity examination. This is strongly supported by the large number of polymorphic DNA bands (i.e. a total of 67 out of 95) produced, which is higher than those observed in other cultivated crops, such as grapevine where 35 polymorphic bands were generated among closely related germplasm in presence of 12 ISSR primers [44]. In addition, the registered collective rate value of resolving power (Rp) is 43.02 with a mean of 6.14 (Table 1). Similar Rp values have been reported in lupin germplasm collection [28], potato cultivars [55], and figs [67]. Moreover, since the $(AG)_{10}G$, $(AG)_{10}C$ and $(AG)_{10}T$ primers presented the highest Rp rates, it is assumed that these oligonucleotides mainly contribute in the accessions' characterization and to examine the genetic diversity in this crop.

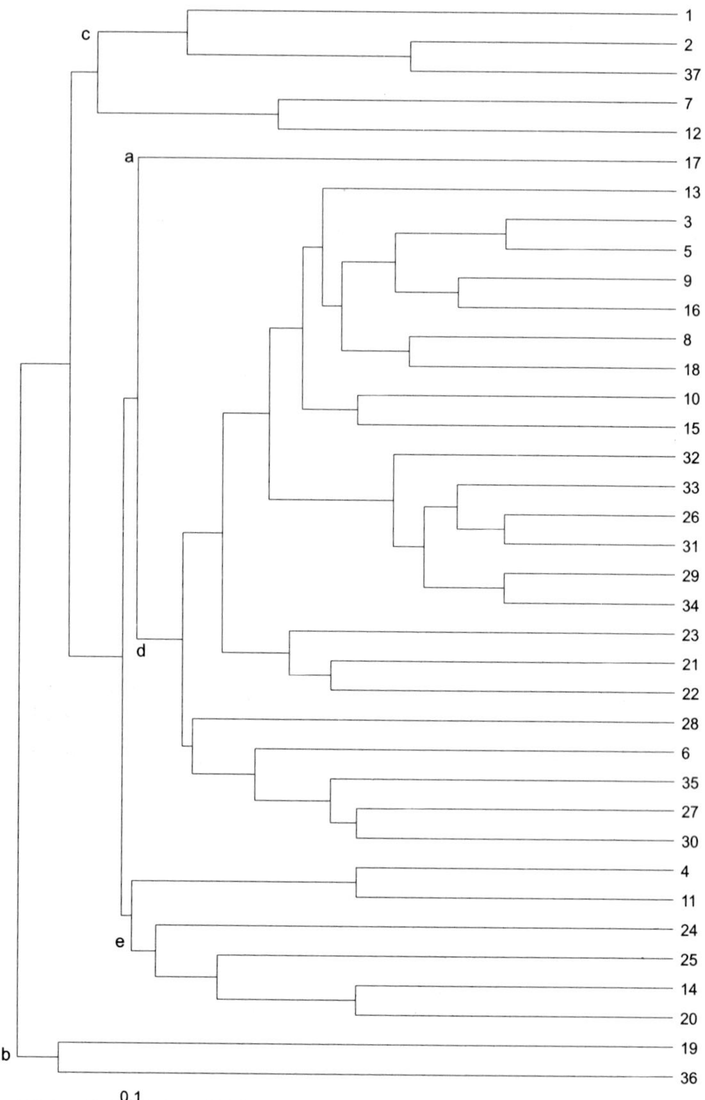

Fig. 4. *UPGMA dendrogram of 37 Tunisian date-palm varieties constructed from Nei & Li genetic distance based on 54 RAPD markers. The scale indicates the relative genetic distance.*

1: Deglet Nour; 2: Okht Deglet; 3: Deglet Hassen; 4: Deglet Bey; 5: Boufagous; 6: Khou Boufagous; 7: Ftimi; 8: Khou Ftimi; 9: Kenta; 10: Kentichi; 11: Goundi; 12: Okht Goundi; 13: Ammari; 14: Zehdi; 15: Hlaoui; 16: Lagou;17: Bouaffar; 18: Om Laghlez; 19: Besser Hlou; 20: Ghars Mettig; 21: Horra; 22: Angou;; 23: Arichti; 24: Irhaymia; 25: Khadhraoui; 26: Darbouzi; 27: Bidh Hmam; 28: Khalt Saad; 29: Khalt Ali Meskine; 30: Tronja; 31: Khad Khadem; 32: Cheknet Hanene; 33: Cheknet Etterzi; 34: Cheknet Ebbay Hamed; 35:Tazerzit Safra; 36: Tazerzit Soda; 37: Gasbi

Table 1. ISSR bands and resolving power (Rp) rates of the used primers in Tunisian date-palm ecotypes

Primer sequence	*Tm Optimum*	*Amplified bands*		*Rp*
		Total	*Polymorphic*	
$(AGG)_6$	55°C	13	10	6.61
$(TGGA)_5$	55°C	0	-	-
$(ACTG)_4$	45°C	6	0	-
$(GACA)_4$	45°C	0	-	-
$(GACAC)_4$	55°C	10	0	-
$(AG)_{10}$	55°C	Smear	-	-
$(AG)_{10}G$	60°C	13	10	9.46
$(AG)_{10}C$	60°C	12	11	7.37
$(AG)_{10}T$	57°C	12	10	7.15
$(CT)_{10}A$	57°C	7	7	2.86
$(CT)_{10}G$	60°C	10	9	4.84
$(CT)_{10}T$	57°C	12	10	4.73
Total		95	67	43.02

The binary data matrix was computed to estimate the genetic distances between accessions. These ranged from 0.23 to 0.98 with a mean of 0.54, suggesting a high degree of genetic diversity at the DNA level (Table 2). The smallest distance value of 0.23 was observed between DF1 and T169, indicating that these two ecotypes are the most similar. The maximum distance value (0.98) is scored between Besser Hlou/Deglet Nour, Angou/Kenta, Bouhattam/Deglet Bey, Lemsi/Zehdi, Lemsi/Khou Ftimi, Lemsi/Horra, Lemsi/T138 Lemsi/T158, Denga/Horra, Gasbi/T138, and Gasbi/T158. This result suggests that the above mentioned varieties are characterized by great divergence.

Phylogenic Relationships as Revealed by ISSR Data

The genetic distance matrix was computed with the Neighbour program using the UPGMA algorithm in order to cluster the accessions according to their genetic similarity and to draw the relationships between the tested accessions. The derived dendrogram supported a varietal clustering made independently of the trees' sex and the ecotypes geographic origin and exhibited two major clusters, labelled **a** and **b**

Table 2. Genetic distance matrix among 34 Tunisian date-palm ecotypes based on ISSR data and estimated from Nei & Li's formula

Ecotype	1	2	3	4	5	6	7	8	9	10	11	12	13	14	15	16	17	18	19	20	21	22	23	24	25	26	27	28	29	30	31	32	33
1 Deglet Nour																																	
2 Boufagous	0.47																																
3 Ftimi	0.37	0.43																															
4 Kenta	0.47	0.28	0.43																														
5 Kintichi	0.61	0.61	0.31	0.61																													
6 Deget bey	0.37	0.31	0.40	0.26	0.50																												
7 Ghars mettig	0.26	0.37	0.40	0.37	0.50	0.29																											
8 Zehdi	0.34	0.40	0.43	0.47	0.61	0.43	0.31																										
9 Arichti	0.54	0.47	0.50	0.40	0.78	0.37	0.37	0.34																									
10 Khouftimi	0.54	0.69	0.37	0.53	0.40	0.57	0.50	0.50	0.53																								
11 Horra	0.43	0.50	0.47	0.73	0.37	0.50	0.53	0.31	0.65	0.37																							
12 Okhet degla	0.31	0.43	0.53	0.43	0.57	0.40	0.34	0.37	0.57	0.5	0.34																						
13 T124	0.47	0.34	0.57	0.53	0.61	0.50	0.37	0.34	0.53	0.61	0.37	0.31																					
14 T138	0.61	0.69	0.73	0.53	0.69	0.57	0.50	0.47	0.61	0.53	0.43	0.43	0.47																				
15 T158	0.53	0.53	0.73	0.53	0.69	0.50	0.43	0.34	0.47	0.61	0.37	0.43	**0.23**	0.28																			
16 T169	0.43	0.65	0.78	0.50	0.82	0.40	0.47	0.65	0.65	0.57	0.61	0.40	0.50	0.50	0.43																		
17 DF1	0.26	0.65	0.61	0.57	0.65	0.47	0.34	0.50	0.57	0.43	0.53	0.47	0.57	0.50	0.50	**0.23**																	
18 DG9	0.28	0.61	0.65	0.47	0.61	0.50	0.37	0.47	0.69	0.40	0.50	0.43	0.47	0.61	0.53	0.31	0.31																
19 Lemsi	0.69	0.87	0.82	0.78	0.78	0.92	0.82	**0.98**	0.87	**0.98**	**0.98**	0.73	0.87	**0.98**	**0.98**	0.73	0.65	0.78															
20 Bouhattam	0.73	0.92	0.69	0.73	0.73	**0.98**	0.69	0.82	0.73	0.82	0.92	0.87	0.73	0.73	0.73	0.61	0.78	0.65	0.43														
21 Smiti	0.69	0.78	0.50	0.78	0.47	0.65	0.57	0.61	0.53	0.69	0.65	0.65	0.78	0.78	0.69	0.65	0.73	0.87	0.53	0.43													
22 Denga	0.57	0.73	0.69	0.73	0.82	0.61	0.61	0.73	0.65	0.92	**0.98**	0.78	0.92	0.82	0.73	0.78	0.78	0.82	0.43	0.34	0.57												
23 Hamra	0.43	0.65	0.69	0.65	0.65	0.69	0.53	0.65	0.65	0.73	0.69	0.69	0.57	0.73	0.65	0.61	0.47	0.50	0.43	0.61	0.50	0.61											
24 Aguiwa	0.65	0.65	0.61	0.65	0.73	**0.98**	0.69	0.82	0.73	0.65	0.69	0.69	0.82	0.73	0.73	0.53	0.69	0.65	0.57	0.28	0.31	0.47	0.61										
25 Kharroubi	0.78	0.87	0.73	0.69	0.53	0.73	0.65	0.87	0.61	0.69	0.73	0.65	0.61	0.69	0.61	0.65	0.82	0.87	0.87	0.43	0.34	0.65	0.73	0.43									
26 Angou	0.69	0.69	0.82	0.98	0.61	0.65	0.57	0.69	0.78	0.61	0.57	0.73	0.61	0.61	0.47	0.57	0.50	0.69	0.87	0.73	0.69	0.57	0.92	0.65	0.61								
27 Besser Hlou	0.98	0.78	0.65	0.78	0.69	0.92	0.73	0.78	0.53	0.61	0.73	0.73	0.69	0.69	0.69	0.57	0.65	0.78	0.61	0.50	0.34	0.73	0.73	0.31	0.47	0.78							
28 Rhaimia	0.65	0.65	0.69	0.50	0.57	0.69	0.69	0.73	0.73	0.73	0.87	0.78	0.92	0.92	0.82	0.47	0.47	0.73	0.50	0.61	0.57	0.53	0.61	0.47	0.57	0.65	0.50						
29 Bidh-hmam	0.57	0.50	0.78	0.57	0.57	0.61	0.53	0.65	0.57	0.65	0.69	0.53	0.57	0.73	0.65	0.47	0.40	0.50	0.73	0.61	0.57	0.61	0.53	0.47	0.50	0.43	0.43	0.40					
30 Gasbi	0.54	0.61	0.73	0.69	0.87	0.65	0.65	0.69	0.61	0.87	0.82	0.82	0.87	0.98	0.98	0.82	0.65	0.69	0.61	0.65	0.78	0.37	0.82	0.50	0.53	0.53	0.69	0.50	0.57				
31 KchdouAhmar	0.61	0.61	0.73	0.53	0.78	0.73	0.73	0.61	0.78	0.78	0.82	0.73	0.78	0.69	0.61	0.57	0.50	0.78	0.61	0.50	0.61	0.92	0.82	0.65	0.61	0.69	0.69	0.65	0.50	0.78			
32 Fhalksebba	0.69	0.61	0.82	0.69	0.78	0.73	0.50	0.61	0.40	0.53	0.65	0.50	0.40	0.61	0.53	0.65	0.65	0.53	0.87	0.73	0.61	0.82	0.65	0.50	0.47	0.47	0.34	0.65	0.31	0.47	0.78		
33 Rakli	0.54	0.61	0.73	0.40	0.61	0.65	0.50	0.61	0.61	0.69	0.73	0.50	0.61	0.87	0.61	0.73	0.82	0.47	0.61	0.50	0.69	0.50	0.73	0.43	0.47	0.61	0.78	0.73	0.57	0.50	0.69	0.53	
34 Fermla	0.47	0.69	0.57	0.53	0.53	0.82	0.57	0.87	0.87	0.69	0.82	0.73	0.61	0.78	0.69	0.57	0.57	0.61	0.40	0.37	0.50	0.65	0.50	0.50	0.69	0.87	0.61	0.43	0.73	0.87	0.47	0.87	0.69

(Fig. 5). This interpretation is well illustrated in the case of the cluster **a** that includes either varieties or male ecotypes. Furthermore, the cluster **b** is composed of varieties originating from the different date-palm oases.

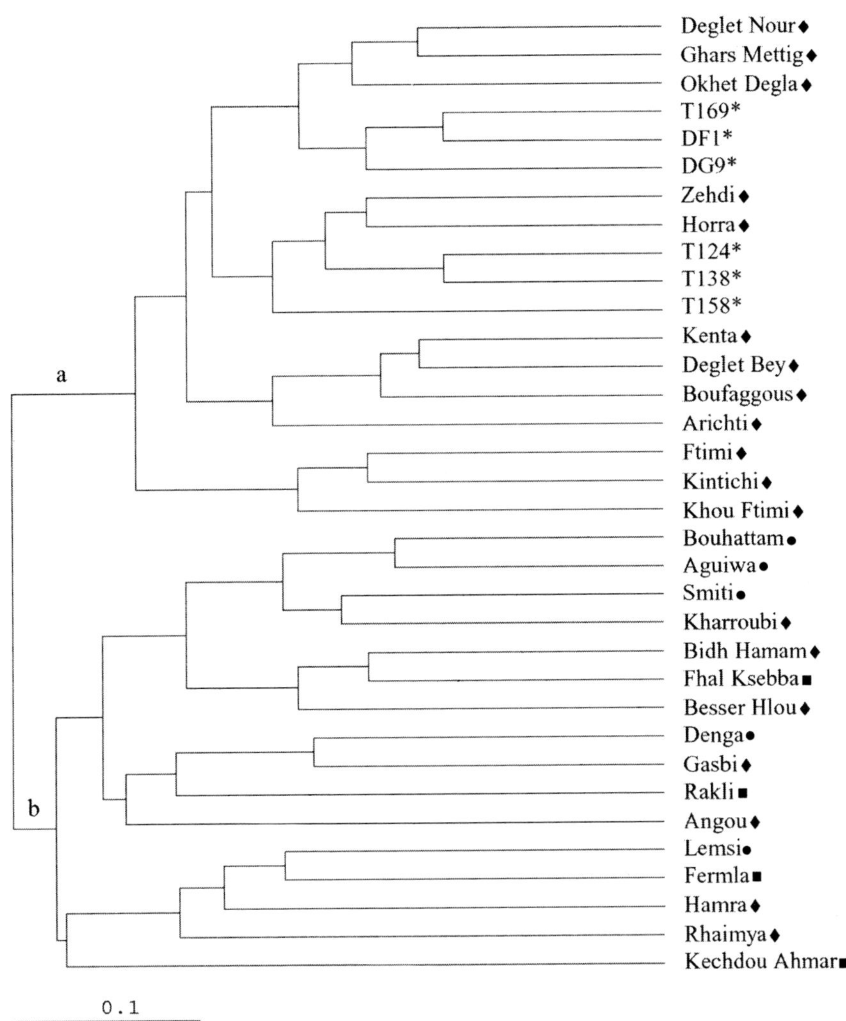

Fig. 5. *UPGMA dendrogram of 34 Tunisian date-palm accessions constructed from Nei & Li genetic distance based on 67 ISSR markers. The scale indicates the relative genetic distance. Accessions originated from Tozeur, Gabès and Kébili oases are labelled (♦), (●) and (■) respectively. Males are labelled with an asterisk.*

Genetic Diversity and Phylogenic Relationships as Inferred by Combined RAPD and ISSR Data

Since slight differences are denoted in the RAPD and the ISSR dendrograms, RAPD data were taken into account together with the ISSRs in order to identify the relationships between a subset of 17 varieties. A total 102 polymorphic bands (54 RAPDs and 48 ISSRs) were used to examine the genetic polymorphisms in this crop. The derived genetic distance matrix based on the Nei and Li's formula and reported in Table 3 exhibited values ranging from 0.29 to 0.82 with a mean of 0.53. The smallest genetic distance of 0.29 is scored between Boufagous and Deglet Bey varieties suggesting their great similarities at the DNA level. However, the Deglet Nour and Besser Hlou, the Okhet Degla and Besser Hlou seem to be the most dissimilar varieties, as the genetic distance registered between them is the highest.

The derived dendrogram illustrated in Fig. 6 based on RAPD markers together with ISSRs, permitted to cluster the varieties studied in two main groupings. The first one is composed of Besser Hlou, Rhaimya, Bidh Hmam, Angou and Gasbi varieties. All the remaining ones are clustered in the second group that exhibited three subgroups. As reported above, this typology clusters the varieties based on their nomenclature and/or fruit parameters. This dendrogram's topology is also almost identical to this, based on the ISSR data either by the number of clusters, or by their composition (data not shown). However, a few differences have been scored between dendrograms based on RAPD data and the combined RAPD/ISSR. This result suggests that the designed methods contribute differently in the discrimination of the date-palm varieties. This assumption is strongly supported since polymorphisms obtained with RAPD and ISSR markers have different underlying sources at the molecular level and may differ in their informativeness for the exploration of genetic diversity and the establishment of relationships between ecotypes [22]. Hence, the determination of the contribution of each technique in the observed diversity is of a great interest in order to determine whatever they contribute in the Tunisian date-palm polymorphisms. This could be made possible by estimation of the correlation indices between the three obtained genetic distances matrix. As reported in Table 4, the Pearson and Spearman coefficients exhibited high positive and significant correlations between RAPD data matrix and the combined data (Pearson's coefficient = 0.2629, rP = 0.00; Sperman's

Table 3. Genetic distance matrix among 17 Tunisian date-palm cultivars based on RAPD and ISSR data and estimated from Nei & Li's formula

Cultivar	*1*	*2*	*3*	*4*	*5*	*6*	*7*	*8*	*9*	*10*	*11*	*12*	*13*	*14*	*15*	*16*
1 DegletNour																
2 OkhetDegla	0.35															
3 Deglet Bey	0.48	0.48														
4 Boufeggous	0.48	0.42	**0.29**													
5 Ftimi	0.47	0.47	0.51	0.45												
6 Khou Ftimi	0.60	0.44	0.48	0.42	0.38											
7 Kenta	0.48	0.48	0.27	0.24	0.48	0.51										
8 Kintichi	0.65	0.51	0.53	0.41	0.36	0.36	0.53									
9 Zehdi	0.51	0.42	0.50	0.44	0.51	0.55	0.47	0.47								
10 BesserHlou	**0.82**	**0.82**	0.64	0.71	0.73	0.62	0.64	0.71	0.67							
11 Ghars Mettig	0.44	0.56	0.33	0.42	0.50	0.53	0.36	0.51	0.31	0.62						
12 Horra	0.64	0.50	0.48	0.39	0.50	0.38	0.58	0.33	0.39	0.65	0.47					
13 Angou	0.71	0.80	0.55	0.55	0.60	0.50	0.73	0.55	0.73	0.69	0.50.	0.44				
14 Arichti	0.62	0.58	0.44	0.47	0.55	0.51	0.47	0.64	0.44	0.67	0.45	0.51	0.51			
15 Rhaimya	0.71	0.64	0.58	0.58	0.67	0.56	0.55	0.58	0.62	0.55	0.60	0.60	0.64	0.69		
16 Bidh Hmam	0.51	0.55	0.44	0.41	0.62	0.55	0.47	0.50	0.56	0.50	0.39	0.51	0.42	0.50	0.39	
17 Gasbi	0.53	0.50	0.62	0.51	0.60	0.64	0.58	0.69	0.65	0.73	0.64	0.67	0.53	0.62	0.50	0.62

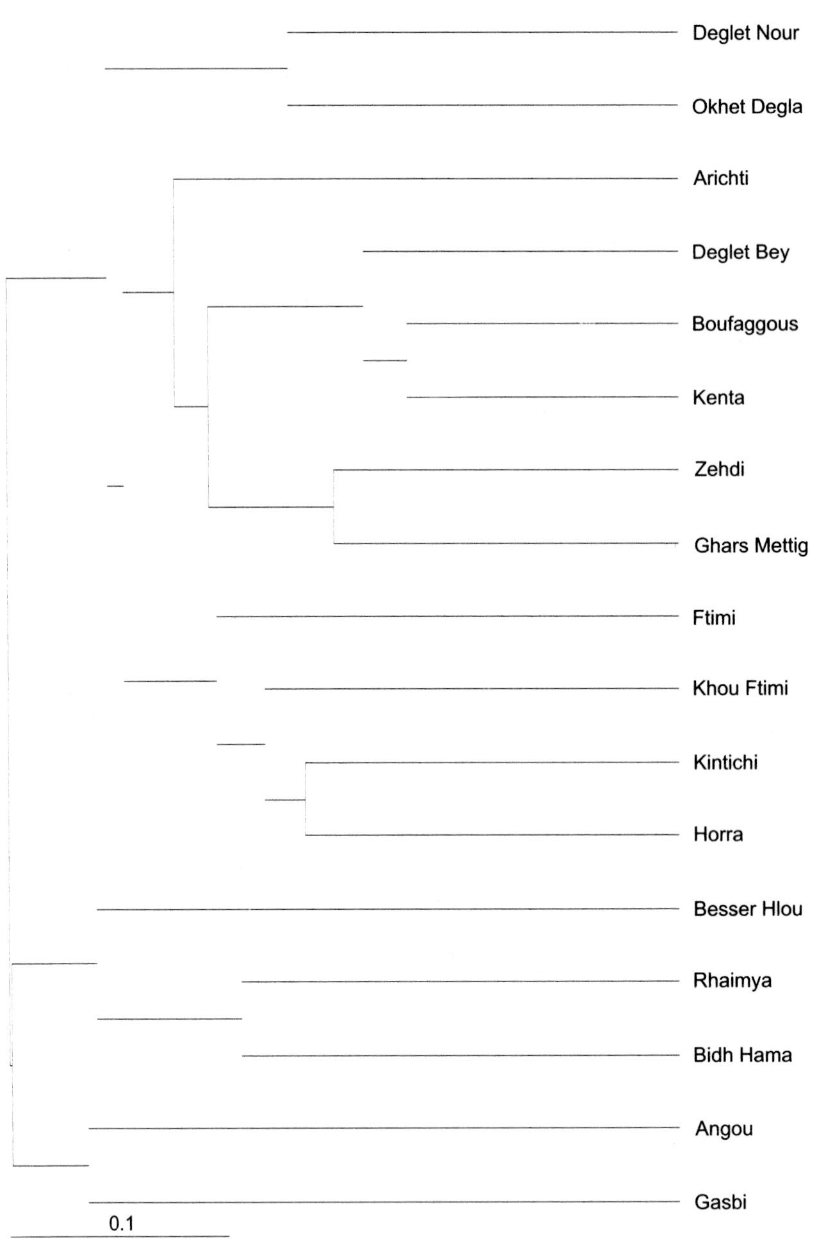

Fig. 6. UPGMA dendrogram of 17 Tunisian date-palm cultivars constructed from Nei and Li's genetic distance estimated on 102 combined RAPD and ISSR data. The scale indicates the genetic distance.

coefficient = 0.1684, rP = 0.00). Negative, but not significant, correlations were obtained between ISSR and RAPD on the one hand and ISSR/RAPD on the other hand, supporting the groupings' differences observed in dendrograms based on RAPDs and the RAPD/ ISSR combined data. Therefore, it may be assumed that ISSRs are more informative in terms of discriminating date-palm varieties rather than the RAPDs. Differences in the distribution of these classes of markers throughout the genome constitute a critical factor in determining their efficiency in the investigation of the genetic variability of this crop. This is in agreement with data reported in other plant species [22, 38, 47]. Moreover, other studies have described a faster evolution rate of the neutral ISSR markers and an association of RAPD bands with functionally important loci [22, 54].

Table 4. Pearson's coefficient (upper diagonal) and Sperman's coefficient (lower diagonal) based on ISSR RAPD and ISSR/RAPD data *: significative value (0.05)

Marker	*ISSR*	*RAPD*	*ISSR/RAPD*
ISSR	1.0000	–0.1754	–0.1754
RAPD	–0.1468	1.0000	0.2629*
ISSR/RAPD	–0.1468	0.1684*	1.0000

Genetic Diversity as Inferred by SSR Data

Out of the 16 primer pairs tested for their ability to generate expected SSR banding patterns in Tunisian date-palms, 14 have successfully established the ecotypes' genotypes. The SSR profiles exhibited more than four different alleles per locus with clearly identifiable homozygous and heterozygous genotypes. A total of 100 alleles with a mean of 7.14 alleles per locus were scored (Table 5). The number of alleles per locus varied from 4 (mPdCIR16) to 10 (mPdCIR78).

H_{exp} values (Table 6) ranged from 0.40 (mPdCIR35) to 0.83 (mPdCIR85) indicating that the Tunisian date-palm collection is characterized by a high degree of genetic diversity. The alleles' number, as well as their frequencies, varied significantly among the Tunisian subgroups. This is well exemplified in the case of the mPdCIR90 locus exhibiting seven alleles in Tozeur oasis, and four out of them have not been evidenced in the remaining subgroups. Moreover, the mean number

Table 5. Summary of 14 microsatellite loci revealed in the Tunisian date-palm genotypes studied

Locus	*Alleles*		*Genotypes*
	Size	*Number*	
mPdCIR10	142 - 181	8	16
mPdCIR15	142 - 157	7	11
mPdCIR16	148 - 156	4	7
mPdCIR25	219 - 250	7	14
mPdCIR32	302 - 318	8	16
mPdCIR35	200 - 214	5	7
mPdCIR50	172 - 222	8	19
mPdCIR57	260 - 288	8	15
mPdCIR63	139 - 171	5	8
mPdCIR70	205 - 227	7	15
mPdCIR78	138 - 173	10	22
mPdCIR85	175 - 197	8	21
mPdCIR90	162 - 193	7	12
mPdCIR93	181 - 202	8	18
Total		100	201

of alleles varied from one group to another (Table 7). However, there is no significant difference in allelic richness among the four groups (P=0.15). This result agrees with multilocus means of expected heterozygosity (H_{exp}) and unbiased heterozygosity (H_{nb}) that did not significantly differ among groups at P>0.05 (Table 7).

As reported in Table 8, it is assumed that for all the loci studied, the H_s and H_t values are nearly similar, suggesting that the maximum variability is locally maintained. This assumption is confirmed by the low values of G_{st}. For instance, the multilocus values of H_s, H_t and G_{st} are 0.6449, 0.6885 and 0.0633, respectively. Moreover, these results suggested that only 7% of the genetic diversity is explained at the inter-group level, while 93% of this variability is maintained at the intra-group level.

In addition, estimation of F_{is} values, according to the formula of Weir and Cockerham [81], indicates that the Tozeur group as well as the male genotypes showed a significant deviation from Hardy-Weinberg Equilibrium (HWE). Results of the exact test [64] and the specific test

Table 6. Expected (H_{exp}) and observed (H_{obs}) Heterozygosity in each group by locus computed using GENETIX

Locus		*Tozeur*	*Kébili*	*Gabès*	*Males*	*All accessions*
mPdCIR10	H_{exp}	0.75	0.42	0.81	0.54	0.73
	H_{obs}	0.67	0.60	1.00	0.57	0.69
mPdCIR15	H_{exp}	0.71	0.64	0.40	0.62	0.68
	H_{obs}	0.70	0.40	0.50	0.43	0.59
mPdCIR16	H_{exp}	0.60	0.66	0.61	0.49	0.60
	H_{obs}	0.67	0.60	0.83	0.57	0.67
mPdCIR25	H_{exp}	0.73	0.58	0.57	0.70	0.72
	H_{obs}	0.83	0.60	0.83	0.14	0.71
mPdCIR32	H_{exp}	0.76	0.82	0.78	0.66	0.77
	H_{obs}	0.67	1.00	0.67	0.86	0.71
mPdCIR35	H_{exp}	0.45	0.32	0.28	0.24	**0.40**
	H_{obs}	0.33	0.00	0.33	0.00	0.26
mPdCIR50	H_{exp}	0.75	0.78	0.74	0.65	0.75
	H_{obs}	0.77	0.60	0.83	0.71	0.73
mPdCIR57	H_{exp}	0.71	0.66	0.74	0.50	0.71
	H_{obs}	0.67	0.20	0.67	0.29	0.57
mPdCIR63	H_{exp}	0.64	0.42	0.28	0.72	0.63
	H_{obs}	0.23	0.60	0.33	0.14	0.26
mPdCIR70	H_{exp}	0.71	0.78	0.72	0.72	0.75
	H_{obs}	0.30	0.20	0.00	0.29	0.26
mPdCIR78	H_{exp}	0.77	0.68	0.72	0.78	0.79
	H_{obs}	0.73	0.80	1.00	0.57	0.73
mPdCIR85	H_{exp}	0.83	0.80	0.78	0.76	**0.83**
	H_{obs}	0.73	0.80	1.00	0.57	0.75
mPdCIR90	H_{exp}	0.69	0.58	0.61	0.49	0.68
	H_{obs}	0.67	0.80	0.83	0.57	0.69
mPdCIR93	H_{exp}	0.80	0.70	0.72	0.72	0.80
	H_{obs}	0.90	0.60	0.83	0.86	0.86

for heterozygote deficiency (U test [64]) strongly supported this finding since statistically significant deficits for the mentioned groups have been registered. However, no deviation from HWE is observed in the two remaining groups (i.e. Gabès and Kébili) (Table 7).

Table 7. Genetic diversity indices for the four groups

Group	H_{exp}	H_{nb}	H_{obs}	F_{is}	*P value*	*Mean Number of alleles/locus*
Tozeur	0.71	0.72	0.63	0.1222	0.0000	6.79
Kébili	0.63	0.70	0.56	0.2258	0.0983	4.00
Gabès	0.63	0.68	0.69	-0.0140	0.7789	4.00
Males	0.62	0.66	0.47	0.3083	0.0001	3.71
All accessions	0.70	0.71	0.61	0.142	0.0000	7.14

Table 8. The total genetic diversity (Ht), the mean genetic diversity within population (Hs) and the genetic differentiation among groups (Gst) estimated using the program Genetix 4.04

Locus	*Hs*	*Ht*	*Gst*
mPdCIR10	0.6294	0.6851	0.0813
mPdCIR15	0.5945	0.6307	0.0574
mPdCIR16	0.5901	0.6039	0.0229
mPdCIR25	0.6456	0.7123	0.0937
mPdCIR32	0.7540	0.7940	0.0504
mPdCIR35	0.3239	0.3346	0.0320
mPdCIR50	0.7299	0.7556	0.0340
mPdCIR57	0.6525	0.7093	0.0801
mPdCIR63	0.5163	0.5734	0.0996
mPdCIR70	0.7333	0.7697	0.0473
mPdCIR78	0.7367	0.8034	0.0831
mPdCIR85	0.7914	0.8374	0.0549
mPdCIR90	0.5933	0.6594	0.1003
mPdCIR93	0.7378	0.7703	0.0422
Multilocus	0.6449	0.6885	0.0633

Phylogenetic Relationships as Revealed by Microsatellite Markers

A derived NJ dendrogram based on Das genetic distance, exhibited three main clusters, each one composed of males as well as cultivars (Fig. 7). The two Deglet Nour and the two Arichti accessions, which originated from different date-palm oases, are strongly clustered and suggest that

Table 9. Pairwise comparisons of the multilocus Fst values

Group	*PopT*	*PopM*	*PopG*	*PopK*
Tozeur	0.0000			
Males	0.0095	0.0000		
Gabès	0.0202	0.0492	0.0000	
Kébili	–0.0219	–0.0144	–0.0193	0.0000

these cultivars correspond to similar genotypes. In addition, the observed clustering topology showed that groupings of accessions are made independently either from their geographic origin or the sex of trees. This result is corroborated by the absence of geographic structure in the plotting of accessions on the two first PCO axes (data not shown). Moreover, pairwise comparisons of the multilocus Fst values scored among the pre-established groups were not significant (P>0.05), indicating that all groups revealed high genetic affinity (Table 9).

CONCLUSION

The objective of this study was to characterize a large number of Tunisian date-palm ecotypes with the help of the RAPD, ISSR and SSR markers in order to investigate their phylogenic relationships. The designed procedures have enabled the survey of the DNA polymorphism in the collection analyzed. On the whole, our results concur with those describing the use of RAPD technique in date-palms starting from Moroccan, Tunisian, Saudi and Iraqi collections [3, 6, 72, 79]. These authors assumed that the studied accessions are clustered independently of their geographic origin and suggested a narrow genetic diversity in this crop. Both analyses have generated a dendrogram topology, which is in agreement with those based on morphometric criteria particularly related to the fruit parameters [59]. This is well exemplified in the case of Boufaggous and Deglet Bey varieties characterized by nearly similar dates (large size and dark colour). In addition, accessions' groupings are not well defined either according to their geographical origin or the sex of trees, since the introduced varieties and the male ecotypes did not significantly diverge from the autochthonous female accessions. Therefore, data derived from the evidenced markers (RAPDs, ISSRs and SSRs) suggest that all the date-palm ecotypes are interrelated in spite of their phenotypic divergence. A common and narrow genetic basis is

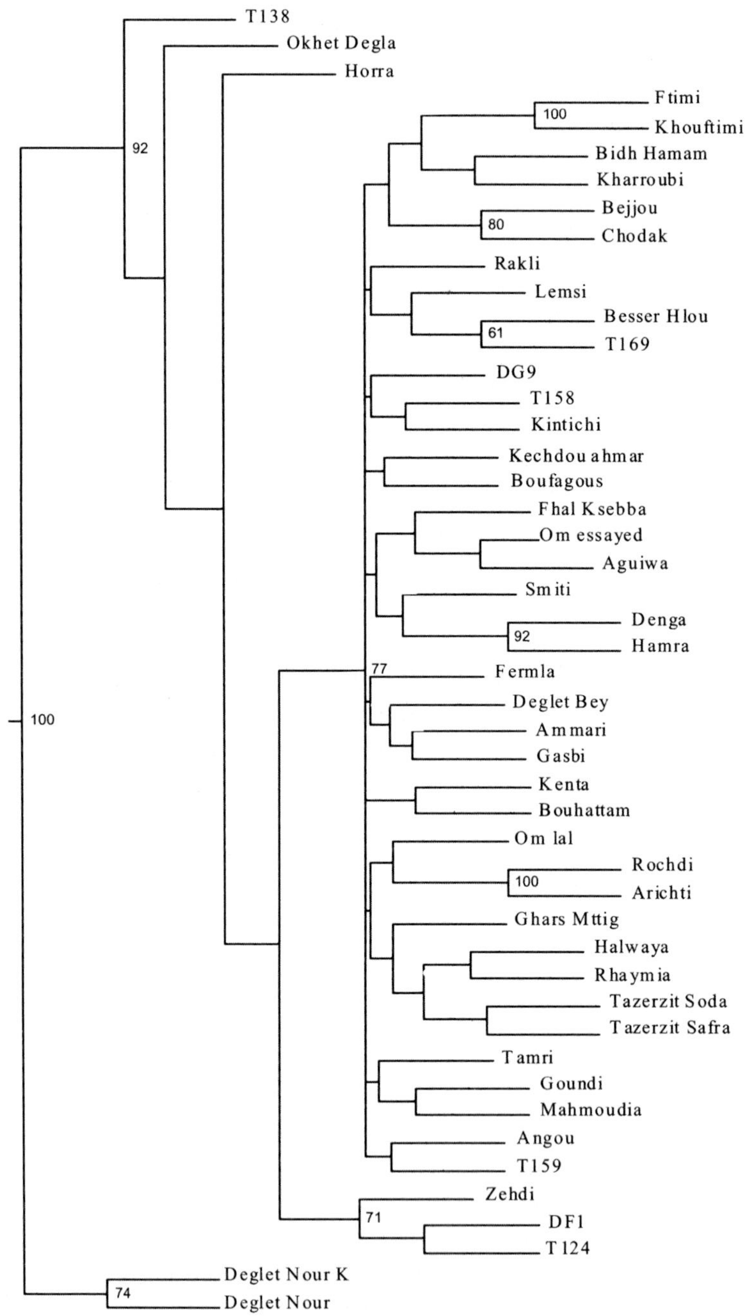

Fig. 7. *Dendrogram of Tunisian date-palm ecotypes based on Das genetic distance estimated from microsatellite data.*

strongly supported by selection applied by farmers based either on the fruit quality or the ecotypes' adaptation to local conditions. Hence, a small part of the genome that encodes interesting agronomic parameters is affected by this selection in the natural populations from which clones originated. Consequently, it may be assumed that the data strongly supported the ancient date-palm's Mesopotamian "Fertile Crescent" domestication origin [30, 45, 83, 86].

Compared to the diversity reported in other fruit crops, the overall polymorphism exhibited in the present study is rather high, suggesting that the designed methods are very effective in assessing the molecular polymorphism of this crop. In addition, the revealed SSR alleles were successfully used to discriminate the studied ecotypes since an identification key has been established on the basis of three microsatellite loci. In this case, the data provided higher percentage of resolution than that scored in date-palms using either isoenzymes or plastid haplotypes [8, 9, 11, 52, 66]. The transfer of SSR analysis to other laboratories over the world would be of great interest to label at a large scale offshoots, any other plant material at early stage and *in vitro* plantlets. A precise fingerprinting of unlimited number of these closely related ecotypes reported across the world could be made possible on the basis of our procedure. Such a strategy could improve cultivars' differentiation and then contribute to their labelling (homonymy and synonymy) since large numbers of ecotypes have been reported in the date-palm' growing countries [1, 7, 43, 61, 71]. Moreover, since offshoots' exchanges are currently occurring, the evidenced SSR markers are greatly recommended to be used as descriptors in the certification and the control of original labels of date-palm material originated from these countries.

A deeper insight of the reported markers could constitute a powerful tool to molecularly characterize the genetic diversity of this crop and trace parentage and genetic relationships within closely related genotypes at the specific level and below. This can be made possible through analysis of natural hybrids between date-palm and its relatives from the *Phoenix* species reported in Algeria, Morocco, Punjab and Senegal [45]. Research is presently being carried out to specify the phylogenic and the culture origins of the date-palms [46] and to shed light on the domestication process in this phytogenetic resource.

Acknowledgement

This work was supported by grants from the Tunisian "Ministère de l'Enseignement Supérieur" and the "Ministère de la Recherche Scientifique, de la Technologie et du Développement des Compétences" and from the IPGRI "Projet FEM-PNUD-IPGRI, RAB 98 G31". The authors would like to thank Professors J.C. Mounolou, F. Quetier, A. Rode and C. Hartmann (Université Paris Sud, France), J.C. Pintaud, T. Couvreur (IRD, Montpellier, France) and N. Billotte (CIRAD, Montpellier, France) for their fruitful collaboration and for stimulating discussions.

References

[1] Ajibade SR, Weeden NF, Chite SM. Inter simple sequence repeat analysis of genetic relationships in the genus Vigna. Euphytica 2000; 111: 45-55.

[2] Al Helal AA. Amylase isoenzymes and protein of date palm (*Phoenix dactylifera* L.) fruit. Bot Bull Academica Sinica 1988; 29: 239-244.

[3] Al-Khalifah NS, Askari E. Molecular phylogeny of date palm (*Phoenix dactylifera* L) cultivars from Saudi Arabia by DNA fingerprinting Theor Appl Genet 2003; 107: 1266-1270.

[4] Alvarez AE, Van de Wiel CCM, Smulders MJM, Vosman B. Use of microsatellites to evaluate genetic diversity and species relationships in the genus *Lycopersicon*. Theor Appl Genet 2001; 103: 1283-1292.

[5] Belkhir K, Goudet J, Chikhi L, Bonhomme F. Genetix (Ver. 4.01), logiciel sous windowsTM pour la génétique des populations 2000, Laboratoire Génome et Population, Université Montpellier II, Montpellier, France.

[6] Ben Abdallah A, Stiti AK, Lepoivre P, Du Jardin P. Identification de cultivars de palmier dattier (*Phoenix dactylifera* L) par l'amplification aléatoire d'ADN (RAPD). Cahiers Agriculture 2000; 9: 103-107.

[7] Ben Khalifa A. Diversity of date palm *Phoenix dactylifera* L. Options Méditerranéennes 1996; 28: 160.

[8] Bendiab K, Baaziz M, Brakez Z, Sedra MyH. Electrophoretic patterns of acid soluble proteins and active isoforms of peroxidases and polyphenoloxidase typifying calli and somatic embryos of two reputed date palm cultivars in Morocco. Euphytica 1994; 76: 159-168.

[9] Bennaceur M, Lanaud C, Chevalier MH, Bounaga N. Genetic diversity of the date palm (*Phoenix dactylifra* L.) from Algeria revealed by enzyme markers. Plant Breed 1991; 107: 56-69.

[10] Billotte N, Marseillac N, Brottier P, et al. Nuclear microsatellite markers for the date-palm (*Phoenix dactylifera* L.): characterization, utility across the genus *Phoenix* and in other palm genera. Molecular Ecology Notes 2004; 4: 256-258.

[11] Booij L, Montfort S, Ferry M. Characterization of thirteen date-palm (*Phoenix dactylifera* L.) cultivars by enzyme electrophoresis using the Phast-System. J Plant Physiol 1995; 145: 62-66.

[12] Bouabidi H, Reynes M, Rouissi MB. Critères de caractérisation des fruits de quelques cultivars de palmier dattier (*Phoenix dactylifera* L) du sud tunisien. Ann. Inst Rech Agr Tunisie 1996; 69: 73-87.

[13] Bruford MW, Wayne RK. Microsatellites and their application to population genetic studies. Curr Opin Genet Dev 1993; 3: 939-943.

[14] Chakraborty R, Jin L. A unified approach to study hypervariable polymorphisms: statistical considerations of determining relatedness and population distances. In: Chakraborty R, Epplen JT, Jeffreys AJ, eds. DNA fingerprinting: State of the Science. Birkhäuser Verlag, 1993: 153-175.

[15] Condit R, Hubbell SP. Abundance and DNA sequence of two-base repeat regions in tropical tree genomes. Genome 1991; 34: 66-71.

[16] Cornicquel B, Mercier L. Date-palm (*Phoenix dactylifera* L.) cultivar identification by RFLP & RAPD. Plant Sc 1994; 101: 163-172.

[17] Cornicquel B, Mercier L. Identification of date-palm (*Phoenix dactylifera* L.) cultivars by RFLP: partial characterization of a cDNA probe that contains a sequence encoding a zinc finger motif. Int J Plant Sc 1996; 158: 152-156.

[18] Dellaporta SL, Wood J, Hicks JB. A plant DNA preparation. Version II. Plant Mol Biol Reporter 1984; 4: 19-21.

[19] Dieringer D, Schlöterer C. Microsatellite analyser (MSA): a platform independent analysis tool for large microsatellite data sets. Molecular Ecology Notes 2003; 2: 1-3.

[20] Djerbi M, ed. Précis de phoéniciculture. FAO, Rome, Italie, 1985.

[21] El Hadrami I, El BellaJ M, El Idrissi A, et al. Biotechnologies végétales et amélioration du palmier-dattier (*Phoenix dactylifera* L.), pivot de l'agriculture oasienne Marocaine. Cahiers Agriculture 1998; 7: 463-468.

[22] Esselman EJ, Jianqiangt LD, Crawford J, et al. Clonal diversity in the rare *Calamagrotis proteri* ssp *insperata* (Poaceae): Comparative results for allozymes and random amplified polymorphic DNA (RAPD) and Inter Simple Sequence Repeat (ISSR) markers. Mol Ecol 1999; 8: 443-451.

[23] Fang DQ, Roose ML. Identification of closely related citrus cultivars with inter-simple sequence repeat markers. Theor Appl Genet 1997; 95: 408-417.

[24] Felsenstein J. Evolutionary trees from DNA sequences: a maximum likelihood approach. J Mol Evol 1981; 17: 368-376.

[25] Felsenstein J. PHYLIP (Phylogeny Interference Package), version 3.5c Distributed by the author, Department of Genetics, University of Washington, Seattle, Washington, USA, 1995.

[26] Garcia de Leon FJ. Marqueurs hypervariables chez le loup (*Dicentrachus labrax* L.) Applications aux programmes d'amélioration génétique à l'étude de populations naturelles. Thèse de Doctorat, université d'Aix-Marseille II, 1995.

[27] Garcia de Leon FJ, Chikhi L, Bonhomme F. Microsatellite polymorphism and population subdivision in natural populations of European sea bass *Dicentrarchus labrax* (Linnaeus, 1758). Mol Ecol 1997; 6: 51-62.

[28] Gilbert JE, Lewis RV, Wilkinson MJ, Caligari PDS. Developing an appropriate strategy to assess genetic variability in plant germplasm collections. Theor Appl Genet 1999; 98: 1125-1131.

[29] Goldstein DB, Pollock DD. Launching microsatellites: a review of mutation processes and methods of phylogenetic inference. Journal of Heredity 1997; 88: 335-342.

[30] Goor A. The history of the date through the ages in the Holy Land. Econ Bot 1967; 21: 320-340.

[31] Goudet J. Fstat version 1.2. A computer program to calculate F-statistics. Journal of Heredity 1995; 86: 485-486.

[32] Gupta M, Chyi YS, Romeo-Severson J, Owen JL. Amplification of DNA markers from evolutionarily diverse genomes using single primers of simple-sequence repeats. Theor Appl Genet 1994; 89: 998-1006.

[33] Gupta PK, Farshney RK. The development and use of microsatellite markers for genetic analysis and plant breeding with emphasis on bread wheat. Euphytica 2000; 113: 163-185.

[34] Haddouch M. Situation actuelle et perspectives de développement du palmier dattier au Maroc. Options Méditerranéenes 1996; 28: 63-79.

[35] Hodkinson TR, Chase MW, Renvoize SA. Characterization of a genetic resource collection for *Miscanthus* (Saccharinae, Andropogoneae, Poaceae) using AFLP and ISSR PCR. Ann Bot 2002; 89: 627-636.

[36] Innan H, Terauchi R, Miyashita NT. Microsatellite polymorphism in natural populations of the wild plant *Arabidopsis thaliana*. Genetics 1997; 146: 1441-1452.

[37] Kresovich S, Szewc-McFadden AK, Bliek SM, McFerson JR. Abundance and characterization of simple-sequence repeats (SSRs) isolated from a size-fractionated genomic library of *Brassica napus* L. (rapeseed). Theor Appl Genet 1995; 91: 206-211.

[38] Lai JA, Yang WC, Hsiao JY. An assessment of genetic relationships in cultivated tea clones and native wild tea in Taiwan using RAPD and ISSR markers. Bot Bull Acad Sc 2001; 42: 93-100.

[39] Langella O. Populations 1.2.28 Software. CNRS UPR9034, France, 2002.

[40] Liu ZW, Biyashev RM, Saghai-Maroof MA. Development of simple sequence repeat DNA markers and their integration into a barley linkage map. Theor Appl Genet 1996; 93: 869-876.

[41] MacHugh DE, Shriver MD, Loftus RT, et al. Microsatellite DNA variation and the evolution, domestication and phylogeography of taurine and zebu cattle (*Bos taurus* and *Bos indicus*). Genetics 1997; 146: 1071-1086.

[42] Mitchell SE, Kresovich S, Jester CA, et al. Application of multiplex PCR and fluorescence-based, semi-automated allele sizing technology for genotyping plant genetics resources. Crop Sci 1997; 37: 617-624.

[43] Mohamed S, Shabana HR, Mawlod BA. Evaluation and identification of Iraqi date: fruit characteristics of fifty cultivars. Date Palm J 1983; 2: 27-56.

[44] Moreno S, Martin JP, Ortiz JM. Inter-simple sequence PCR for characterization of closely related grapevine germplasm. Euphytica 1998; 101: 117-125.

[45] Munier P, ed. Le palmier-dattier. Paris, France: Maisonneuve & Larose, 1973.

[46] Munier P. Origine de la culture sur palmier-dattier et sa propagation en Afrique. Fruits 1981; 36 (7): 437-450.

[47] Nagaoka T, Ogihara O. Applicability of inter-simple sequence repeat polymorphism in wheat for use as DNA markers in comparison to RFLP and RAPD markers. Theor Appl Genet 1997; 98: 86-92.

[48] Nei M, Li WH. Mathematical model for studying genetical variation in terms of restriction endonucleases. Proc Natl Acad Sci USA 1979; 74: 5267-5273.

[49] Nei M, ed. Molecular evolutionary genetics, New York: Columbia University Press, 1987.

[50] Nixon RW, Carpenter B. Growing date in the United States. US Dep Bull. 1978; 207: 63.

[51] Ould Mohamed Salem A, Trifi M, Sakka H, et al. Genetic inheritance of four enzymes in date palm (*Phoenix dactylifera* L). Genet Res Crop Evol 2001a; 48: 361-368.

[52] Ould Mohamed Salem A, Trifi M, Salhi-Hannachi A, et al. Genetic variability analysis of Tunisian date-palm (*Phoenix dactylifera* L.) cultivars. Journal of Genetics & Breeding, 2001b; 55: 269-278.

[53] Page RDM. TREEVWIEW: http//:taxonomy.zoology.gla.uk./rod.html. An application to display phylogenetic trees on personal computers. Comput Appl Biosci 1996; 12: 357-358.

[54] Penner GA. RAPD analysis of plant genomes. In: Jauhar PP, ed. Methods of genome analysis in plants. Boca Raton: CRC Press, 1996: 251-268.

[55] Prevost A, Wilkinson MJ. A new system of comparing PCR primers applied to ISSR fingerprinting of potato cultivars. Theor Appl Genet 1999; 98: 107-112.

[56] Raymond M, Rousset F. GENEPOP (ver1.2): A population genetics software for exact test and ecumenicism. Journal of Heredity 1995; 86: 248-249.

[57] Reddy KD, Nagaraju J, Abraham EG. Genetic characterisation of the silkworm Bombyx mori by simple sequence repeats (SSR)-anchored PCR. Heredity 1999; 83: 681-687.

[58] Reynes M, Bouabidi H, Pionbo G, Risterrucci AM. Caractérisation des principales variétés de dattes cultivées dans la région du Djerid en Tunisie. Fruits 1994; 49: 189-198.

[59] Rhouma A, ed. Le palmier dattier en Tunisie I: Le patrimoine génétique. Tunis Tunisie: Arabesque, 1994.

[60] Rhouma A, ed. Le palmier dattier en Tunisie II: Le patrimoine génétique. IPGRI, 2005.

[61] Rizvi M, Davis J. Structural features of the date market in Sind Pakistan. Date Palm J 1983; A: 103-122.

[62] Röder MS, Plaschke J, König SU, et al. Abundance, variability and chromosomal location of microsatellites in wheat. Mol Gen Genet 1995; 246: 327-333.

[63] Röder MS, Korzun V, Wendehake K, et al. A microsatellite map of wheat. Genetics 1998; 149: 2007-2023.

[64] Rousset F, Raymond M. Testing heterozygote excess and deficiency. Genetics 1995; 140: 1413-1419.

[65] Saiki RK, Scharf S, Mullis KB, et al. Enzymatic amplification of beta-globin genomic sequences and restriction site analysis for diagnosis of suckle cell anemia. Science 1988; 239: 1350-1354.

[66] Sakka H, Zehdi S, Ould Mohamed Salem A, et al. Genetic polymorphism of plastid DNA in Tunisian date-palm germplasm (*Phoenix dactylifera* L.) detected with PCR-RFLP. Genet Res Crop Evol 2004; 51: 479-487.

[67] Salhi-Hannachi A, Trifi M, Zehdi S, et al. Inter-simple sequence repeat fingerprintings to assess genetic diversity in Tunisian fig (*Ficus carica* L.) germplasm. Genet Res Crop Evol. 2004; 51: 269-275.

[68] Sambrook J, Fritsch EF, Maniatis T, eds. Molecular Cloning: a Laboratory Manual, Second Edition. Cold Spring Harbor: New York: Cold Spring Harbor Laboratory Press, 1989.

[69] SAS, SAS user's guide: SAS/STAT SAS BASIC Version 607, Fourth edition SAS incl Box 8000 Cary NC 27512-8000 Cary NC: SAS institute Inc: 1990.

[70] Schug MD, Hutter CM, Wetterstrand KA, et al. The mutation rates of di, tri and tetra-nucleotide repeats in *Drosophila melanogaster*. Mol Biol Evol 1998; 15: 1751-1760.

[71] Sedra MyH. La palmeraie Marocaine: composition, caractéristique et potentialités. Options Méditerranéennes. 1996; A-28: 163.

[72] Sedra MyH, Lashermes H, Trouslot P, et al. Identification and genetic diversity analysis of date-palm (*Phoenix dactylifera* L.) varieties from Morocco using RAPD markers. Euphytica 1998; 103: 75-82.

[73] Senior ML, Heun M. Mapping maize microsatellites and polymerase chain reaction confirmation of the targeted repeats using a CT primer. Genome 1998; 36: 884-889.

[74] Sneath PHA, Sokal RR, eds. Numerical taxonomy. San Francisco, USA: Freeman, 1973.

[75] Stepansky A, Kovalski I, Perl-Treves R. Intraspecific classification of melons (*Cucumus melo* L.) in view of their phenotypic and molecular variation. Plant Syst Evol 1999; 217: 313-333.

[76] Taramino G, Tingey S. Simple sequence repeats for germplasm analysis and mapping in maize. Genome 1996; 39: 277-287.

[77] Tautz D, Renz M. Simple sequences are ubiquitous repetitive components of eukaryotic genomes. Nucleic Acids Res 1984; 12: 4127-4138.

[78] Testolin R, Marrazzo T, Cipriani G, et al. Microsatellite DNA in peach (*Prunus persica* L. Batsch) and its use in fingerprinting and testing the genetic origin of cultivars. Genome 2000; 43: 512-520.

[79] Trifi M, Rhouma A, Marrakchi M. Phylogenetic relationships in Tunisian date-palm (*Phoenix dactylifera* L) germplasm collection using DNA amplification fingerprinting. Agronomie 2000; 20: 665-671.

[80] Triki MA, Zouba A, Khoualdia O, et al. "Maladie des feuilles cassantes" or brittle leaf disease of date palms in Tunisia: biotic or abiotic disease. J Plant Pathology 2003; 85 (2): 71-79.

[81] Weir BS, Cockerham C. Estimating F-statistics for the analysis of population structure. Evolution 1984; 38: 1358-1370.

[82] Williams GK, Kubelik AR, Livak KJ, et al. DNA polymorphisms amplified by arbitrary primers are useful as genetic markers. Nucl Acids Res 1990; 18: 6531-6535.

[83] Wrigley G. Date-palm (*Phoenix dactylifera* L.). In: Smartt J, Simmonds NW, eds. The evolution of crop plants. Essex, United Kingdom: Longman, 1995: 399-403.

[84] Zehdi S, Trifi M, Billote N, et al. Genetic diversity of Tunisian date palms (*Phoenix dactylifera* L.) revealed by nuclear microsatellite polymorphism. Hereditas 2004; 141: 278-287.

[85] Zietkiewicz E, Rafalski A, Labuda D. Genome fingerprinting by simple sequence repeat (SSR)-anchored polymerase chain reaction amplification. Genomics, 1994; 20: 176-183.

[86] Zohary D, Spiegl-Roy P. Beginning of fruit growing in the Old Word. Science 1975; 187: 319-327.

CHAPTER

4

Oil Palm

ZUZANA PRICE[1], SEAN MAYES[2], NORBERT BILLOTE[3], FARAH HAFEEZ[1], FREDERIC DUMORTIER[4] and DON MAC-DONALD[1]*

[1]Department of Genetics, Cambridge University, Cambridge, UK

[2]Nottingham University, Division of Biosciences, Sutton Boninghton Campus, Loughborough, Leicester LE12 5RD, UK

[3]CIRAD (CIRAD-CP) TA 80/03 Avenue Agropolis 34398, Montpellier Cedex, France

[4]DAMI, OPRS, New Britain Palm Oil Ltd., P.O. Box 165, Kimbe, West New Britain Province, Papua New Guinea

ABSTRACT

Palm breeders have made significant increases in the genetic yield potential of oil palm since breeding began systematically at the beginning of the 19th century, with an estimated four-fold increase in yields during this period [22]. This represents a doubling of yields due to genetic improvement and a further doubling due to improved agronomic practices.

One of the major breeding developments was the recognition of the genetic control of shell-thickness by a single gene [3]. This alone increased yield by 25% with a switch from planting thick shelled (dura) fruited trees to the thinner shelled hybrid (tenera). This switch has also had a fundamental effect on the way in which oil palm is bred and improved, as all commercial material must now be of the tenera shell-type and is essentially a hybrid between the dura mother palm and the pisifera pollen palm.

[*]*Address for correspondence:* Zuzana Price, Cambridge University, Department of Genetics, Downing Site, Downing Street, Cambridge CB2 2EH, UK. E-mail z.price@gen.cam.ac.uk

It will be a major challenge to continue to generate significant improvements in the genetic material used for palm oil production in the coming years (both for economic and environmental benefits).

In the first section of this article, the background to the current success is examined, both in terms of the biology and breeding of oil palm. The second part of the article deals with how biotechnology has and will effect the genetic improvement of oil palm.

Key Words: *Oil palm, breeding, molecular markers, crop improvement*

INTRODUCTION

Botanical Classification and Phylogeny

Palms

Palms are woody monocotyledons belonging to the family *Arecaceae* (an alternative name to *Palmae*), in the order *Arecales* [61]. They are a natural group of plants with fossil records dating from the late Cretaceous, and with a characteristic appearance. The present evidence also suggests that palms probably evolved very early in the history of the monocotyledons. Phylogenetic analysis of monocot relationships based on plastid *atp*B (chloroplast gene encoding the β chain of ATP synthetase), *rbcL* (the large subunit of RUBISCO), *mat*K (gene located within the intron of chloroplast gene *trnK)*, and *ndh*F (chloroplast gene), mitochondrial *atp*A (gene encoding the α chain of ATP synthetase) and nuclear 18S and 26S rDNA (ribosomal DNA) [12] placed *Arecales* as a sister-group to the rest of Commelinids, which includes orders *Commeliniales*, *Zingiberales* and *Poales*. However, *Arecales* (single family *Arecaceae,* the palms) were not included in the Commelinids although Dahlgren *et al.* [24] noted that the palms shared some characteristics with those families and the connection of *Arecaceae* to *Pandanaceae* and *Cyclantacae* based on similarities in habit and inflorescence indicate parallelism rather than close phylogenetic relationship.

There are no morphological synapomorphies (shared, derived state) for the palms overall. However, phylogenetic studies of *rbc*L [11, 28]), 18S rRNA [46, 108], and the chloroplast gene *rps*4 [83], all support monophyly of the palm family [114]. The latest phylogenetic analysis of *Arecaceae* [47] added sequence data of 51 genera of Arecoid Line for plastid genes (*atp*B, *rbc*L and *ndh*F) and the plastid intergenic spacers (*trn*T and *trn*Q-*rps*16). Furthermore, Hahn [48] added the nuclear DNA

(18S rDNA) and chloroplast DNA (cpDNA; *atp*B and *rbc*L) sequence for 65 palm genera. The detailed extensive array of molecular phylogenetic studies of palms has identified four major groups corresponding to: (1) Calamoideae, (2) Nypoideae, (3) Coryphoideae+ Caryoteae, and (4) Arecoideae. The Arecoid line is the largest of the four groups with approximately 60% of the genera in the family. The analysis of biogeographic patterns present in Arecoid line suggests that the group is of Gondwana origin [47].

Palms are evolving slowly at the sequence level. Studies of *rbc*L and comparative studies of chloroplast DNA RFLPs revealed that palm plastid genomes have an approximately five-fold slower rate of synonymous substitution compared to other monocots, especially grasses [39, 118]. This difference in rates of divergence has been confirmed with the nuclear *Adh* gene ([40, 82]. Studies of variation in palm mitochondrial genes have been focused on the gene *atp*A [32]. Results showed that the nuclear genes evolve faster than the chloroplast genes, and the chloroplast genes evolve faster than the mitochondrial genes as shown previously by Wolfe *et al.* [119]. The relative rates of divergence within groups seem consistent, but palms evolve at synonymous sites more slowly than grasses.

The patterns of non-synonymous substitutions are different; only *rbc*L and *Adh*1 evolve significantly faster in grasses compared to palms. The rates of nucleotide substitution in the nuclear ribosomal small subunit (18S nrDNA) are significantly lower than that seen in other monocots, and comparable to those of the plastid gene *rbc*L [108].

In the palm family there is little variation in chromosome number, but the genome size can vary significantly. Differences in chromosome numbers are unusual and polyploidy is also rare in palms [113].

The Elaeidinae

The oil palm (*Elaeis guineensis* Jacq.) belongs to the subfamily Arecoideae, tribe Cocoeae and subtribe Elaeidinae. The analysis of biogeographic patterns present in tribe Cocoeae suggests that the tribe is of Gondwana origin and primary diversification in this group may have coincided with continental breakup [47]. The subtribe Elaeidinae includes only the genus *Elaeis* (from the Greek *elaia*, for the olive tree) and *Barcella*, and is always recovered as monophyletic [47]. The genus *Barcella* has no commercial use at present. The genus *Elaeis* consists of only two species: (1) the African oil palm - *Elaeis guineensis* Jacq., and (2) the Latin American oil palm - *Elaeis oleifera* Cortez [22].

Mitochondrial DNA has been used to assess the phylogeny of the subtribe *Elaeidinae* to which the African and Latin American oil palm and the genus *Barcella* belong [2]. The authors analyzed 288 representative accessions of *Elaeis oleifera* and 38 of *Elaeis guineensis*, by performing RFLP with four mitochondrial probes. The RFLP analysis identified more mitotypes in *E. oleifera* compared to *E. guineensis*, and also confirmed that the divergence between the two species was very low.

Elaeis guineensis – the African oil palm

There are no subspecies in *Elaeis guineensis* Jacq. However, there is a range of breeding populations of restricted origin (BPROs) such as Pobe, Yangambi, Deli Dura, AVROS and others, documented by Rosenquist [100], which play an important role in many breeding programmes. One of the most important BPROs is the Deli Dura, which is believed to have descended from four palms that were planted in 1848 in the Bogor Botanical Gardens, Indonesia. Concern has already been raised about the limited within population genetic diversity for some BPROs such as Deli Dura. (ref. section 1.6).

Elaeis species have 16 pairs of chromosomes [71]. On the basis of their length, the chromosomes of *Elaeis guineensis* were divided into three groups [71]. The size of the haploid *E. guineensis* genome has been estimated to be 1.7×10^9 bp (basepairs) (2C=3.7 pg (picograms)) [97].

Geographical Distribution

Although the two *Elaeis* species occur on separate continents and have different growth habits, they are very similar. However, while they can hybridize to produce some fertile offsprings, the differences between them are sufficient to treat them as separate species. Barcellos *et al.* [2] have proposed the centre of origin of the genus *Elaeis* to be Latin America, based on their results which revealed more variability in *E. oleifera* than in *E. guineensis* accessions examined by them, and on botanical evidence.

However, the fossil, historical and linguistic evidence (particularly from Brazil) for the African origin of *E. guineensis* is strong. Zeven [122] reported finding fossil pollen similar to *E. guinensis* from Miocene and earlier layers in the Niger delta and there have been similar results obtained [30, 31] [96]. Zeven suggested that *Elaeis sp.* originated in Africa and spread to South America via the Tertiary bridges, which are believed to have connected Africa and America. Both results from Africa and South America are consistent with a initial *Elaeis* species, which underwent geographical speciation with the breakup of Gondwana Land, some 60 million years ago, without the need for a specific mechanism to prevent cross-fertilization of the derived species (although fertility is a major issue with the interspecific F1) [53].

The first historical evidence of oil palm cultivation in Africa comes from the Portuguese explorer Ca' Da Mosto (1434-1460; [23] although the species was recorded in detail by Jacquin [58].

It is generally accepted that the present geographical distribution of oil palm is the reflection of favourable climate and of human farming activities with palms almost certainly being spread by the migration of man [22].

Biology of Oil Palm

Inflorescence, flower and fruit structure

Separate male and female inflorescences arise in the leaf axils among the leaves, and the infructescence is large and densely packed with fruit primordia. Approximately two inflorescences are initiated per month and take up to three years to develop into a mature male or female inflorescence. The oil palm is monoecious, alternately producing male and female inflorescences in a cycle of around six months, and is thus naturally out-crossing. Detailed studies of the flowers have, however, shown that each flower primordium is a potential producer of both female and male organs, though one or the other almost always remains rudimentary [50]. The female inflorescence is a panicle consisting of a variable number of rachillae that carry 5 to 30 floral triads, each composed of one female flower accompanied by 2 non-functional male flowers. Although each female flower has a tricarpellate ovary, only one carpel usually develops to give rise to a seed. The ovary is accompanied by two rudimentary androecia. In male inflorescences each male rachilla is composed of 400-1,500 staminate flowers. In very rare cases (tissue culture stress, or very young plants) both androecium and gynoecium develop to give rise to a hermaphrodite flower. Inflorescence abnormalities are by no means uncommon in oil palm. The sex of the inflorescence is influenced by the external conditions for about two years before anthesis and also by a genetic component [20]. The cycle between male and female inflorescences can be biased towards male inflorescences under harsh external conditions, such as drought, and towards female inflorescences under favourable external conditions.

It was originally believed that oil palm, because of its abundant pollen and a reduced flower structure, was wind pollinated. In fact, the introduction of the pollinating insect *Elaiedobius kamerunicus* into Malaysia from Africa showed that insect pollination plays an essential part in the fruit set, especially in wet conditions [51].

The seed (kernel) and the pulp (mesocarp) of the fruit are very rich in oil. In the internal fruit structure the most important differences are to be found in the thickness of the shell (endocarp). There three known fruit forms are: (1) dura, (2) pisifera and, (3) tenera, which were determined by Beinaret and Vanderveryen [3] to be due to a single gene. The importance of this gene is well understood, because only plants of intermediate type (tenera) are grown commercially. In the pisifera form (shsh) there is no shell (endocarp) as such, only a fibrous ring. In the dura form (ShSh) there is a thick endocarp, and in the intermediate tenera form (Shsh) the endocarp is thinner and the fibrous ring is present too. The thickness of the endocarp varies considerably in the dura and tenera forms, the distributions even overlap, so the ultimate criterion for distinguishing dura from tenera is the presence of the fibrous ring in the tenera. There are no fruit types as such recognised in *Elaeis oleifera*, all fruit appear to be of dura form.

The external appearance of a fruit varies when ripened. There are four major fruit types known: (1) *nigrescens,* (2) *virescens,* (3) *albescens and* (4) *poissoni.* The fruit is normally dark and is called *nigrescens.* A relatively uncommon type is *virescens.* Its green colour is due to the absence of anthocyanin in the exocarp of the fruit and the colour changes at maturity to orange. The white colour of the mesocarp (*albescens*) is caused by the absence of, or low level of, carotenoids in the mesocarp. *Poissoni* is an abnormal fruit type which is often referred to as 'mantled', or as 'a fruit with supplementary carpels'.

Oil Palm Products

Oil is the main commercial product of the oil palm. The oil is extracted from the fruit mesocarp (palm oil) and nut kernels (kernel oil). Palm oil, extracted from the fibrous flesh of fruits (mesocarp) after they have been hot-squeezed, has oil content from 40–70%. Prime oil, commercially known as palm kernel oil, is extracted from the seeds.

About 90% of world's palm oil is used for edible purposes [102]. Numerous studies have shown the association between diet and the incidence of coronary disease (CHD). Palmitic acid (44%) is the major saturated fatty acid in palm oil and this is balanced by almost 39% monounsaturated oleic acid and 11% polyunsaturated linoleic acid. The remaining acids are largely stearic (5%) and myristic (1%). This composition is significantly different from palm kernel oil (obtained as a co-product during the processing of oil palm fruits), which is almost 85% saturated, short chain ($C_{12} - C_{14}$), lipids. Nutritional studies showed that diets with a high proportion of palm oil are as healthy as any other diet, because the fat component is equivalent to that of other edible oils [22].

The fatty acid composition of palm oil (≈1:1 saturated to unsaturated fatty acids) is such that the oil is semi-solid at normal room temperature (22°C), which favours its use as the solid-fat component for margarine. More liquid oils need to be treated by hydrogenation to make them solid, leading to trans-fatty acids [115]. Palm oil is particularly suitable for deep-frying because of low content of polyunsaturated linoleic acid and a higher level of saturated fatty acids [102], which are less susceptible to oxidation. In addition, palm oil contains natural antioxidants such as tocopherols and carotenoids. During the last few years interest has been evinced in red palm oil as a source of vitamin A in human nutrition [109, 121].

Palm oil based oleochemicals, diesel and biodegradable plastics

Only about 10% of palm oil is used for non-food products such as oleochemicals e.g. sodium salts (soaps) and glycerol [70]. Furthermore, fatty acids and methyl esters from palm oil can be used as substitutes for diesel [15]. Alcohol can be also produced by fermentation of carbohydrates [22]. Biodegradable plastics such as polyhydroxybutyrate (PHB) could be synthesized from acetyl coenzyme A, the precursor for fatty acid-synthesis by transforming oil palm [55, 74].

Other products from oil palm

One of the major 'waste' products from the oil palm mill is the empty fruit bunches which are left after the fruit is steam-stripped from the fresh fruit bunches. These are ideally used is as fuel, and occasionally as mulch on young field planted palms [22]. After oil extraction, the remaining palm kernel material can be pressed to form fodder cake for cattle. Kernel cake from screw press machines has 8-13% residual oil which can make a useful contribution in animal diets [22]. The wood from palm trunks is relatively soft and is only really useful for forming pressed wooden objects [65, 67].

Oil Palm Crop Improvement

Advances in the crop improvement of oil palm

A more scientific approach to crop improvement began towards the middle of the 20th century, when Beinaret and Vanderveryen proved that the tenera form was generated by crossing dura and pisifera forms. This has been the single most important step in the genetic improvement of oil palm yields. The thick-shelled dura form is homozygous for one allele (ShSh), but has a yield disadvantage of about 25% compared to the thin-shelled heterozygous *tenera* form (Shsh) (commercially grown type), while the shell-less *pisifera* is homozygous for the alternative allele (shsh), but is often female-sterile, and cannot be grown as a crop.

The oil palm is monoecious and is naturally cross-pollinated. Although both parents have male and female flowers, for seed production the dura form is used as a female parent and pisifera has taken the role of the male parent.

Progress to date in oil palm breeding to increase oil yields has been spectacular, with a four-fold increase in yield over the last century [19]. Half of this has been ascribed to improvement in the genetic material. Within Deli Dura, a comparison of unselected material (grown under the best agronomic practices) derived from Bogor Botanical Gardens, Indonesia, and material after four generations of selection, suggested a 50% increase in yield, largely from an increase in mesocarp/fruit [19]. Hardon *et al.* [49] has estimated that improvement in yield per generation has been 10-15%, although this only equates to approximately 1% per year. In general, two approaches have been adopted for oil palm breeding. The basic approaches are Family and Individual Selection (FIS), and Reciprocal Recurrent Selection (RRS). FIS identifies the best families and then selects the next generation of parents from within these, using mainly phenotypic values. Such an approach resembles animal breeding [33]. RRS aims to develop separate and complementary pools of Pisifera and Dura which exploit hybrid vigour when crossed. These are progeny tested against each other before further development, allowing the generation of breeding values. This is the method favoured by maize breeders.

Rosenquist [100] suggested that some of the disadvantages of RRS were the tendency to produce inbreeding within the maternal and the paternal pools of material, as well as the more limited numbers of palms in the base population. In practice, many programmes are a combination of the two approaches, as oil palm has the advantage over maize of being perennial, with palms having good breeding values making repeated contributions to breeding material, and the female sterility of many pisifera sources (such as AVROS) makes progeny testing necessary.

However, recent results from Dami dura, suggest that inbreeding within the maternal pool is no longer an over-riding concern [26]. This might raise a doubt whether many of the deleterious recessive genes have already been eliminated through the RRS programme.

It is difficult to see how such impressive increases in oil yield can be maintained in future generations, without a major contribution from biotechnology.

Molecular Markers in Oil Palm Breeding

The primary objective of oil palm research is to increase profit per hectare from plantations; [17]. Breeding and selection of *Elaeis guineensis* began in the early 1920s and since then considerable improvements have been made both in yield and quality characters [51, 100]. The potential yield of the crop may be as high as 17 tons of oil per hectare [17]. The long breeding cycle and the variation still encountered suggest that there exists considerable scope for improvement in yield [17], disease resistance [36] and oil composition. Molecular markers represent one way in which it may be possible to select within material earlier in the breeding cycle to reduce the generational times.

Fingerprinting and linkage studies

Linkage mapping is frequently performed using polymorphic DNA markers such as isozymes [41, 42] and RFLPs (Restriction Fragment Length Polymorphism; [77]. RFLPs may be generated by gain or loss of restriction sites and/or indels between restriction sites. Jack *et al.* [56] and Mayes *et al.* [77] reported the potential in oil palm for marker identification and application and the use of highly informative oil palm RFLP markers for genotype characterization Mayes *et al.* [77, 78] and Jack *et al.* [57] reported construction of a RFLP map for oil palm and subsequent identification of a marker linked to shell thickness. There were 24 linkage groups identified (resolved to 21 by Rance *et al.*, [95]) although oil palm has only 16 chromosomes. 40 RFLP markers have also been used for estimating genetic similarity within oil palm breeding parents such as Deli, AVROS etc. [79].

RFLPs are gradually being replaced by less laborious and more polymorphic marker systems which are based on PCR, such as SSRs (Simple Sequence Repeat; [4, 110]) and AFLPs (Amplified Fragment Length Polymorphism; [116]). These systems are generally used for saturating already existing RFLP maps. Kulartne *et al.* [68] used AFLP markers for studying the diversity within different populations collected by Malaysian Palm Oil Board (MPOB). Purba *et al.* [92] investigated the genetic relationships between genotypes from different *E. guineensis* populations used in IOPRI (Indonesian Oil Palm Research Institute) breeding programme with the help of AFLP markers. The findings indicated that the crosses among African sub-populations were more potential for breeding than the crosses between the African and the Deli populations currently used in the reciprocal recurrent selection. An AFLP map with 20 linkage groups was reported by Chua *et al.* [16]. Microsatellites (or SSRs) are small arrays of tandem repeats that are simple in sequence (e.g. $[CA]_{10}$). SSRs are thought to have been produced by mutation, unequal crossing-over and DNA slippage. Microsatellites are neutrally evolving, co-dominant markers with high levels of genetic diversity and show Mendelian inheritance.

About 400 microsatellite markers (SSRs) were recently developed in the *E. guineensis* species by CIRAD (Centre de coopération internationale en recherche agronomique pour le développement), employing a microsatellite-enriched library building procedure from a hybridization-based capture methodology using biotin-labelled microsatellite oligoprobes and streptavidin-coated magnetic beads [4]. The SSR polymorphism was characterized in the *E. guineensis* and in the closely related species *E. oleifera*, in which utility of the SSR markers was observed, as well as on a subset of 16 other palm species to which some oil palm SSRs were potentially transferable [5]. Transferability of some date palm and peach palm SSRs in the *Elaeis guineensis* species was also observed [6, 7]. A reference linkage map of oil palm was developed in the control cross LM2T × DA10D, using 944 loci (255 SSR, 688 AFLP, locus *Sh*) distributed on 16 linkage groups representing the 16 chromosome pairs of the oil palm ([8], http://tropgenedb.cirad.fr/oilpalm/publications.html). Two AFLP markers were located on this map at 7 cM and 11 cM on each side of the *Sh* locus controlling the variety type of the fruit in oil palm, using bulk segregant analysis and linkage mapping methods. A further 103 SSRs were also developed by Mayes and co-workers ([90] PIPOC (International Palm Oil Congress); http://www.gen.cam.ac.uk).

A range of other marker systems is available, such as RAPDs (Random Amplified Polymorphic DNA; [117] and ISSR (Inter Simple Sequence Repeats, [123]. RAPDs use random sequence 10 nucleotide primers, and amplification of products depends on the presence of the complementary nucleotide sequence in the opposite orientation on each DNA strand within a stretch of less than approximately 3 Kbp (kilobase pairs). They are rapid, simple but often not reproducible and not transferable between laboratories. In ISSR, primers anchored at the 3' end that anneal to microsatellites are used. Shah [104] assessed the utility of RAPD markers for determination of genetic variation in oil palm and Moretzsohn [81] produced a RAPD linkage map of the shell thickness locus in oil palm. Rajanaidu *et al.* [93] used RAPDs and RFLPs to estimate genetic diversity and compare different populations collected by the MPOB.

There have been a number of Long-Terminal-Repeat Retrotransposon (LTR-RTN)-based marker systems utilizing retrotransposons, such as SSAP (Sequence Specific Amplified Polymorphism), IRAP (Inter Retrotransposon Amplified Polymorphism) and REMAP (Retrotransposon Microsatellite Amplified Polymorphism) developed in plants [91]. In the case of LTR RTN-based markers, the unique biological process of retrotransposition generates polymorphisms, which is an irreversible process resulting in insertions of RTNs into new sites without the loss of the parental copies. The consequences of retrotransposition range from the alteration of a few hundred base pairs to a few kilobases at the site of insertion. By contrast, marker systems employing simple sequence repeats (SSRs) are based on random small-scale changes (i.e. from one up to a few tens of nucleotides). Price *et al.* ([91], http://www.gen.cam.ac.uk) developed a multilocus IRAP marker system based on *copia* - like retrotransposons. The authors also reviewed marker methods based on retrotransposons and concluded that there was scope for using these methods in oil palm breeding and diversity analysis as an alternative to AFLP.

Potential applications of markers—simple traits

Development of markers would shorten the process of selection, especially at the nursery stage. Currently, crossing and growth in the nursery stage take approximately two years. After two years, seedlings are field planted (typically 143 palms per hectare). Although the seedlings start to produce fruit after 2.5 years, 5 years of recorded field data is necessary to properly assess the quality of a potential breeding palm. Marker-assisted selection (MAS) would accelerate the speed of the process. Some of the important areas of interest are shell-thickness, *virescens* and crown disease. A marker for shell thickness could have a potentially high commercial value in breeding programmes. The value would be in determining whether selected palms are *pisifera* or sterile *tenera* before progeny testing. This marker could also assess purity of commercial *tenera* seed lots [77]. Hartley [50] states that *virescens* is possibly monofactorial and dominant. Exploitation of this gene, if a segregating population were available, could make spotting ripe bunches much easier and thus contribute to a higher yield, through decreasing the loss of loose fruit.

The development and establishment of technologies, such as MAS [80], which would allow selection of individuals to be based on the genetic marker information, would represent a major step in the oil palm breeding.

Markers would also assist breeding programmes by: (1) helping to maintain diversity within the populations used for breeding, (2) identifying outcrosses in breeding programmes, and (3) allowing controlled introduction of foreign material into breeding programmes. MAS would be used in conjunction with the existing selection based on General Combining Ability (GCA), Specific Combining Ability (SCA), and other breeding values.

It has been noted by Corley and Tinker [22] that the progress made in the breeding programme depends both on the amount of variation present in the population before the selection starts, and on the heritability of the characteristics to be selected by the breeder. It is worth pointing out that the majority of characters measured by oil palm breeders are likely to be polygenic and these include bunch yield and its components; oil and kernel to bunch and their components; and carotene content. Although the major effect on bunch composition is the shell thickness gene, the attempts made to find a marker for this gene were not entirely successful (the linkage was not sufficiently close) [5, 78, 81], until the recent development of the Billotte map (Link2palm EU Framework 6 programme). Recently there has been an interest in high carotene content [93], for its increased nutritional value. It has also been shown that in the crosses derived from the Nigerian material the carotene content ranged from 180 to 2,500 ppm (parts per million).

The systematic approach of extensive phenotypic markers and GCA analysis would allow the association of markers and phenotypic characters and enable quantitative trait loci (QTL) analysis.

Potential applications of markers—complex traits

Marker based methods could provide a means of using QTL (Quantitative Trait Loci) analysis to target regions of a genetic map and measuring their effects. Linkage mapping of QTL depends on detecting the linkage disequilibrium between marker regions involved and the trait genes themselves. Localization of a QTL depends more on the population size than on the density of markers [66], as well as on the heritability of the trait studied. Firstly, the problem with the QTL mapping in oil palm is that population sizes are small and the basis of the genetic heritability of many quantitative traits is yet to be determined. The number of individuals in any one controlled cross is often limited (<90). Secondly, QTL are more easily identified for inbred lines but they are much more difficult to identify in out-crossing species where there is much more background variation. Thirdly, for parents to provide linkage information, the hybrid F1 must be heterozygous at both a marker and a linked QTL because only in this case can marker-trait associations be made in the progeny.

The availability of a large number of published SSR markers and dominant marker systems such as AFLP, REMAP and IRAP, and the relative ease with which those systems can be converted to automated marker typing, should allow rapid identification of markers linked to agronomically important traits and subsequent QTL analysis.

The first attempts at QTL analysis in oil palm were reported by Rance *et al.* [95]. The authors investigated the underlying genetic basis of quantitative traits (QTL) in oil palm and identified six marker regions associated with QTL effects; RFLP markers were identified linked to yield, oil/bunch and its components, and vegetative characters.

The recent development of the CIRAD genetic map enables a detailed QTL analysis as the populations reach maturity. Already, an initial analysis has identified a QTL for palm height (palm2LINK).

Potential applications of markers—disease resistance

The most serious diseases are *Fusarium* wilt (*Fusarium oxysporum* f.sp.*elaeidis*) in several parts of Africa and Latin America; *Ganoderma* in Asia, dry basal rot (*Ceratocystis*) in West Africa and fatal yellowing and sudden wither on new plantations in Latin America [43]. Unsuccessful attempts at finding RFLP markers linked to *Fusarium* wilt resistance have been made by Buchanan [10]. The present availability of highly polymorphic markers such as microsatellites makes this task much more feasible. *Ganoderma* trunk rot or basal stem rot has been a problem in some areas of Malaysia and Indonesia for the last 40 years, and in recent years it has also been the subject of much research in those countries. Most of this work has been summarized by Flood *et al.* [37], which also includes a general review of the current state of this disease in Asia by Ariffin [1]. de Franqueville *et al.* [25] showed that there were significant differences between families in *Ganoderma* incidence and thus demonstrated the feasibility of breeding for resistance and of potential marker application. There has been extensive research on fatal yellowing (*Thielaviopsis paradoxa*), much of it is reviewed by Gomez *et al.* [43]. To combat this disease, markers could be used to search for disease resistance factors within *Elaeis guineensis* material and in interspecific hybrids with *Elaeis oleifera*.

Genome Organization

BAC (Bacterial Artificial Chromosome) libraries

Plant genomes are remarkably large and dynamic and contain up to 80% of repetitive DNA [103].

BAC libraries can be constructed: (1) to represent the majority of the genome possible (restriction endonucleases such as *Hind*III and *Eco*RI, which cut frequently and show no significant sensitivity to methylation of the genomic DNA), or (2) to try to target the coding regions (rare cutting methylation sensitive restriction endonucleases, such as *Mlu*I and *Not*I). Large insert clone libraries, such as BACs [106] are essential tools. A partial BAC *Hind* III library was reported by Singh *et al.* [107] with the average size inserts of 40 kb and a complete *Hind*III oil palm BAC library has been made recently by CIRAD [87]. Complete libraries have the advantage that they contain all of the clonable sequences in the genome, but with the disadvantage that this requires very large numbers of clones and significant infrastructure and resources for development, maintenance and usage of those clones. An alternative approach is to develop 'targeted' BAC clones. These use methylation patterns within the genomic DNA to produce restriction enzyme cuts where there is no methylation present; lack of methylation is often indicative of coding and coding-associated regions [73]. The construction of an initial test BAC library for oil palm using the methylation-sensitive rare cutting restriction endonuclease, *Mlu*I has allowed this possibility to be examined (Hafeez and Mayes, unpublished). While the average insert size is relatively low (80 Kbp), hybridization of specific sequence probes to colonies [89] and analysis of 600 *Mlu*I BAC end-sequences, compared to *Hind*III derived BAC clones and end-sequence, confirm significant enrichment for low-copy sequences and the exclusion of high copy-number classes of retrotransposon from *Mlu*I clones.

Understanding something about the structure of the oil palm genome could be a major advantage in the development of targeted markers for MAS.

Uses of Conserved Synteny

Perhaps the nearest relative of oil palm from within those major crops, which have been studied in depth, could be cereals. It is estimated that the cereal group and oil palm diverged some of 100 million years ago [12]. Closer relatives of oil palm on which marker work has been conducted, are date palm (*Phoenix dactylus*) and coconut palm (*Cocos nucifera*), and Billotte *et al.* [6] have demonstrated that SSR markers developed in one species could have been used in another. Indeed, one of the targets of the recent EU Framework 6 INCO-DEV Link2palm proposal was to cross map markers in coconut and oil palm. While this work is underway and will provide useful information, oil palm is probably the most developed of these three species in terms of molecular tools and will only make limited gains from the other two species.

An initial study testing RFLPs which were used in cross mapping between a segment of Chromosome 9 in rice with chromosome 5 in wheat [38] gave some evidence for conservation of gene order between cereals and oil palm (3 out of 5 markers mapped in rice/wheat also showed linkage in oil palm [Hafeez, unpublished], however distances between markers were considerably greater in oil palm and currently with the lack of sequence data, this approach is unlikely to make a major impact in the next few years.

Vegetative Propagation of Oil Palm

The reasons for developing methods for vegetative production of oil palm are many, however, the drawback is that it has only one vegetative meristem and cannot be propagated by taking cuttings. The ability to rapidly propagate elite genotypes has immense potential for a species with a selection cycle of 10–16 years. Yield advantages over seedling populations were predicted in the order of 30%. In spite of the fact that attempts at propagating oil palm by tissue culture started in the 1960s, the discovery of abnormal flowering and severe bunch failure caused a major setback, just as commercial exploitation had begun [18]. Presently the application of clonal plant production in oil palm is still limited due to the occurrence of these somaclonal variants. It has been shown that the frequency of abnormal flowering varies between the clones, with some clearly being more susceptible [27]. Furthermore, Eeuwens *et al.* [29] showed that the tissue culture medium largely affected the embryoids during multiplication: short transfer intervals between mediums and low level of auxin result in a high level of cytokinin, which increases the risk of somaclonal variation. Better protocols are being developed alongside molecular work to try to understand the basis of the change in floral morphology and methods of exploitation have also been adapted. The use of tissue culture to produce 'clonal seed' is seen as an intermediate step towards full commercial exploitation. Clonal seed has the advantage that one of the parents is an elite clone, while the other is a seed-derived palm (as dura palms are limiting for the production of commercial seed, the dura is often the clonal palm). This has the advantage that any recessive somaclonal mutations should be masked by the contribution of the seedling gamete.

It has been widely assumed that abnormal flowering is an 'epigenetic' phenomenon. A definition of epigenetics, as postulated by Russo *et al.* [101], says "epigenetics is a study of mitotically and/or meiotically heritable changes in gene function that cannot be explained by the changes in DNA sequence". Cytosine methylation is one of the more prevalent and intriguing mechanisms for generating epigenetic change. Whereas symmetric CpG nucleotides are the major target for methylation in animals, methylcytosine in plants and fungi can occur at C residues at symmetric (CpG; CpNpG) and asymmetric CpXpX (where X is any base other than G) sites [35, 45]. Methylation is required for the normal development of animals and plants [64] and functions as a global repressor of gene expression [9]. There is increasing evidence that reduced DNA methylation can result in abnormal plant development [14, 34]. Interestingly, the "mantled" somaclonal variation in oil palm has been shown to be correlated with DNA hypomethylation, thus indicating that normal and abnormal plants differ in the degree of methylation of nuclear DNA [59, 75]. Furthermore, Kaeppler *et al.* [63], in the review of somaclonal variation in plants, suggested that variation in methylation levels as a result of tissue culture could possibly be a cause for the abnormal flowering.

It has been suggested that methylation evolved as a genomic defence against invasive DNA, including TEs (Transposable Elements) and viruses [120]. Although most plant TEs are not transcriptionally active (i.e. they are neutral), they can be reactivated under the conditions of abiotic or biotic stress, so-called 'genomic stress' [69]. However, only *Tnt*1 and *Tto*1 have been observed to be actually transposing [44, 54,

88]. Further application of a marker system based on retrotransposons to the oil palm somaclonal variants of the same ploidy exhibiting differences in the genome size would perhaps help in understanding the potential role of retrotransposons in the generation of somaclonal floral variants. It is worth noting that so far molecular markers have been unsuccessful in drawing out differences between normal and mantled palms, which would be sufficiently repeatable and efficient enough to be useful as a screening method (proteins – [72]; cytokinins – [62]; DNA markers – [13, 84, 105]; messenger RNA [60, 94, 98, 99, 111, 112].

Transformation Technology

The first evidence of transient expression of a reporter gene (glucurodinase; GUS) in oil palm tissues delivered by microprojectile bombardment was reported in 1993 [76] and since then significant effort has gone into developing potential transformation systems for oil palm, with major focus being oil composition [86], abscission of fruit [52] and higher resistance to disease [10]. The potential of this approach to oil palm improvement was examined and reported by Corley and Stratford [21] and initial transformation of oil palm with a marker gene has been reported [85]. Corley and Stratford [21] estimate that production of a transformant line in sufficient numbers for field planting could be 15 years, even after transformation has been achieved. Given the very mixed results using transformation in annual crops (and the sensitivity in the EU (European Union) over transgenic products) it could be a couple of decades before we see major implementation of this technology for oil palm.

Conclusion

The future of genetic improvement in oil palm

The impressive progress made over the last 90 years will be difficult to match in the future without a substantial contribution from marker-assisted selection and transformation and tissue culture approaches.

These offer the potential to short-circuit the long breeding and selection cycles currently needed for the genetic improvement and multiplication of oil palm, as well as offering novel solutions to genetic and agronomic problems through transformation.

Molecular genetics

Recent advances in creation of generally accessible, co-dominant SSR markers by the EU Link2Palm programme (http://www.neiker.net/link2palm/OilP/DefOIL.htm) will provide resources for generating genetic maps that can be integrated and used for dissecting the genetic basis of some of the key traits in oil palm. Generating molecular markers for direct MAS will be one consequence of this, but possibly, equally important will be gaining knowledge of the genetic basis and inter-relations for a number of agronomic traits. This may enable modification of breeding approaches for improving their efficiency without the intrinsic use of markers and their associated costs.

Also, the creation and characterization of the first oil palm EST clone database should enable the development of the first slide-based microarrays and facilitate the first use of the potentially extremely powerful transcriptomics approach in oil palm (http://www.mpob.gov.my/).

Limited sequence currently exists for oil palm, but this is likely to expand rapidly and a complete genome-sequencing programme for this important oil crop can only be a matter of time.

Models and comparative genetics

Using model systems to investigate oil palm has great potential, despite the considerable genetic distance between oil palm and any of the information-rich model systems. Whether the direct use of conserved gene order between species such as rice and oil palm will be feasible is unclear, although the ability to characterize genes in model systems such as *Arabidopsis thaliana* and to use information from conserved biochemical pathways will be invaluable.

Important areas where comparative genetics may make an impact include control of cell abscission in fruit, reduction in the problem of loose fruit collection, engineering of oil quality and composition through transgenic expression of homologous or heterologous oil biosynthesis genes, or even approaches to reduce height increment without a reduction in yield, as has occurred for many cereal crops with semi-dwarfing genes.

Transgenics and tissue culture

The intrinsic potential of clonal propagation should be realized in the coming years, once concerns over abnormal flowering subside, and this should facilitate the generation of transgenic oil palm expressing specific traits to improve a number of characters and develop novel traits such as materials for bioplastics, polyhydroxybutyrates (PHBs) and polyhydroxyalkanoates (PHAs) (http://minihelix.mit.edu/malaysia/research/me1.htm).

Perhaps one of the most important applications of transgenic technology will be the approach to reduce pests and diseases. For a disease like *Ganoderma*, where there is variation in genetic susceptibility but the inheritance appears complex, transgenic approaches may be critical and would certainly justify the research investment and, not the least, the 15-20 year timescale needed to produce and test a transgenic oil palm.

References

[1] Ariffin D, Idris AS, Gurmit S. Status of *Ganoderma* in oil palm. Flood J, Bridge PD, Holderness M, eds. 49-68,City, Country,: CABI Publishing, 2000.

[2] Barcellos E, Second G, Kahn F, Amblard P, Lebrun P, Sequin M. Molecular markers applied to the analysis of genetic diversity and to the biogeography of *Elaeis* (Palmae). In: Memoirs of The New York Botanical Garden, 1999; 83:191-201.

[3] Beinaret A, Vanderweyen R. Contribution à l'étude génétique et biométrique des variétés d'*Elaeis guineensis* Jacq. Publs INEAC Sér Sci, 1941; 27.

[4] Billotte N, Lagoda PJL, Risterucci AM, Baurens FC. Microsatellite enriched-libraries: applied methodology for the development of SSR markers in tropical crops. Fruits 1999; 54: 277-288.

[5] Billotte N, Risterucci AM, Barcelos E, et al. Development, characterisation, and across-taxa utility of oil palm (*Elaeis guineensis* Jacq.) microsatellite markers. Genome 2001; 44: 413-425.

[6] Billotte N, Marseillac N, Brottier P, et al. Nuclear microsatellite markers for the date palm (Phoenix dactylifera L.): characterization and utility across the genus *Phoenix* and in other palm genera. Mol Ecol Notes 2004a; 4: 256–258.

[7] Billotte N, Couvreur T, Marseillac N, et al. A new set of microsatellite markers for the peach palm (*Bactris gasipaes* Kunth); characterization and across-taxa utility within the tribe Cocoeae. Mol Ecol Notes 2004b; 4(2): 256-258.

[8] Billotte N, Marseillac N, Risterucci AM, et al. Microsatellite-based High Density linkage map in oil palm (*Elaeis guineensis* Jacq.). Theor Appl Genet (2005); 110(4): 754-765

[9] Bird, A. DNA methylation patterns and epigenetic memory. Genes and Dev; 2002; 16: 6-16.

[10] Buchanan, AG. Molecular genetic analysis of Fusarium wilt resistance in oil palm. Thesis, University of Bath, Bath, UK, 1999.

[11] Chase MW, Stevenson DW, Wilkin P, Rudall, PJ. Monocot systematics: a combined analysis. In: Rudall PJ, Cribb PJ, Cutler DF, Humphries CJ. eds. Monocotyledons: systematics and evolution. Royal Botanical Garden, Kew, United Kingdom, 1995.

[12] Chase MW. Monocot relationships: an overview. Amer J Bot 2004; 91: 1645-1655.

[13] Cheah SC, Siti Nor Akmar A, Ooi LCL, RahimahAR, MadonM. Detection of DNA variability in the oil palm using RFLP probes. In: Proc 1991 PORIM Int Palm Oil Conf, Kuala Lumpur, Malaysia: Palm Oil Res. Inst, 1993; 144-150.

[14] Chen RZ, Pettersson U, Beard C, et al. DNA hypomethylation leads to elevated mutation rates. Nature 1998; 395: 89-93.

[15] Choo Y M, Cheah KY. Biofuel. In: Basiron Y, Jalani BS, Chan KW, (eds). Advances in oil palm research. Vol. 2. Kuala Lumpur, Malaysia: Malaysia: Oil Palm Board, 2000; 806-844.

[16] Chua KL, Singh R, Cheah SC. Construction of oil palm (Elaeis guineensis) linkage maps using AFLP markers. In: Proc. 2001 Int Palm Oil Congr, Kuala Lumpur, Malaysia: Malaysian Palm Oil Board, 2001; 461-465.

[17] Corley RH V. Potential productivity of tropical perennial crops. Exp Aric 1983; 19: 217-237.

[18] Corley RHV, Lee C H, Law I H, Cundall E. Abnormal flower development in oil palm clones. In: Proc 1987 Int Oil Palm Conf. Kuala Lumpur, Malaysia, Palm Oil Res. Inst 1986; 173-185.

[19] Corley R H V, Lee C H. The physiological basis for genetic improvement of oil palm in Malaysia. Euphytica 1992; 60: 179-184.

[20] Corley R H V, Donough C R. Effects of defoliation on sex differentiation in oil palm clones. Expl Agric 1995; 31(17): 177-189.

[21] Corley H R V, Stratford R. Biotechnology and oil palm: opportunities and future impact. In: Proc 1998 Int Oil palm Conf '*Commodity of the past and the future*' Medan, Indonesia: Indonesian Oil Palm Res Inst. 1998; 80-91.

[22] Corley HRV, Tinker PB. The oil palm. Fourth edition. United Kingdom, Blackwell Science Ltd, 2003.

[23] Crone GR. The voyages of Cadamosto and other documents on Western Africa in the second half of the fifteenth century. Hakluyt Society 1937; Series II 80.

[24] Dahlgren RMT, Cliffort HT, Yeo PF. The families of monocotyledons: structure, evolution and taxonomy. Berlin, Germany: Springer, 1985.

[25] de Franqueville H, Asmady H, Jacquemard JC. Indications on sources of oil palm (Elaeis guneensis Jacq.) genetic resistance and susceptibility to *Ganoderma* sp., the cause of basal stem rot. In: Proc 2001 Int. Palm Oil Congr Kuala Lumpur, Malaysia: Malaysian Palm Oil Board 2001; 420-431.

[26] Dumortier F. Utilisation of oil palm genetic resources at DAMI oil palm research station. Paper presented at Int Symp 'Oil palm genetic resources and Utilization', 8-10 June 2000, Kuala Lumpur, Malaysia: Malaysian Oil Palm Board, 2000.

[27] Durand-Gasselin T, Duval Y, Baudoin L, et al. Description and degree of mantled flowering abnormality in oil palm (*Elaeis guineensis* Jacq.) clones produced using the

Orston-CIRAD procedure. In: Rao V, Henson IE, Rajanaidu N, eds. Recent developments in oil palm tissue culture and biotechnology, Kuala Lumpur, Malaysia: Palm Oil Res Inst, 1995; 48-63.

[28] Duval M R, Clegg M T, Chase M W, et al. Phylogenetic hypotheses for the monocotyledons constructed from *rbc*L sequence data. Ann Missouri Bot Gard; 1993; 80: 607-619.

[29] Eeuwens C J, Lord S, Donough C R, et al. Effects of tissue culture conditions during embryoid multiplications on the incidence of 'mantled' flowering in clonally propagated oil palm. Pl Cell Tissue Organ Culture 2002; 70: 311-323.

[30] Elenga H, Schwartz D, Vincens A. Pollen evidence of late Quaternary vegetation and inferred climatic change in the Congo. Paleography, Paleoclimat Paleoecol. 1994; 109: 345-346.

[31] Ergo A B. New evidence for African origin of Elaeis guineensis Jacq. by the discovery of the fossil seeds in Uganda. Ann Gembloux 1997; 102: 191-201.

[32] Eyre-Walker A, Gaut BS. Correlated rates of synonymous site evolution across plant genomes. Mol Biol Evol 1997; 14: 455-460.

[33] Falconer D S. Introduction to quantitative genetics. 2nd edn. London: Longman, 1981.

[34] Finnegan E, Peacock W J, Dennis E S. Reduced DNA methylation in *Arabidopsis thaliana* results in abnormal plant development. Proc Natl Acad Sci USA 1996; 93: 8449-8454.

[35] Finnegan E J, Genger R K, Peacock W J, Dennis E S. DNA methylation in plants. Annu Rev Plant Physiol Plant Mol Biol 1998; 49: 223-47.

[36] Flood J, Cooper RM, Lees P E. An investigation into the pathogenity of four isolates of *Fusarium oxysporum* from South America, Africa and Malaysia to clonal oil palm. J Phytopathol 1989; 124: 80-88.

[37] FloodJ, Bridge PD,Holderness M. *Ganoderma* diseases of perennial crops, Wallingford, UK: CABI Publishing, 2000. p101-112

[38] Foote T, Roberts M, Kurata N et al. Detailed comparative mapping of cereal chromosome regions corresponding to the Ph1 locus in wheat. Genetics 1997; 147: 801-807.

[39] Gaut B S, Muse S V, Clark W D, CleggM T. Relative rates of nucleotide substitutions at the *rbc*L locus in monocoyledonous plants. J Mol Evol 1992; 35: 292-303.

[40] Gaut B S, Muse S V, Clark W D, CleggM T. Substitution rates comparisons between grasses and palms: Synonymous rate differences at the nuclear gene *Adh* and parallel rate differences at the plastid gene *rbc*L. Proc Nat Acad Sci USA 1996; 93: 10274-10279.

[41] Ghesquiere M. Enzyme polymorphism in oil palm (*Elaeis guineensis* Jacq).1. Genetic-control of 9 enzyme-systems. Oleagineaux 1984; 39: 561-574.

[42] Ghesquiere M. Enzyme polymorphism in oil palm (*Elaeis guineensis* Jacq). 2. Variability and genetic – structure of 7 origins of oil palm. Oleagineaux 1985; 40: 529-537.

[43] Gomez P L. Ayala L, Munevar F. Characteristics and management of bud rot, a disease of oil palm. In: Pushparajah E, (ed). Proc Int Planters Conf 'Plantation tree crops in the new millennium: the way ahead'..) Kuala Lumpur, Malaysia: Incorp Soc Planters, 2000; 545-533.

[44] Grandbastien M A, Spielman A, Caboche M. *Tnt1*, a mobile retroviral-like transposable element of tobacco isolated by plant cell genetics. Nature 1989; 337: 376-380.

[45] Gruenbaum Y, Naveh-Many T, Cedar H, Razin A. Sequence specificity of methylation in higher plant DNA. Nature 1981; 292: 860-862.

[46] Hahn W J, Kress J W, Zimmer E A. 18S nrDNA sequence phylogenetics of the monocots. Amer J Bot 1996; 83: 211-212.

[47] Hahn W J. A phylogenetic analysis of the Arecoid Line of palms based on plastid DNA sequence data. Mol Phylogen Evol 2002a; 23: 189-204.

[48] Hahn W J. A molecular phylogenetic study of the *Palmae* (*Arecaceae*) based on *atp*B, rbcL, and 18S nrDNA sequences. Syst Biol 2002b; 51: 92-112.

[49] Hardon J J, CorleyRH V, Lee C H. Breeding and selecting the oil palm. In: Abbot AJ, Atkin RH, eds. Improving vegetatively propagated crops London: Academic Press, 1987; 63-81.

[50] Hartley C W S. The Oil Palm. Second edition. London: Longmans 1977.

[51] Hartley C W S. The Oil Palm. Third edition. London: Longmans, 1988.

[52] Henderson J, Davies HA, Heyes SJ,Osborne DJ. The study of a monocotyledon abscission zone using microscopic, chemical, enzymatic and solid state ^{13}C CP/MAS NMR analyses. Phytochem 2001; 56: 131-139.

[53] Heywood HV. Flowering plants of the world. London: B T Batsford Ltd, 1993.

[54] Hirochika H. Activation of tobacco retrotransposons during tissue-culture. EMBO J 1993; 12: 2521-2528.

[55] Houmiel K L, Slater S, Broyles D, et al. Poly(beta-hydroxybutyrate) production in oilseed leukoplasts of *Brassica napus*Planta. 1999; 209(4): 547-555.

[56] Jack P L, Dimitrijevic T A F, Mayes S. Assessment of nuclear mitochondrial and chloroplast RFLP markers in oil palm (*Elaeis guineensis* Jacq.). Theor Appl Genet 1995; 90: 643-649.

[57] Jack P L, James C, Price Z, et al. Application of DNA markers in oil palm breeding. In: Proc 1998 Int Oil Palm Conf, Medan, Indonesia: Indonesian Oil Palm Res Inst 1998; 315-324.

[58] Jacquin NJ. *Selectarum stirpium*. Americanarum historia, 1793.

[59] Jaligot E, Rival A, Beulé T, et al. Somaclonal variation in oil palm (*Elaeis guineensis* Jacq.): the DNA methylation hypothesis. Plant Cell Reports 2000; 19: 684-690.

[60] Jaligot E, Beulé T, Rival A. Methylation-sensitive RFLPs: characterisation of two oil palm markers showing somaclonal variation- associated polymorphism. Theo Appl Genet 2002; 104: 1263-1269.

[61] Jones D L. Palms throughout the world. Australia: Reed Books, 1994.

[62] Jones LH, Hanke DE, Euwens CJ. An evaluation of the role of cytokinins in the development of abnormal inflorescences in oil palms (*Elaeis guineensis* Jacq.) regenerated from tissue culture. J Pl Growth Reg 1995; 14: 135-142.

[63] Kaeppler SM, Kaeppler H F, Rhee Y. Epigenetic aspects of somaclonal variation in plants. Pl Molec Biol 2000; 43: 179-188.

[64] Kakutani T, Jeddeloh JA, Flowers SK, et al. Developmental abnormalities and epimutations associated with DNA hypomethylation mutations. Proc Nat Acad Sci USA 1996; 93: 12406-12411.

[65] Kamarudin N, Walker A K, Basri Wahid M, et al. Population studies of Oryctes rhinoceros in an oil palm replant using pheromone traps. In: Preprints, 1999 PORIM Int Palm Oil Conf Kuala Lumpur, Malaysia: Palm Oil Research Inst 1999; 477-496.

[66] Kearsey M J, Pooni HS. The genetical analysis of quantitative traits. ? Chapman and Hall, 1996, London, UK

[67] Koh M P, Rahim S, Mohd Nor MY, et al. Manufacture of building materials from oil palm biomass. In: Gurmit Singh *et al.* eds. Oil palm and the environment – a Malaysian perspective, Kuala Lumpur, Malaysia: Malaysian Oil Palm Growers Council, 1999; 199-211.

[68] Kulartne R S, Shah F H, Rajanaidu N. Investigation of genetic diversity in African natural oil palm populations and Deli dura using AFLP markers. Paper presented at Int Symp. 'Oil palm genetic resources and Utilization', 8-10 June, Kuala Lumpur, Malaysia: Malaysian Oil Palm Board, 2000.

[69] Kumar A, Bennetzen J L. Plant retrotransposons. Annual Review Genet, 1999; 33(1): 479-532.

[70] Kunton A and Hamirin K. Soaps from palm products. In: Advances of oil palm research, Vol.2. (Ed. by Y Basiron, B S Jalani and Chan K W); p 1102-1140, Malaysian Oil Palm Board, Kulala Lumpur, Malaysia

[71] Madon M, Clyde M.. Cytological analysis of *Elaeis guineensis*. Elaeis. 1995; 17(2): 124-134.

[72] Marmey P, Besse I, Verdeil J L. A proteic marker found to differentiate 2 types of calli of the same clones in oil palm. C R Acad Sci, Paris, Ser. III, 1991; 313: 333-338.

[73] Martienssen R. Transposons, DNA methylation and gene control. Trends In Genetics 1998; 14 (7): 263-264.

[74] Masani Mat Usus A, Ho C L, Parveez G K A. Construction of PHB gene expression vectors for the production of biodegradable plastics in transgenic oil palm. In: Proc 2001 Int Palm Oil Congr Kuala Lumpur, Malaysia: Malaysian Oil Palm Board, 2001; 674-711.

[75] Matthes M, Singh R, Cheah S C, Karp A. Variation in oil palm (*Elaeis guineensis* Jacq.) tissue culture-derived regenerants revealed by AFLPs with methylation-sensitive enzymes. Theor Appl Genet 2001; 102: 971-979.

[76] Mayes, S., Batty, N., Jack, P.L., and Corley R.H.V. Applications of biotechnology to oil palm breeding.Proceedings 1993 PIPOC Int Palm Oil Cong. Kuala Lumpur, Malaysia: 1993; 885-898

[77] Mayes S, James CM, Horner SF, et al. The application of restriction fragment polymorphism to genetic fingerprinting in oil palm (*Elaeis guineensis* Jacq.). Mol Breeding 1995; 2: 175-180.

[78] Mayes S, Jack PL, Marshall DF, Corley HRV. Construction of a RFLP map genetic linkage map for oil palm (*Elaeis guineensis* Jacq.). Genome 1997; 40: 116-122.

[79] Mayes S, Jack P L, Corley R H. The use of molecular markers to investigate the genetic structure of an oil palm breeding programme. Heredity 2000; 85(3): 288-93.

[80] Mohan M, Nair S, Bhagwat A, et al. Genome mapping, molecular markers, and marker assisted selection in crop plants. Mol Breeding 1997; 3: 87-103.

[81] Moretzsohn M C, Nunes C D M, Ferriera M E,Grattapaglia, D. RAPD linkage mapping of the shell thickness locus in oil palm (*Elaeis guineensis* Jacq.). Theor Appl Genet 2000; 100: 63-70.

[82] Morton B R, Gaut B S, Clegg MT. Evolution of alcohol dehydrogenase genes in palm and grass families. Proc Natl Acad Sci USA 1996; 93: 11735-11739.

[83] Nadot S, Bittar G, Carter L, et al. A phylogenetic analysis of monocotyledons based on the chloroplast gene rps4, using parsimony and new numerical phenetic methods. Mol Phyloge Evol 1995; 4: 257-282.

[84] Paranjothy K, Ong L M, Sharifah S. DNA and protein changes in relation to clonal abnormalities. In: Rao V, Henson IE, Rajanaidu N. eds. Recent developments in oil palm tissue culture and biotechnologyKuala Lumpur, Malaysia: Palm Oil Res. Inst, 1995; 86-97.

[85] Parveez G K A, Masri M M, Zainal, A., et al. Transgenic oil palm: production and projection. Biochem Soc Trans 2000; 28: 969–972.

[86] Parveez G K A, Chowdhury, M K U, and Saleh, N M. Current status of genetic engineering of oil bearing crops. Asia Pasific J Molec Biol Biotechnol 1994; 2:174-192

[87] Piffanelli P, Lagoda P, Clément D, Thibivilliers S, Vilarinhos A D, Sabau X, Billotte N, Séguin M, Chalhoub B and Glaszmann J C. Bactrop: A BAC-based platform for physical mapping of tropical species. [Poster]. In : Plant, Animal and Microbe Genomes 10th Conference, January 12 - 16, 2002, San Diego, California

[88] Pouteau S, Grandbastien MA, Boccara M. Microbial elicitors of plant defence response activate transcription of a retrotransposon. Plant J 1994; 5: 535-542.

[89] Price, Z; Dumortier, F; MacDonald, D.W., and Mayes, S. Characterisation of *copia*-like retrotransposons in oil palm (*Elaeis guineensis* Jacq.) Theor Appl Genet 2002; 104 p 860-867

[90] Price Z, Dumortier F, Mayes S. The Development and Initial Application of DNA - Based Genetic Markers to the new Britain Palm Oil Ltd (PNG) Breeding Programme. Proceedings 2003 PIPOC IntPalm Oil Cong. Kuala Lumpur, Malaysia: 2003; 885-898.

[91] Price Z, Schulman A, Mayes, S. Development of new marker systems: oil palm. Plant Genet Res: Charact and Util 2004; 1(2/3): 105-115.

[92] Purba A R, Noyer J L, Baudouin, L, et al. A new aspect of genetic diversity of Indonesian oil palm (*Elaeis guineensis* Jacq.) revealed by isoenzyme and AFLP markers and its consequences to breeding. Theor Appl Genet 2000; 101: 956-961.

[93] Rajanaidu N, Maizura I, Cheah S C. Screening of oil palm natural populations using RAPD and RFLP molecular markers. Paper presented at Int Symp 'Oil palm genetic resources and Utilization', 8-10 June, Kuala Lumpur, Malaysia: Malaysian Oil Palm Board, 2000.

[94] Rajinder S, Cheah S C, Madon M, et al. Genomic strategies for enhancing the value of oil palm. In: Proc 2001 Int Palm Oil CongrKuala Lumpur, Malaysia: Malaysian Palm Oil Board, 2001; 3-17.

[95] Rance K A, Mayes S, Price Z, et al. Quantitative trait loci for yield components in oil palm (*Elaeis guineensis* Jacq.). Theor Appl Genet 2001; 103(8): 1302-1310.

[96] Raynaud F I, Maley J, Wirrmann D. Vegetation and climate in the forest of S-W Cameroon since 4770 years BP. Pollen analysis of sediments from Lake Ossa. C. R Acad Sci Ser IIA. Sciences de la terre et des plantes 1996; 332:479.

[97] Rival A, Beule T, Barre P, et al. Comparative flow cytometric estimation of nuclear DNA content in oil palm (*Elaeis guineensis* Jacq) tissue cultures and seed derived plants. Plant Cell Reports 1997; 16: 884-887.

[98] Rival A, Tregear J, Verdeil J L, et al. Molecular search for mRNA and genomic markers of the oil palm 'mantled' somaclonal variants in oil palm (*Elaeis guineensis*, Jacq.). Acta Hort 1998; 461: 165-171.

[99] Rival A, Tregear J, Jaligot E, et al. Oil palm biotechnology at CIRAD. In Proc 2001 Int Palm Oil Congr Kuala Lumpur, Malaysia: Malaysian Palm Oil Board, 2001; 51-82.

[100] Rosenquist E A. The genetic base of oil palm breeding populations. Proc Palm Oil Res Inst Malaysia 1985; 10: 10-27.

[101] Russo V E A, Martienssen R A, Riggs A D. Epigenetic mechanisms of gene regulation. Cold Spring Harbor, NY: Cold Sprin Harbor Laboratory Press, 1996.

[102] Sambanthamurthi R, Sundram K, Tan Y. Chemistry and biochemistry of palm oil.Prog Lipid Res 2000; 39: 507-58.

[103] San Miguel P, Tikhonov A, Jin Y K, et al. Nested retrotransposons in the intergenic regions of the maize genome. Science 1996; 274: 765-768.

[104] Shah F H, Rashid O, Simons A J, Dunsdon A. The utility of RAPD markers for determination of genetic variation in oil palm (*Elaeis guineensis* Jacq.). Theor Appl Genet 1994; 89: 713-718 [105] Sharifah S S A, Singh R, Cheah SC. Molecular dissection of the floral clonal abnormality in oil palm. In: Preprints, 1999 PORIM Int Palm Oil Conf Malaysia, Kuala Lumpur: Palm Oil Research Inst, 1999; 477-496.

[106] Shizuya H, Birren B, Kim U-J, et al. Cloning and stable maintenance of 300-kilobase-pair fragments of human DNA in Escherichia coli using an F-factor-based vector. Proc Natl Acad Sci USA 1992; 89: 8794-8797.

[107] Singh R, Lessard P A, Guan T S, Panandam J M, Sinskey, A and Cheah S C. Preliminary attempts at the construction of large insert DNA libraries for oil palm (*Elaeis guineensis* Jacq.). J. Oil Palm Res (2003);15(1): 12-20.

[108] Soltis D E, Soltis P S, Nickrent D L, et al. Angiosperm phylogeny inferred from 18S ribosomal DNA sequences. Ann Missouri Bot Gard 1997; 18: 1-49.

[109] Sundram K, Sambanthamurthi R, Tan Y A. Palm fruit chemistry and nutrition.Asia Pac J Clin Nutr 2003; 12: 355-62.

[110] Tautz D, Renz M. Simple sequences are ubiquitous component of eucaryotic genomes. Nucl Acids Res 1984; 12: 4127-4138.

[111] Toruan-Mathius N, Harris N, Ginting G. Use of the biomolecular techniques in studies of abnormalities in oil palm clones. In: Proc 1998 Int Oil Palm Conf. '*Commodity of the past and the future*' Medan, Indonesia: Indonesian Oil Palm Res Inst. 1998; 115-126.

[112] Tregear J W, Morcillo F, Richaud F, et al. Characterization of a defensin gene expressed in oil palm inflorescences: induction during tissue culture and possible association with epigenetic somaclonal variation events. J Exp Bot 2002 53: 1387-1396.

[113] Uhl N W, Dransfield J. Genera palmarum. Ithaca, NY: International palm society and L. H Bailey Hortorium, 1987.

[114] Uhl N W, Dransfield J, Davis J I, et al. Phylogenetic relationship among palms: Cladistic analyses of morphological and chloroplast DNA restriction site variation. Monocot systematics: a combined analysis. In: Rudall PJ, Cribb DF, Cutler DF, Humphries CJ, (eds). Monocotyledons: systematics and evolution.; Royal Botanical Garden Kew , United Kingdom,1995.

[115] van Duijn G. Technical aspects of trans-reduction in margarines. Oil Corps Gras Lipides 2000; 7: 95-98.

[116] Vos P, Hogers R, Bleeker M, et al. AFLP: a new technique for DNA fingerprinting. Nucl Acids Res 1995; 23 (21): 4407-4414.

[117] Williams J, Kubelik A, Livak K, et al. DNA polymorphisms amplified by arbitrary primers are useful as genetic markers. Nucleic Acids Res 1990; 18: 631-6535.

[118] Wilson M A, Gaut B, Clegg M T. Chloroplast DNA evolves slowly in the palm family (*Arecaceae*). Mol Biol Evol 1990; 7: 303-314.

[119] Wolfe K H, Gouy M, Li W H, Sharp P M. Rates of nucleotide substitution vary greatly among plant, mitochondrial, chloroplast and nuclear DNAs. Proc Natl Acad Sci USA 1987; 84: 9054-9058.

[120] Yoder J A, Walsh C P, Bestor T H. Cytosine methylation and the ecology of intragenic parasites. Trends in Genet 1997; 13 (8): 335-340. [121] Zagre N M, Delpeuch F, Traissac P, Delisle H. Red palm oil as a source of vitamin A for mothers and children: impact of a pilot project in Burkina Faso. Public Health Nutr 2003; 6: 733-742.

[122] Zeven A C. On the origins of the oil palm (*Elaeis guineensis* Jacq). Grana Palynol 1964; 5: 121-123.

[123] Zietkiewicz E, Rafalski A, Labuda A. Genome fingerprinting by simple sequence repeat (SSR)-anchored polymerase chain reaction amplification. Genomics 1994; 20: 176-183.

CHAPTER

5

Genome Complexity of *Allium*

ELLEN B. PEFFLEY

Texas Tech University, Department of Plant & Soil Science, Lubbock, Texas, USA

ABSTRACT

The *Allium* genome is one of the largest among monocotyledonous plants. Understanding relationships between the Alliaceae taxa facilitates in determining the strategies to select superior germplasm used in crop improvement programs. Species hybridization broadens genetic variation. Molecular-marker technology allows portions of genomes to be targeted and exploited when designing new crops with desirable traits found in wild or related species. In order to place *Allium* species in proper taxa, the *Allium* genus has been investigated at molecular and DNA levels so that genetic differences that separate and identify the species are revealed. The objective of this monograph is to present a survey of recent biochemical, molecular, and cytogenetic approaches that have been used on alliaceae taxa so as to elucidate genome organization in *Allium*.

Key Words: Alliaceae, molecular markers, genome organization, chemotaxonomy, DNA technology

Abbreviations: AFLP = amplified fragment length polymorphism, CAPs = cleaved amplified polymorphic sequences, CMS = cytoplasmic male sterility, ESTs = expressed sequence tags, FISH = fluorescence in situ hybridization, GISH = genomic in situ hybridization, Indels = short

Address for correspondence: Texas Tech University, Department of Plant & Soil Science, PO Box 42122 (post), Corner of 15th and Boston, Ag Science 101 (UPS), Lubbock, Texas 79409, USA. E-mail: ellen.peffley@ttu.edu

insertion = deletion event, ITSs = internal transcribed spacers of nuclear ribosomal DNA, RAPDs = random amplified polymorphic DNA markers, RFLPs = restriction fragment length polymorphism, SCAR = sequence-characterized amplified region, SNPs = single nucleotide polymorphisms, SSRs = simple sequence repeats

INTRODUCTION

Taxonomically *Allium* is a complicated genus, placed ambiguously into the families Liliaceae [80] and Amaryllidaceae by the International Code of Botanical Nomenclature. *Allium* is the most important genus of Alliaceae, nevertheless classification is problematic and not explicit [36]. Watson and Dallwitz [125] place Alliaceae into Subclass Monocotyledonae, Superorder Liliiflorae, and Order Asparagales. Others [31] follow the hierarchy of Takhtajan [113] and place Alliaceae into Class Liliopsida, Subclass Liliidae, Superorder Liliianae, and Order Amaryllidales with four subgenera: (1) *Melanocrommyum*, (2) *Rhizirideum s. lat.*, (3) *Amerallium*, and (4) *Allium*. While members of the Alliaceae are morphologically varied, they share common characteristics. They are mucilaginous herbs, produce essential oils, flavonols (kaempferol and quercetin) but alkaloids are absent; they are bulbaceous or rhizomatous; mesophytic or xerophytic; leaves are aromatic (onion-scented, with allylic sulphides), simple, not evergreen, and may be alternate, distichous (leek), spiral, flat, rolled, angular, sessile or petiolate, or sheathing [125]. Early classification by morphology ([20] referenced by [78]) divided *Allium* species into seven groups. A later subdivision ([124], cited by [31]) placed cultivated *Allium* species into four sections: (1) Cepa (bulb onion), (2) Phyllodolon (Japanese bunching onion), (3) Porrum (garlic and leek), and (4) Rhizirideum (chive). Subsequent classification based on morphology, crossability, and karyotype by Traub [114] divided them among the four sections: (1) Allium, (2) Cepa, (3) Fistulosa, and (4) Rhizirideum.

Of the eight independent global centers of origin established by Nikolai Vavilov, *Allium* species are found in four [123]. From Center of Origin I, Chinese, came *A. chinense* Don. (*A. odorum* L.), *A. fistulosum* L., *A. macrostemon* Bge. *A. pekinense* Prokh.; from Center III, Inner-Asian (northwestern India, Afghanistan, and Uzbekistan) came *A. cepa* L.

(sensu lato), *A. pskemense* Fedtsch., *A. vavilovii* Vved., *A. sativum* L., and *A. longicuspis* E. Regel.; from Center IV, Asia Minor, are *A. cepa* (secondary origin), *A. porrum* L., and *A. ampeloprasum* L.; and from Center VI, Abyssinian (Egypt and Somalia), *A. ascalonicum* L. arrived from Southwest Asia (Pakistan and Iran). Estimations of the number of species within the *Allium* genus vary, from 600 [125] to some 750 species [109] belonging to three subgenera and 18 sections [114]. *Alliums* are cultivated as ornamentals [51] and have been domesticated as food crops. Although most *Alliums* are edible and consumed by indigenous populations, some of the most important species are cultivated as food crops, which include *A. ampeloprasum* syn. *A. porrum* (leek), *A. cepa* (bulb onion), *A. fistulosum* (Japanese bunching or Welsh onion), *A. sativum* (garlic), and *A. schoenoprasum* (chives).

These and other *Allium* species have been investigated at molecular and DNA levels in an attempt to place them in proper taxa. The objective of this monograph is to present a survey of recent biochemical, molecular, and cytogenetic approaches that have been used on to elucidate genome organization in *Allium*. The intention is not to provide an exhaustive treatment of classification (for this refer to others [109, 31]), but rather to demonstrate the methods that have been used, singly and in combination, to understand the organization of the *Allium* genome.

SPECIES HYBRIDIZATION

Hybridization plays a significant role in the production of novel phenotypes for the evolution and speciation of plants [126]. Crosses between *Allium* taxa have been made to attain novel types and to introgress desirable characters from one species into another. Interspecific hybridization within the genus has been of keen interest for many decades. The earliest of the crosses was between *Allium cepa* and *A. fistulosum* [24, 25, 66, 73]. Due to sterility barriers *A. cepa* x *A. fistulosum* interspecific derivatives are difficult to obtain [7, 15, 16, 24, 25, 67, 73, 84, 85, 86, 116], yet introgressants have been recovered in advanced *A. cepa* x *A. fistulosum* backcross populations [88].

Non-Mendelian inheritance of genes in interspecific hybrids and derivatives, manifested by full or partial sterility, results from pre- or post-zygotic barriers. Various strategies to expose these barriers include biochemical, cytogenetic, and DNA assays. In *Allium*, early mechanisms hypothesized that sterility of F_1 generations and preferential genome transmission in advanced generations included stylar-incongruity [118] and nuclear-cytoplasmic incompatibility [7, 115, 117]. Pre-zygotic barriers observed in crosses of *A. cepa* x *A. ampeloprasum* [89] and *A. cepa* x *A. sphaerocephalon* [52] have been overcome with judicious breeding efforts. In *A. cepa* x *A. fistulosum* F_2 and backcrosses, the nuclear genome most closely related to the maternal cytoplasmic genome is preferentially recovered in advanced generations [16, 116]; with persistent crossing, recombinants with paternal and maternal genomes have been recovered [41, 88]. Post-zygotic barriers selectively eliminate zygotes in plants causing distorted segregation in progeny populations. This distortion has been attributed to disharmony in interaction between parental genomes in the embryo and the endosperm [35]. A recent systematic examination of interspecific derivatives by Mangum and Peffley [74] describes a third mechanism which more fully explains distorted gene segregation and fecundity of advanced generations of *A. cepa* x *A. fistulosum* interspecific hybrids, i.e. central cell nuclear-cytoplasmic incongruity. The latter provides evidence for the underlying mechanism of incongruity in the central cell and predicts the success realized when genomes used for 'bridging' can be attributed to the resulting balanced and healthy endosperm formation. A success story using a bridge cross involves three genomes, *A. roylei*, *A. fistulosum* and *A. cepa*. Early studies reported these species to be closely related [64, 121] and that crosses between *A. fistulosum* and *A. roylei* resulted in a large extent of recombination [54]. Khrustaleva and Kik [55] used the *A. roylei* genome as a bridge between *A. fistulosum* and *A. cepa*. They observed that recovery of individuals with three genomes was possible because *A. roylei* genes could circumvent or restore the nucleo-cytoplasmic imbalance that led to the sterility shown by the (*A. fistulosum* x *A. cepa*) x *A. cepa* backcross of Ulloa et al. [117]. Information attained by using the underlying mechanisms was utilized using classical cytogenetic analyses of pure and derived (i.e. of species hybridization [97]) karyotypes and genomic hybridization, revealing recombinant chromosomes. This investigation exemplifies the current state of *Allium* research where progressively a synergistic blending of techniques as genomes is probed.

BIOCHEMICAL AND MOLECULAR APPROACHES FOR CLASSIFICATION

The following section is a survey of salient molecular, cytogenetic, and chemotaxonomic approaches used in systematics of different taxa of Alliaceae. The literature is sorted according to the marker system used, listed alphabetically by taxa, followed by reference number. Where more than one approach has been used, the reference may be repeated or categorized by the most relevant approach; where more than one taxa has been investigated, the references may either be repeated, or sorted by taxa first appearing in the title.

Molecular markers have been used to elucidate geographic distribution patterns, patterns of genetic diversity, track introgression, identify hybrids, sort taxa, and detect maternal or paternal ancestry, construct genetic maps, genome analyses [61, 62], and in comparative studies [11]. Nuclear markers, which may be codominant (isozyme and microsatellites) or dominant [RAPDs (random amplified polymorphic DNA markers), RFLPs (restriction fragment length polymorphism), AFLPs (amplified fragment length polymorphism), and ITSs (internal transcribed spacers of nuclear ribosomal DNA)] are useful to track paternal or maternal inheritance. Mitochondrial markers have been associated with cytoplasmic male sterility (CMS) and used to distinguish male fertile from male sterile cytoplasm. Genomic changes may be revealed by cytogenetic chromosomal analyses and changes in nuclear repetitive DNA further elaborated with GISH (genomic in situ hybridization).

DNA CONTENT IN *ALLIUM*

Among vascular plants there is approximately a hundred-fold variation in nuclear DNA content [10, 94]. *Alliums* are one of the several monocotyledonous plants (Table 1). Cultivated onion has nuclear DNA which is 40 times greater than that of rice, and in the case of lack it is 60 times greater. (0.8pg/2C) [4].

Nuclear DNA amounts of *Allium* vary widely by species [94] and are as low as 15.2 pg/2C (*A. sibiricum*) [49] and as great as 94.96 pg (picogram) in the octaploid *A. nutans* [6]. In a comparative study of nuclear DNA content of cultivated plant species across families, members of the Liliaceae and Alliaceae families have some of the largest genomes, *Tulipa*, *Allium ampeloprasum* and *A. cepa*, 63.6, 50.27, and

Table 1. Nuclear DNA content of some important monocotyledonous plant species determined by flow cytometry [from 4]

Scientific Name	*Common Name*	*Family*	*Nuclear DNA Content pg/2C*[a]	*mbp*[b]*/1C*
Aegilops squarrosa	Goatgrass	Poaceae	8.34	4,024
Allium ampeloprasum	Leek	Alliaceae	50.27	24,255
Allium cepa	Onion	Alliaceae	32.74	15,797
Asparagus officinalis	Asparagus	Liliaceae	2.71	1,308
Avena sativa	Oats	Poaceae	23.45	11,315
Hordeum vulgare	Barley	Poaceae	10.10	4,873
Oryza longistaminata	African rice	Poaceae	0.78	376
Oryza sativa ssp. Indica	Rice	Poaceae	0.87-0.96	419-463
Oryza sativa ssp. Japonica	Rice	Poaceae	0.86-0.91	415-439
Oryza sativa ssp. Javanica	Rice	Poaceae	0.88	424
Triticum aestivum (2n=6X)	Wheat	Poaceae	33.09	15,966
Triticum monococcum	Einkorn wheat	Poaceae	11.92	5,751
Tulipa sp.	Garden tulip	Liliaceae	51.2-63.6	24,704-30,687
Zea mays	Corn	Poaceae	4.75-5.63	2,292-2,716

[a]Value for each cultivar determined by two or more measurements of at least 2,000 nuclei.

[b]1 picogram (pg) = 965 million base pairs (Mbp) (Bennett and Smith 1976)

32.74 pg/2C, respectively; diploid onion (*A. cepa*) carries the same amount as hexaploid wheat (*Triticum aestivum*) (33.09 pg/2C) [4, 75]. Accurate estimations of nuclear genomes are critical for plant genome studies. Precise quantification of DNA will aid in detecting nuclear DNA that has been modified by ploidy changes, introgressed segments, or additions or deletions of chromosomes or chromosome segments [95]. Establishing a plant standard in calculating the DNA content of plants may reduce a source of error, which affects estimations of DNA [48]. Johnston et al. [48] found general congruence of DNA contents when estimates as determined by flow cytometry were compared with those of standard Feulgen microspectrophotometry and a suggestion by them to launch an international, coordinated effort to set an agreed-upon plant calibration standard may lead to an agreement within the Allium community of researchers on baseline quantities of DNA.

Over 85% of the total genomic DNA in cereals consists of repetitive sequences, whereas genes constitute about 1%. Repetitive DNA sequences represent the most variable part of the eukaryotic genome and can have major effects on genome organization and function [11]. Earlier cytological techniques such as C-banding [5, 50] and giemsa banding [87, 100] have been supplanted with more refined and precise cellular and molecular analyses, and gene localization in situ at the chromosomal level has greatly influenced the study of biodiversity. In situ hybridization in *Allium* is a powerful and accurate tool when mapping DNA onto chromosomes, including, but not limited to single copy [8, 19], or highly repetitive DNA sequences [65], satellite DNA [8], chromosomes/ chromosome segments flow-sorted or microdissection [33], single gene location [96], total genomic DNA which provides a general picture of genomic divergence within taxa [99, 101], and alien chromosome segments [41]. Genomic *in situ* hybridization with total genomic DNA as a probe provides unique information about similarities between repetitive DNA from related species, as well as the physical location of conserved sequences on chromosomes; this method can yield a generalized picture of genomic divergence inside taxonomic groups. Telomeric repeat sequences (or lack of) can even more precisely predict phylogenetic groupings [2]. Molecular documentation serves as a useful aid in tracing genetic diversity and phylogenetic studies [61].

MOLECULAR MARKER SYSTEMS

Isozyme Analysis

Allium cepa [41, 70, 73, 77, 84, 98]
Allium fistulosum [42, 70, 77, 84, 88]
Allium oschaninii [70]
Allium sativum [44, 71]
Allium vavilovii [72]
Allium subg. *Allium* [61]
Allium subg. *Amerallium* [61]
Allium subg. *Bromatorrhiza* [61]
Allium subg. *Caloscordum* [61]
Allium subg. *Melanocrommyum* [61]
Allium subg. *Rhizirideum* [61]

DNA Analysis

i. Restriction fragment length polymorphism (RFLP) of chloroplast (cp) DNA

  Allium altaicum [30]
  Allium cepa var. *ascalonicum* [3]
  Allium fistulosum [30]
  Allium subg. *Melanocrommyum* [82]
  Allium x *wakegi* [3]
  Allium (29) *spp*. [78]

mitochondrial (mt) DNA

  Allium ampeloprasum [27, 38]
  Allium cepa [26, 99]
  Allium commutatum [57]
  Allium porrum [57]
  Allium schoenoprasum [27]

genomic DNA

  Allium cepa [60]
  Allium subg. *Allium* [61]
  Allium subg. *Amerallium* [61]
  Allium subg. *Bromatorrhiza* [61]
  Allium subg. *Caloscordum* [61]
  Allium subg. *Melanocrommyum* [61]
  Allium subg. *Rhizirideum* [61]

ITS sequence analyses

  Allium L. subg. *Melanocrommyum* [22, 23, 82]
  Order Asparagales [63]

Nuclear DNA

  Allium roylei [121, 122]

cDNA

  Allium cepa [79]

ii. Random amplified polymorphic DNA (RAPD)

  Allium aaseae [107]
  Allium altaicum [30]
  Allium ampeloprasum [89]
  Allium cepa [18, 60, 89, 112]

Allium cepa var. *ascalonicum* [3]
Allium fistulosum [30]
Allium kermesinum Rchb. [112]
Allium oschaninii [62]
Allium sativum [44, 70, 127]
Allium x *wakegi* [3]
Allium vavilovii [62]
Allium subg. *Allium* [61]
Allium subg. *Amerallium* [61]
Allium subg. *Bromatorrhiza* [61]
Allium subg. *Caloscordum* [61]
Allium subg. *Melanocrommyum* [29, 61]
Allium subg. *Rhizirideum* [61]

iii. Amplified fragment length polymorphism (AFLP)
Allium cepa [119, 120]
Allium fistulosum [120]
Allium porrum [105]
Allium subg. *Rhizirideum* [122]
Allium roylei [119, 120]
Allium sativum [44, 45]

iv. PCR-based markers
CMS
Allium cepa [26, 89, 90]
Allium section *cepa* [68]
Allium ampeloprasum [39, 90]
Anthocyanidin synthase
Allium cepa [59]
RT-PCR
Allium cepa [58]
CAPs (nrDNA)
Allium L. subg. *Melanocrommyum* (Webb et Berth.) Rouy [22]
Allium giganteum Regel. [21]
Indels
Allium cepa [47]

SCAR

Allium cepa (shallot) [77]

Allium fistulosum [77]

SNPs

Allium cepa [47, 76]

Allium cepa (shallot) [77]

Allium sativum [127]

SSRs

Allium cepa [47, 76]

Allium sativum [127]

v. Repetitive DNA and microsatellites

Allium fistulosum [46, 108]

vi. Expressed sequence tags (ESTs)

Allium cepa [76, 79]

vii. Telomeric repeats

Order Asparagales, Alliaceae [2]

Allium altaicum [17]

Allium cepa [2, 17, 32, 91]

Allium chevsuricum [32]

Allium fistulosum [91]

Allium x *proliferum* (Moench) Schrad. [91]

Alliaceae family genera *Allium*, *Nothoscordum*, *Tulbaghia* [1]

Allium globosum [32]

Allium sativum [32]

Cytogenetic Chromosomal Analyses

i. In situ hybridization (ISH)

Allium cepa [41]

Allium fistulosum [41, 46]

a. Fluorescence in situ hybridization (FISH)

Allium altaicum [17]

Allium cepa [17, 32, 56, 91, 102]

Allium chevsuricum [32]

Allium fistulosum [32, 91]

Allium globosum [32]

Allium x *proliferum* (Moench) Schrad. [91, 102]

Allium schoenoprasum [103]
Allium sativum [32]
Alliaceae family genera *Allium*, *Nothoscordum*, *Tulbaghia* [1]

b. Genomic in situ hybridization (GISH)
Allium ampeloprasum [90]
Allium cepa [9, 62, 90, 101, 110]
Allium cepa var. *viviparum* [92]
Allium fistulosum [9, 110]
Allium oschaninii [31, 62]
Allium vavilovii [31, 62]
Allium wakegi [103]
Allium subg. *Melanocrommyum* [29]

ii. Chromosomal analyses

a. Bromodeoxyuridine (BrdU)
Allium fistulosum [34]

b. C-banding
Allium altyncolicum [28]
Allium fistulosum [46]

c. Flow cytometry/karyotype
Allium albidum Fisch. ex Bieb. [52]
Allium angulosum [52]
Allium carolinianum DC [52]
Allium cepa [52]
Allium chevsuricam Tscholok. [52]
Allium flavellum Vved [52]
Allium hymenorrhizum Ldb. [52]
Allium jodanthum Vved. [52]
Allium karelinii Poljak. [52]
Allium lineare L.s.l. [52]
Allium obliquum L. [52]
Allium rubens Schad. ex Willd. [52]
Allium saxatile M. *Bieb.* [52]
Allium sphaerocephaon [52]
Allium victorialis L. [52]

CHEMOTAXONOMIC RELATIONSHIPS

Health benefits of *Allium* species for their phytochemical value and application as phytopharmaceuticals is gaining significance [39, 40, 43]. Recent advances in chemotaxonomic relationships have identified constituents unique to species [13, 14]. Application of chemotaxonomy in crop improvement allows for selection of germplasm to design novel phenotypes with desired phytochemical constituency. Aroma profiles in *Allium* have been found to be taxa specific [12] and, thus, are useful for chemotaxonomical classification [111]. Sulphur-containing compounds responsible for the characteristic onion aroma and classes of flavonoids responsible for scale and leaf pigmentation are mainly accountable for the health properties found in *Alliums*. As more information is gained about these classes of compounds, they become candidates for single gene insertion and targets for manipulation [58]. The major flavonol found in onion (*Allium cepa*), quercetin, could become a selection trait in breeding programs because of: (1) potential benefits to human health, (2) it is largely unaffected by the environment [69, 83], and (3) is genetically inherited [106].

Chemical Constituents

i. Aroma, sulfur-containing compounds, cystein sulphoxides (CS)
   Allium seven chemotypes of 43 species [110]
   Allium cepa [53, 112]
   Allium globosum [53]
   Allium kermesinum Rchb. [112]
   Allium obliquum [53]
ii. Flavonoids separated into groups by detection of flavonols and flavones
   Allium cepa Aggregatum Group [104]
   Allium fistulosum [104]
   Allium Section *Allium* (37 species) [37]
   Allium Ampeloprasum Group [37]
   Allium Guttatum Group [37]
   Allium Rotundum Group [37]
   Allium Sphaerocephalon Group [37]
iii. Anthocyanidin synthase
   Allium cepa [58]

CONCLUSION

Given its wide variability, *Allium* continues to be an interesting crop, at cytogenetic, chemical, and molecular levels. As it has one of the largest genomes in the plant kingdom and is taxonomically complex, it poses unique challenges and opportunities for elucidating genome organization. Significant progress has been made in recent years in revealing genomic organization at the chromosomal and DNA levels. The emerging technology of proteomics may provide even greater insight into the genomes and possibilities of protein manipulation. Marker-assisted breeding is routine in many *Allium* improvement programs and with the wide array of target species and plethora of investigative methodologies, great strides are being made towards better understanding of this complex genus.

Acknowledgement

Deep gratitude is expressed to my mentor and friend, Dr Joe Corgan, for introducing me to the world of *Alliums*.

[1] Adams SP, Leitch IJ, Bennett MD, Leitch AR. *Aloe* L.- a second plant family without $(TTTAGGG)_n$ telomeres. Chromosoma 2000; 109: 201-205.

[2] Adams SP, Hartman TPV, Lim KY, et al. Loss and recovery of *Arabidopsis*-type telomere repeat sequences 5'-$(TTTAGGG)_n$-3' in the evolution of a major radiation of flowering plants. Proc R Soc Lond 2001; 268: 1541-1546.

[3] Arifin NS, Ozaki Y, Okubo H. Genetic diversity in Indonesian shallot (*Allium cepa* var. *ascalonicum*) and *Allium* x *wakegi* revealed by RAPD markers and origin of *A.* x *wakegi* identified by RFLP analyses of amplified chloroplast genes. Euphytica 2000; 111: 23-31.

[4] Arumuganathan K, Earle ED. Nuclear DNA content of some important plant species. Plant Mol Biol Reporter 1991; 9(3): 208-218.

[5] Badr A, Elkington TT. Giemsa C-banded karyotypes and taxonomic relationships of some North American *Allium* species and their relationship to Old World species (*Liliaceae*). Plant Syst and Evol 1984; 144: 17-24.

[6] Baranyi M, Greilhuber J. Genome size in *Allium*: in quest of reproducible data. Annals of Botany 1999; 83(6): 687-695.

[7] Bark OH, Havey MJ, Corgan JN. Restriction fragment length polymorphism (RFLP) analysis of progeny from an *Allium fistulosum* x *A. cepa* hybrid. Chromosoma 1994; 92: 185-192.

[8] Barnes SR, James AM, Jamieson G. The organization, nucleotide sequence, and chromosomal distribution of a satellite DNA from *Allium cepa*. Chromosoma 1985; 92: 185-192.

[9] Barthes L, Ricroch A. Interspecific chromosomal rearrangements in monosomic addition lines of *Allium*. Genome 2001; 44: 929-935.

[10] Bennett MD, Smith JB. Nuclear DNA amounts: angiosperms. Phil Trans R Soc Lond B 1976; 274: 227-274.

[11] Bennetzen JL. Comparative sequence analysis of plant nuclear genomes: microlinearity and its many exceptions. The Plant Cell 2000; 12: 1021-1029.

[12] Bilyk A, Cooper PL, Sapers GM. Varietal differences in distribution of quercetin and kaempferol in onion (*Allium cepa* L.) tissue. Journal of Agr Food Chem 1984; 32: 274-276.

[13] Block E, Naganathan S, Putman D, Zhao SH. *Allium* chemistry: HPLC analysis of thiosulfinates from onion, garlic, wild garlic (ransoms), leek, scallion, shallot, elephant (great-headed) garlic, chive, and Chinese chive. Uniquely high allyl to methyl ratios in some garlic samples. Journal of Agr Food Chem 1992; 40(12): 2418-2430.

[14] Block E, Putman D, Zhao SH. *Allium* chemistry: GC-MS analysis of thiosulfinates and related compounds from onion, leek, scallion, shallot, chive, and Chinese chive. Journal of Agri Food Chemistry 1992; 40(12): 2431-2438.

[15] Cryder C. A study of the associations of heritable traits in progeny from the interspecific backcross (*Allium fistulosum* x *Allium cepa* L.) x *Allium cepa* L. Ph.D. Dissertation, New Mexico State University, Las Cruces, New Mexico. 1988.

[16] Cryder et al. Isozyme analysis of progeny derived from (*Allium fistulosum* x *Allium cepa*) x *Allium cepa*. Theor Appl Genet 1991; 82: 337-345.

[17] Cuñado N, Sánchez-Morán E, Barrios J, Santos JL. Searching for telomeric sequences in two *Allium* species. Genome 2001; 44: 640-643.

[18] D'Ennequin MLT, Panaud O, Robert T, Richroch A. Assessment of genetic relationships among sexual and asexual forms of *Allium cepa* using morphological traits and RAPD markers. Heredity 1997; 78: 403-409.

[19] de Jong JH, Fransz P, Zabel P. High resolution FISH in plants- techniques and applications. Trends in Plant Science 1999; 4(7): 258-263.

[20] Don G. 1827, a monograph of the genus *Allium*. Edinburgh, an advance reprint of Mem. Wernerian Nat Hist Soc 1832; 6: 1-102.

[21] Dubouzet JG, Shinoda K, Murata N. Interspecific hybridization of *Allium gigantum* Regel: production and early verification of putative hybrids. Theor Appl Genet 1998; 96: 385-388.

[22] Dubouzet JG, Shinoda K. Phylogeny of *Allium L.* subg. *Melanocrommyum* (Webb et Berth.) Rouy based on DNA sequence analysis of the internal transcribed spacer region of nrDNA. Theor Appl Genet 1998; 7: 541-549.

[23] Dubouzet JG, Shinoda K. Relationships among Old and New World *Alliums* based on DNA according to ITS DNA sequence analysis. Theor Appl Genet 1999; 98: 422-433.

[24] Emsweller SL, Jones HA. An interspecific hybrid in *Allium*. Hilgardia 1935; 9(5): 265-273.

[25] Emsweller SL, Jones HA. Meiosis in *Allium fistulosum*, *Allium cepa*, and their hybrid. Hilgardia 1935; 9: 277-288.

[26] Engelke T, Terefe T, Tatlioglu T. A PCR-based marker system monitoring CMS-(S), CMS-(T) and (N)-cytoplasm in the onion (*Allium cepa* L.) Theor Appl Genet 2003; 107: 162-167.

[27] Engelke T, Agbicodo E, Tatlioglu T. Mitochondrial genome variation in *Allium ampeloprasum* and its wild relatives. Euphytica 2004; 137: 181-191.

[28] Friesen N, Borisjuk N, Mes THM, et al. Allotetraploid origin of *Allium altyncolicum* (sect. *Schoenoprasum*) as investigated by karyological and molecular markers. Plant Syst Evol 1997; 206: 317-335.

[29] Friesen N, Fritsch RM, Bachmann K. Hybrid origin of some ornamentals of *Allium* subgenus *Melanocrommyum* verified with GISH and RAPD. Theor Appl Genet 1997; 95: 1229-1238.

[30] Friesen N, Pollner S, Bachmann K, Blattner F. RAPDs and noncoding chloroplast DNA reveal a single origin of the cultivated *Allium fistulosum* from A. *Altaicum* (*Alliaceae*). Am J Bot 1999; 86(4): 554-562.

[31] Fritsch RM, Friesen N. Evolution, Domestication and Taxonomy. In: Allium Crop Science: Recent Advances. New York: CAB International, 2002; 5-30.

[32] Fuchs J, Brandes A, Schubert I. Telomere sequence localization and karyotype evolution in higher plants. Pl Syst Evol 1995; 196: 227-241.

[33] Fuchs J, Houben A, Schubert I. Chromosome 'painting' in plants - a feasible technique? Chromosoma 1996; 104: 315-320.

[34] Fujishige I, Taniguchi K. Sequence of DNA replication in *Allium fistulosum* chromosomes during S-phase. Chromosome research 1998; 6(8): 611-619.

[35] Grant WF, Genetics of flowering plants. New York: Columbia University Press, 1975.

[36] Hanelt P. Taxonomy, evolution, and history. In: Rabowitch HD, Brewster JL, eds. Onions and allied crops. Botany, physiology, and genetics. Boca Raton, FL: CRC Press, 1990; 1: 1-26.

[37] Harborne JB, Williams CA. Notes on flavonoid survey. In: A review of *Allium* Section *Allium*. Whitstable, Kent, UK: Whitstable Litho Ltd., 1996; 41-44.

[38] Havey MJ, Lopes Leite D. Toward the identification of cytoplasmic male sterility in leek: evaluation of organellar DNA diversity among cultivated accessions of *Allium ampeloprasum*. J Am Soc Hort Sci 1999; 124: 626-629.

[39] Hertog MGL, Hollman PCH, Katan MB. Content of potentially anticarcinogenic flavonoids of 28 vegetables and 9 fruits commonly consumed in the Netherlands. Journal of Agr Food Chem 1992; 40: 2379-2383.

[40] Hertog MGL, Hollman PCH. Potential health effects of the dietary flavonol quercetin. Eur J of Clinical Nutr 1996; 50: 63-71.

[41] Hou A, Peffley EB. Recombinant chromosomes of advanced backcross plants between A. *cepa* and A. *fistulosum* L. revealed by in situ hybridization. Theor Appl Genet 2000; 100(8): 1190-1196.

[42] Hou A, Geoffriau E, Peffley EB. Esterase isozymes are useful to track intogression between *Allium fistulosum* L. and A. *cepa* L. Euphytica 2001; 122: (1): 1-8.

[43] Hughes BG, Lawson LD. Antimicrobial effects of *Allium sativum* L. (Garlic), *Allium ampeloprasum* L. (Elephant garlic) and *Allium cepa* L. (Onion) garlic compounds and commercial garlic supplement products. Phytotherapy Res 1991; 5: 154-158.

[44] Ipek M, Ipek A, Simon PW. Comparison of AFLPs, RAPD markers, and isozymes for diversity assessment of garlic and detection of putative duplicates in germplasm collections. J Am Soc Hort Sci 2003; 128(2): 246-252.

[45] Ipek M, Ipek A, Almquist SG, Simon PW. Demonstration of linkage and development of the first low-density genetic map of garlic, based on AFLP markers. Theor Appl Genet 2005; 110: 228-236.

[46] Irifune K, Hirai K, Zheng J, et al. Nucleotide sequence of a highly repeated DNA sequence and its chromosomal localization in *Allium fistulosum*. Theor Appl Genet 1995; 90: 312-316.

[47] Jakse J, Martin W, McCallum J, Havey MJ. Single nucleotide polymorphisms, indels, and simple sequence repeats for onion cultivar identification. J Amer Soc Hort Sci 2005; 130(6): 912-917.

[48] Johnston JS, Bennett MD, Rayburn AL, et al. Reference standards for determination of DNA content of plant nuclei. Am J Bot 1999; 86(5): 609-613.

[49] Jones RN, Rees H. Nuclear DNA variation in *Allium*. Heredity 1968; 23(4): 591-605.

[50] Kalkman ER. Analysis of the C-banded karyotype of *Allium cepa* L. standard system of nomenclature and polymorphism. Genetica 1984; 65: 141-148.

[51] Kamenetsky R, Fritsch RM. Ornamental Alliums. In: Robinowtich HD, Currah L, eds. Allium Crop Science: Recent Advances. New York: CAB International, 2002; 459-491.

[52] Keller ERJ, Schubert I, Fuchs J, Meister A. Interspecific crosses of onion with distant *Allium* species and characterization of the presumed hybrids by means of flow cytometry, karyotype analysis and genomic *in situ* hybridization. Theor Appl Genet 1996; 92: 417-424.

[53] Keusgen M, Schulz H, Glodek J, et al. Characterization of some *Allium* hybrids by aroma precursors, aroma profiles, and alliinase activity. J Agric Food Chem 2002; 50: 2884-2890.

[54] Khrustaleva LI, Kik C. Cytogenetical studies in the bridge cross *Allium cepa* x (A. *fistulosum* x A. *roylei*). Theor Appl Genet 1998; 96: 8-14.

[55] Khrustaleva LI, Kik C. Introgression of *Allium fistulosum* into A. *cepa* mediated by A. roylei. Theor Appl Genet 2000; 100: 17-26.

[56] Khrustaleva Li, Kik C. Localization of single-copy T-DNA insertion in transgenic shallots (*Allium cepa*) by using ultra-sensitive FISH with tyramide signal amplification. The Plant J for Cell Mol Biol 2001; 25(6): 699-707.

[57] Kik C, Samoylov AM, Verbeek WHJ, Ramsdonk LWD. Mitochondrial DNA variation and crossability of leek (*Allium porrum*) and its wild relatives from the *Allium ampeloprasum* complex. Theor Appl Genet 1997; 94: 465-471.

[58] Kim S, Jones R, Yoo KS, Pike LM. Gold color in onions (*Allium cepa*): a natural mutation of the chalcone isomerase gene resulting in a premature stop codon 2004; Mol Gen Genomics 272: 411-419.

[59] Kim S, Yoo KS, Pike LM. Development of a co-dominant, PCR-based marker for allelic selection of the pink trait in onions (*Allium cepa*), based on the insertion mutation in the promoter of the anthocyanidin synthase gene 2005; Theor Appl Genet 110: 628-633.

[60] King JJ, Bradeen JM, Bark O, et al. A low-density genetic map of onion reveals a role for tandem duplication in the evolution of an extremely large diploid genome. Theor Appl Genet 1997; 96: 52-62.

[61] Klaas M. Applications and impact of molecular markers on evolutionary and diversity studies in the genus *Allium*. Plant Breeding 1998; 117: 297-308.

[62] Klaas M, Friesen N. Molecular markers in *Allium*. In: Rabinowitch HD, Currah L, eds. *Allium* Crop Science. New York, NY: Cab International, 2002; 159-185.

[63] Kuhl JC, Catanach A, Cheung F, et al. A unique set of 11,008 onion expressed sequence tags reveals expressed sequence and genomic differences between the monocot orders Asparagales and Poales. Plant cell 2004; 16(1): 114-125.

[64] Labini RM, Elkington TT. Nuclear DNA variation in the genus *Allium* L. (*Liliaceae*). Heredity 1987; 59: 119-128.

[65] Leitch IJ, Leitch AR, Heslop-Harrison JS. Physical mapping of plant DNA sequences by simultaneous *in situ* hybridization of two differently labeled fluorescent probes. Genome 1991; 34: 329-333.

[66] Levan A. Die zytologie von *Allium cepa* x *A. fistulosum*. Hereditas 1936; 21: 195-214.

[67] Levan A. The cytology of the species hybrid *Allium cepa* x *A. fistulosum* and its polyploid derivatives. Hereditas 1941; 27: 253-272.

[68] Lilly JW, Havey MJ. Sequence analysis of a chloroplast intergenic spacer for phylogenetic estimates in *Allium* section *Cepa* and a PCR-based polymorphism detecting mixtures of male-fertile and male-sterile cytoplasmic onion. Theor Appl Genet 2001; 102: 78-82.

[69] Lombard K, Geoffriau E, Peffley EB. Total quercetin content in onion: Survey of cultivars grown at different locations. Hort Technol 2004; 14(4): 1-4.

[70] Maaß HI. Studies on triploid viviparous onions and their origin. Genet Res Crop Evol 1997; 44(2): 95-99.

[71] Maaß HI, Klaas M. Infraspecific differentiation of garlic (*Allium sativum* L.) by isozyme and RAPD markers. Theor Appl Genet 1995; 91: 89-97.

[72] Maaß HI. About the origin of the French grey shallot. Genetic Resources and Crop Evolution 1996; 43: 291-292.

[73] Maeda T. Chiasma studies in *Allium fistulosum*, *Allium cepa* and their F_1, F_2, and backcross hybrids. Japan J Genet 1937; 13: 146-159.

[74] Mangum PD, Peffley EB. Central cell nuclear-cytoplasmic incongruity: a mechanism for segregation distortion in advanced backcross and selfed generations of (*Allium cepa* L. x *Allium fistulosum* L.) x *A. cepa* interspecific hybrid derivatives. Cytogenetic and Genome Research. Plant Cytogenetics 2005; 109(1-3): 400-407.

[75] Marie D, Brown S. A cytometric exercise in plant DNA histograms, with 2C values for 70 species. Biol Cell 1993; 78: 41-51.

[76] Martin WJ, McCallum J, Shigyo M, Jakse J, Kuhl JC, Yamane N, Pither-Joyce M, Gokce AF, Sink KC, Town CD, Havey MJ. Genetic mapping of expressed sequences in onion and in silico comparison with rice show scant colinearity. Mol Gen Genomics 2005; 274: 197-204

[77] Masuzaki, S, Shigyo M, Yamauchi N. Direct comparison between genomic constitution and flavonoid contents in Allium multiple alien addition lines reveals chromosomal locations of genes related to biosynthesis from dihydrokaempferol to quercetin glucosides in scaly leaf of shallot (*Allium cepa* L.) Theor Appl Genet 2006; 112 (4): 607-617

[78] Mathew B. A Review of *Allium* section *Allium*. Whitstable, Kent, UK: Whitstable Litho Ltd., 1996; 176.

[79] McCallum J, Leite D, Pither-Joyce M, Havey MJ. Expressed sequence markers for genetic analysis of bulb onion (*Allium cepa* L.) Theor Appl Genet 2001; 103: 979-991.

[80] McCollum GD. Onions and allies. In: Simmonds NW, ed. Evolution of Crop Plants. London: Longman, 1976; 186-190.

[81] Mes THM, Friesen N, Fritsch RM, et al. Criteria for sampling in *Allium* (*Alliaceae*) based on chloroplast DNA PCR-RFLPs. Systematic Botany 1998; 22: 701-712.

[82] Mes THM. Fritsch RM, Pollner S, Bachmann K. Evolution of the chloroplast genome and polymorphic ITS regions in *Allium* subg. *Melanocrommyum*. Genome 1999; 42: 237-247.

[83] Patil BS, Pike LM, Yoo KS. Variation in the quercetin content in different colored onions (*Allium cepa* L.). J Am Hort Sci 1995; 120: 909-913.

[84] Peffley EB, Corgan JN, Horak K, Tanksley SD. Electrophoretic analysis of *Allium* alien addition lines. Theor Appl Genet 1985; 71: 176-184.

[85] Peffley EB. Evidence for chromosomal differentiation of *Allium fistulosum* and *A. cepa*. J Am Soc Hort Sci 1986; 111: 126-129.

[86] Peffley EB, Currah L. The chromosomal locations of enzyme-coding genes in *Adh-1* and *Pgm-1* in *Allium fistulosum*. Theor Appl Genet 1988; 75: 945-949.

[87] Peffley EB, deVries JN. Giemsa G-banding in *Allium*. Biotechnique and Histochemistry 1993; 68(2): 83-86.

[88] Peffley EB, Hou A. Bulb-type onion introgressants possessing *Allium fistulosum* L. genes recovered from interspecific hybrid backcrosses between *A. cepa* L. x *A. fistulosum* L. Theor Appl Genet 2000; 100: 528-534.

[89] Peterka H, Budahn H, Schrader O. Interspecific hybrids between onion (*Allium cepa* L.) with S-cytoplasm and leek (*Allium ampeloprasum* L.). Theor Appl Genet 1997; 94: 383-389.

[90] Peterka H, Budahn H, Schrader O, Havey M. Transfer of male-sterility-inducing cytoplasm from onion to leek (*Allium ampeloprasum*). Theor Appl Genet 2002; 105: 173-181.

[91] Pich U, Fuchs J, Schubert I. How do Alliaceae stabilize their chromosome ends in the absence of TTTAGGG sequences? Chromosome Research 1996; 4: 207-213.

[92] Puizina J, Javornik B, Bohanec B, et al. Random amplified polymorphic DNA analysis, genome size, and genomic *in situ* hybridization of triploid viviparous onions. Genome 1999; 42(6): 1208-1216.

[93] Rabinowitch HD, Brewster JC. Onions and allied crops. Vol 1. Boca Raton, FL: CRC 1990, 273 p.

[94] Ranjekar PK, Pallotta D, Lafontaine JG. Analysis of plant genomes v. comparative study of molecular properties of DNAs of seven *Allium* species. Biochemical Genetics 1978; 16(9/10): 957-970.

[95] Ricroch A, Brown SC. DNA base composition of *Allium* genomes with different chromosome numbers. Gene 1997; 205(1/2): 255-260.

[96] Ricroch A, Peffley EB, Baker RJ. Chromosomal location of rDNA in *Allium*: in situ hybridization using biotin- and fluorescein-labeled probe. Theor Appl Genet 1992; 83: 413-418.

[97] Rieseberg LH, Peterson PM, Soltis DE, Annable CR. Genetic divergence and isozyme number variations among four varieties of *Allium douglasii* (Alliaceae) Amer J Bot 1987; 74; 11:1614-1624.

[98] Rouamba A, Sandmeier M, Sarr A, Ricroch A. Allozyme variation within and among populations of onion (*Allium cepa* L.) from West Africa. Theor Appl Genet. 2001: 103: 855-861.

[99] Sato Y. PCR amplification of CMS-specific mitochondrial nucleotide sequences to identify cytoplasmic genotypes of onion (*Allium cepa* L.). Theor Appl Genet 1998; 96: 367-370.

[100] Schubert I, Ohle H, Hanelt P. Phylogenetic conclusions from Giemsa banding and NOR staining in top onions (*Liliaceae*). Plant Syst Evol 1983. 143(4): 245-256.

[101] Schwarzacher T, Heslop-Harrison P. Practical In situ hybridization. Oxford, BIOS Scientific Publishers, 2000; 203.

[102] Shibata F, Hizume M. Evolution of 5S rDNA units and their chromosomal localization in *Allium cepa* and *Allium schoenoprasum* revealed by microdissection and FISH. Theor Appl Genet 2002; 105: 167-172.

[103] Shibata F, Hizume M. The identification and analysis of the sequences that allow the detection of *Allium cepa* chromosomes by GISH in the allodiploid A. *wakegi*. Chromosoma 2002; 111: 184-191.

[104] Shigyo M, Tashiro Y, Iino M, et al. Chromosomal locations of genes related to flavonoids and anthocyanin production in leaf sheath of shallot (*Allium cepa* L. Aggregatum group). Genes Genet Syst 1997. 72: 149-152.

[105] Smilde WD, van Heusden AW, Kik C. AFLPs in leek (*Allium porrum*) are not inherited in large linkage blocks. Euphytica 1999; F110: 127-132.

[106] Smith C, Peffley EB, Lombard KA, Liu W. Genetic Analysis of Quercetin in the 'Lady Raider' Onion (*Allium cepa* L. 'Lady Raider'). Texas Journal of Agri and Natl Res 2003; 13: 24-28.

[107] Smith J, Vuong Pham T. Genetic diversity of the narrow endemic *Allium aaseae* (*Alliaceae*). Am J Bot 1996; 83(6): 717-726.

[108] Song YS, Suwabe K, Wako T, et al. Development of microsatellite markers in bunching onion (*Allium fistulosum* L.). Breeding Science 2004; 54(4): 361-365.

[109] Stearn WT. How many species of *Allium* are known? Kew magazine 1992; 9: 180-182.

[110] Stevenson M, Armstrong SJ, Ford-Lloyd BV, Jones GH. Comparative analysis of crossover exchanges and chiasmata in *Allium cepa* x *fistulosum* after genomic *in situ* hybridization (GISH). Chromosome Research 1998; 6(7): 567-574.

[111] Storsberg J, Schulz H, Keller ERJ. Chemotaxonomic classification of some *Allium* wild species on the basis of their volatile sulphur compounds. J App Bot 2003; 77: 160-162.

[112] Storsberg J, Schulz H, Keusgen M, et al. Chemical characterization of interspecific hybrids between *Allium cepa* L. and *Allium kermesinum* Rchb. Journal of Agri Food Chem 2004; 52(17): 5499-5505.

[113] Takhtajan A. Diversity and Classification of Flowering Plants. New York, NY: Colombia University Press, 1997; 643.

[114] Traub H. The subgenera, sections and subsections of *Allium* L. Plant Life 1968; 24: 147-163.

[115] Ulloa GM. A cytogenetic, isozyme and morphological study on interspecific progenies from *Allium fistulosum* x *Allium cepa* crosses. Ph.D. Dissertation, Las Cruces, NM, New Mexico State University, 1993.

[116] Ulloa GM, Corgan JN, Dunford M. Chromosome characteristics and behavior differences in *Allium fistulosum* L., A. *cepa* L., their F_1 hybrid, and selected backcross progeny. Theor Appl Genet 1994; 89: 567-57.

[117] Ulloa GM, Corgan JN, Dunford M. Evidence for nuclear-cytoplasmic incompatibility between *Allium fistulosum* and A. *cepa*. Theor Appl Genet 1995; 90: 746-754.

[118] van der Valk PC, Kik C, Verstappen F, et al. Pre- and post-fertilization barriers to backcrossing the interspecific hybrid between *Allium fistulosum* L. and A. *cepa* L. with A. *cepa*. Euphytica 1991; 53: 201-209.

[119] van Heusden AW, van Ooijen JW, Vrielink-van Ginkel R, et al. A genetic map of an interspecific cross in *Allium* based on amplified fragment length polymorphism (AFLP) markers. Theor Appl Genet 2000; 100: 118-126.

[120] van Heusden AW, Shigyo M, Tashiro Y, et al. AFLP linkage group assignment to the chromosomes of *Allium cepa* L. via monosomic addition lines. Theoretical and applied genetics 2000; 100: 480-486.

[121] van Raamsdonk LWD, Smiech MP, Sandbrink JM. Introgression explains incongruence between nuclear and chloroplast DNA-based phylogenies in *Allium* section *Cepa*. Botanical Journal of the Linnean Soc 1997; 123: 91-108.

[122] van Raamsdonk LWD, Vrielink-van Ginkel M, Kik C. Phylogeny reconstruction and hybrid analysis in *Allium* subgenus *Rhizirideum*. Theoretical and Applied Genetics 2000; 100: 1000-1009.

[123] Vavilov NI. (English translation). *Origin and Geography of Cultivated Plants*. Cambridge, UK: Cambridge University Press, 1992; 498.

[124] Vvedensky AI, Kovalevskaya SS. Rod 151, (7) Allium L.- Luk zhua (kaz.) piez (tadzh.). Opredelitel rastenij srednej azii. Kriticheskij konspekt flory, vol. ii. Izdatel'stvo 'FAN' Uzbekskoj SSR, Tashkent 1971; 39-898, 311-328.

[125] Watson L, Dallwitz MJ. The Families of Flowering Plants: Description, Illustrations, Identification and Information Retrieval. Version: 14th December 2000. http://biodiversity.uno.edu/delta/.

[126] Wissemann, V. Hybridization and the evolution of the nrITS spacer region. In: Plant Genome: Biodiversity and Evolution Vol 1, eds. Sharma AK and A Sharma. Science Publishers, Inc. Enfield NH, USA 2003; pp.57-66.

[127] Zwedie Y, Havey MJ, Prince JP, Jenderek MM. The first genetic linkages among expressed regions of the garlic genome. J Amer Soc Hort Sci 2005; 130(4): 569-574.

CHAPTER

6

Molecular Phylogenetics of Chinese *Cymbidium* (Orchidaceae) Based on nrITS Sequence and RAPD Data

XIU-LIN YE, MING-YONG ZHANG, CAI-YUN SUN, CHENG-YE LIANG and KUAI-FEI XIA

South China Botanical Garden, The Chinese Academy of Sciences, Guangzhou, China

ABSTRACT

The sequences of internal transcribed spacers (ITS) of nuclear ribosomal DNA nr(DNA) and random amplified polymorphic DNA (RAPD) data were used to evaluate the genetic diversity and phylogenetic relationships of *Cymbidium* of China. Variant of ITS sequence is low among genus *Cymbidium* and could only partly define the phylogenetic relationships of *Cymbidium* although van den Berg et al. [21] found the sequence variant levels of *Cymbidium* to be higher than other orchids. The phylogenetic tree generated from ITS sequences is only partially congruent with the current taxonomic classification of the genus at the subgenus level, but better congruent with the traditional section classification, except *C. dayanum,* which was placed with the member subsection *Cyperorchis*. The removal of *C. dayanum*, subgenus *Cyperochis* was a monophyletic group; and subgenus *Jensoa* also appeared paraphyletic, with *C. lancifolium* being the sister-group to the remaining genus; species of subgenus *Cymbidium* appeared polyphyletic, being split into several clades and intermixed with the main subgenus *Cyperochis* and subgenus *Jensoa* clades, respectively. The unweighted pain-group methods with arithmetical averages (UPGMA) dendrogram based on RAPD data further confirmed that *C. dayanum*

Address for correspondence: Xiu-Lin Ye and Ming-Yong Zhang, South China Botanical Garden, The Chinese Academy of Sciences, Guangzhou 510650, P.R. China. E-mail: zhangmy2005@yahoo.com.cn

was located in subgenus *Cyperorchis*, and high polyphyly of subgenus *Cymbidium*. However, some apparent incongruence could be noted between the phylogenetic relationships constructed from the ITS and RAPD data. The RAPD data also showed that the cultivars of *C. goeringii* and *C. eburneum* had higher genetic diversity than other species, which could have originated from the long history of cultivation and selection, and geographic isolation.

Key Words: *Cymbidium,* Orchidaceae, phylogeny, ITS, RAPD

Abbreviations: nrITS = Internal transcribed spacers of nuclear ribosomal DNA RAPD = Random amplified polymorphic DNA.

INTRODUCTION—DISTRIBUTION PATTERNS

Cymbidium is one genus of the family Orchidaceae, which is of immense importance in horticulture. There are about 48 species of *Cymbidium* (Orchidaceae) in the world, and about 29 species and varieties in China [5, 6]. It is widely distributed in southeast Asia, from northwest India to Japan, and south Australia, with the species diversity center in northeast India, southwest China, Indo-China and Malaysia. *Cymbidium* is most abundant in southwest China, followed by the southeast China, especially in Yunnan Province, where this species flourishes [23] (Table 1).

C. *goeringii* Rchb.f. and C. *faberi* Rolfe are most widely distributed, which often grow in the cold areas such as south of Ganshu Province, south of Qingling mountain of Shanxi Province, Heinan, Anhui, Hubei, Hunan, Jiangxi, Zhejiang, Jiangshu, Taiwan, Fujian, Guangxi, Sichuan, Guizhou, Yunnan, Xizhuang Province, and northeast of Guangdong Province. C. *kanran* Makio., C. *floribundum* Lindl., and C. *lancifolium* Hook. are found in south China, including Hubei, Hunan, Jiangxi, Fujian, Zhejiang, Taiwan, Guangdong, Guangxi, Sichuan, Guizhou, and Yunnan. C. *ensifolium* (L.) Sw. is distributed in south of Zhejiang, Jiangxi, Fujian, Taiwan, Hunan, Guangdong, Guangxi, Hainan, Sichuan, Guizhou, and Yunnan Province; C. *sinense* Jackson ex Andr. Willd. in Taiwan, Fujian, south of Sichuan, Jiangxi, Anhui, Guangdong, Hainan, Guangxi, Guizhou, and Yunnan. C. *hookerianum* Rchb.f., C. *iridioides* D.Don and C. *tracyanum* L.Castle are found in Sichuan, Guizhou, Yunnan, Xizhang, and Guangxi Province. C. *insigne* Rolfe only is found in Hainan Province [21].

Table 1. The distribution of *Cymbidium* in provinces of China

Species	*Distribution in provinces of China*																		*Altitude above sea level ('M')*
	SX	*He*	*GS*	*AH*	*HB*	*Hu*	*ZJ*	*JX*	*JS*	*TW*	*FJ*	*GX*	*GD*	*Ha*	*SC*	*YN*	*XZ*	*GZ*	
C. *aloifolium*												√	√			√		√	100-1,100
C. *bicolor*												√	√	√		√		√	200-1,600
C. *dayanum*										√	√	√	√	√	√	√			300-1,600
C. *floribundum*					√	√	√	√		√	√	√	√			√		√	100-3,300
C. *Suavissimum*																√		√	700-1,100
C. *tracyanum*																√	√	√	1,200-1,900
C. *iridioides*															√	√	√		900-2,800
C. *erythraeum*															√	√	√		1,400-2,800
C. *hookerianum*												√			√	√	√	√	1,100-2,700
C. *wilsonii*																√			2,000-2,400
C. *lowianum*																√			1,300-1,900
C. *insigne*														√					1,700-1,850
C. *Wenshanense*																√			
C. *eburneum*												√		√		√			unclear
C. *mastersii*																√			1,600-1,800
C. *elegans*															√	√		√	1,700-2,800
C. *cochleare*										√	√								300-1,000
C. *tigrinum*																√			
C. *ensifolium*				√		√	√	√		√	√	√	√	√	√	√		√	600-1,800

Table 1 contd.

Table 1 contd.

C. *sinense*				√				√		√	√	√	√	√	√	√		√	300-2,000
C. *defoliatum*															√	√		√	
C. *nanulum*														√		√		√	unclear
C. *kanran*				√	√	√	√	√		√	√	√	√	√	√	√		√	400-2,400
C. *cyperifolium*												√	√	√		√		√	900-1,600
C. *faberi*	√	√	√	√	√	√	√	√		√	√	√	√		√	√		√	700-3,000
C. *Goeringii*	√	√	√	√	√	√	√	√	√	√	√	√	√		√	√		√	300-2,200
C. *qiubeinse*																√		√	700-1,800
C. *lancifolium*						√	√			√	√	√	√		√	√		√	300-2,200
C. *macrorhiizon*															√	√		√	700-1,500

SX: Shanxi; He: Henan; GS: Ganshu; AH: Anhni; HB: Hubei; Hu: Hunan; ZJ: Zhejiang ; JX: Jiangxi; JS: Jiangshu; TW: Taiwan; FJ: Fujian; GX: Guangxi; GD: Guangdong; Ha: Hainan; SC: Sichuan; YN: Yunnan; XZ: Xizhang; GZ: Guizhou

Cymbidium species usually occur in the zone of 100-2,000 m above sea level. Looking from the growing latitude, the *Cymbidium* distribute at the north latitude 24°-34°. From the uprightness, different areas yield the same *Cymbidium* species but different sea levels and different *Cymbidium* species can grow on the same altitude. Two to three species of *Cymbidium* can grow in the same sea level, for instance, C. *Goeringii* and C. *sinense* often grow together (Fig. 1).

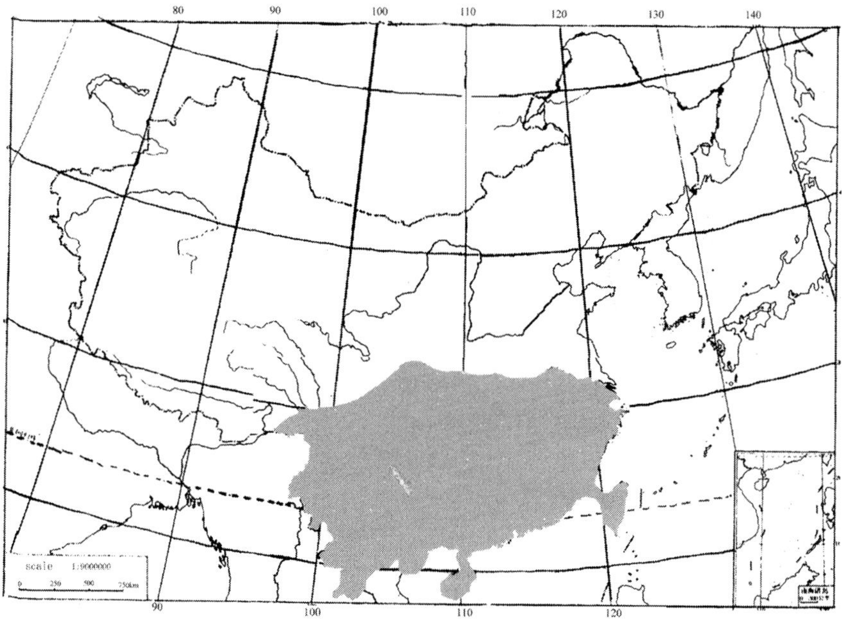

Fig. 1. *The distribution map of* Cymbidium *in China.* *Cymbidium grows in the shadow zone of China.*

THE TRADITIONAL TAXONOMY IN *CYMBIDIUM*

Du Puy and Cribb [9] have reviewed four main classifications of the genus *Cymbidium* (Table 2). In Dressler's [8] framework of Orchidaceae, *Cymbidium* is placed in Cymbidieae of Vandioideae, which contains all the sympodial vandoid orchids mostly with two pollinia. The generic delimitation of *Cymbidium* has been controversial, largely due to the different emphasis of the character variation of pollinia and lip (Table 2). Schlechter [15] completed revision of this group of plants on the basis of the modern infrageneric classification of *Cymbidium* although two genera, *Cymbidium* and *Cyperorchis*, were recognized in his system. He emphasized the fusion of the base of lip and the base of column as the

Table 2. Comparison of traditional supraspecific classifications proposed for *Cymbidium* (modified from Du Puy and Cribb [9])

Blume	*Schlechter*	*Seth & Cribb*	*Du Puy & Cribb*
Cymbidium Sw.	*Cymbidium* Sw.	*Cymbidium*	*Cymbidium*
		subgen. *Cymbidium*	subgen. *Cymbidium*
	sect. *Eucymbidium*	sect. *Cymbidium*	sect. *Cymbidium*
			sect. *Borneense*
	sect. *Himantophyllum*	sect. *Himantophyllum*	sect. *Himantophyllum*
	sect. *Austrocymbidium*	sect. *Austrocymbidium*	sect. *Austrocymbidium*
		sect. *Floribundum*	sect. *Floribundum*
		sect. *Suavissimum*	
	sect. *Bigibbarium*	sect. *Bigibbarium*	sect. *Bigibbarium*
	Cyperorchis Bl.	subgen. *Cyperorchis*	subgen. *Cyperorchis*
Iridorchis Bl.	sect. *Iridorchis*	sect. *Iridorchis*	sect. *Iridorchis*
		sect. *Eburnea*	sect. *Eburnea*
	sect. *Annamaea*	sect. *Annamaea*	sect. *Annamaea*
Cyperorchis Bl.	sect. Eucyperorchis	sect. *Cyperorchis*	sect. *Cyperorchis*
	sect. *Parishiella*	sect. *Parishiella*	sect. *Parishiella*
	Cymbidium Sw.	subgen. *Jensoa*	subgen. *Jensoa*
	sect. *Jensoa*	sect. *Jensoa*	sect. *Jensoa*
	sect. *Maxillarianthe*	sect. *Maxillarianthe*	sect. *Maxillarianthe*
	sect. *Geocymbidium*	sect. *Geocymbidium*	sect. *Geocymbidium*
	sect. *Macrorhizon*	sect. *Pachyrhizanthe*	sect. *Pachyrhizanthe*

unique distinguishing feature of *Cyperorchis*, rather than emphasizing the pollinarium and pollinium shapes as did Blume [2, 3, 4], Reichenbach [13] and Hooker [10]. Schlechter [15] also proposed several sections for both genera. Hunt [11] reduced *Cyperorchis* to within *Cymbidium* and maintained Schlechter's sectional divisions. Seth and Cribb [16] started to use the subgenus concept and three subgenera were proposed, mainly based on the number of pollinia and state of fusion between lip and column: subgenus *Cymbidium* with two pollinia and free lip, subgenus *Cyperorchis* with two pollinia and fusion of lip and column-base, and subgenus *Jensoa* with four pollinia and free lip. This system was slightly modified by Du Puy and Cribb [9].

In this chapter analysis of phylogenetic relationship in *Cymbidium* of China has been undertaken based on the sequences of the internal transcribed spacer region (ITS) of nuclear ribosomal DNA (nrDNA) and RAPD. ITS sequences [1, 21, 27] and RAPD [12, 25, 26] have been widely used to infer phylogenetic relationships among closely related genera and species. RAPD plays an important role in the research of genetic diversity and the relationships of species. The phylogenetic relationships formulated were evaluated against the existing supraspecific classification proposed by Du Puy and Cribb [9].

ANALYSIS OF PHYLOGENETIC RELATIONSHIPS IN *CYMBIDIUM* (ORCHIDACEAE) BASED ON nrITS SEQUENCE DATA

A total of 30 taxa of *Cymbidium* from around China were sampled (voucher specimens deposited in South China Botanical Garden (SCBG)). Their distribution, source, and the GenBank accession numbers are shown in Table 3. Du Puy and Cribb's [9] classification of *Cymbidium* was followed for the purpose of this phylogenetic analysis. Two species from tribe Cymbidieae, *Eulophia graminea*, *Geodorum densiflorum*, and one from tribe Orchideae, *Amitostigma pinguiculum*, were designated as outgroups. All samples for ITS were collected from the wild in China and cultivated in SCBG.

Total DNA was extracted from fresh leaves, following the 2 x CTAB protocol [7] . Nuclear ribosomal internal transcribed spacers (ITS1, 5.8S, ITS2 region) were amplified using the forward primer ITS5 [22] and the reverse primer CA26 [24]. Amplified double stranded DNA fragments were purified using QIAquick Gel Extraction Kit (QIAGEN). Automated sequencing was performed using fluorescent dye-labeled

Table 3. The species and cultivars of Cymbidium in this study

No. of sample	*Species or cultivars*	*Source; voucher specimen*	*GenBank accession number*
(1)	C. *aloifolium* (L.) Sw	Yunnan; Sun 98-2	AF284695
(2)	C. *aloifolium* (L.) Sw 'xishanwenbanlan'	Yunnan; Sun 98-1	
(3)	C. *aloifolium* (L.) Sw 'xiaguanwenbanlan'	Yunnan; Sun 98-3	
(4)	C. *bicolor* Lindl.subsp.*obtusum* Du Puyet Cribb	Yunnan; Sun 98-4	AF284696
(5)	C. *dayanum* Rchb.f.	Yunnan; Sun 98-5	AF284697
(6)	C. *floribundum* Lindl.	Yunnan; Sun 98-6	AF284698
(7)	C. *floribundum* Lindl. 'tailanta'	Yunnan; Sun 98-10	
	C. *pumilum* Rolfe	Yunan; Sun 98-7	AF284699
(8)	C. *suavissimum* Sander ex C.Curtis	Yunnan; Sun 98-8	AF284700
(9)	C. *tracyanum* L. Castle	Yunnan; Sun 98-9	AF2284701
(10)	C. *hookerianum* Rchb.f.	Yunnan; Sun 98-51	AF284702
(11)	C. *wilsonii* (Rolfe ex Cook) Rolfe	Yunnan; Sun 98-11	AF284703
(12)	C. *insigne* Rolfe	Zhejiang; Sun 98-12	AF284704
(13)	C. *iridioides* D. Don	Yunnan; Sun 98-13	AF284705
(14)	C. *erythraeum* Lindl.	Yunnan; Sun 98-14	AF284706
(15)	C. *lowianum* (Rchb.f) Rchb.f.	Yunnan; Ye 15	AF284707
(16)	C. *wenshanense* Y.S.Wu et F.Y. Liu	Yunnan; Sun 98-16	AF284708
(17)	C. *elegans* Lindl.	Yunnan; Sun 98-19	AF284709
(18)	C. *mastersii* Griff. ex Lindl.	Yunnan; Sun 98-21	AF284710
(19)	C. *mastersii* Griff. ex Lindl. 'maguanlan'	Yunnan; Sun 98-18	

Table 1 contd.

Table 1 contd.

(20)	C. *eburneum* Lindl.	Yunnan; Sun 98-22	AF284711
(21)	C. *eburneum* Lindl. 'xiangyabai'	Hainan; Sun 98-32	AF284712
(22)	C. *tigrinum* Parish ex Hook.	Yunnan; Sun 98-31	AF284713
	C. *ensifolium* (L) Sw.	Guangdong; Ye 32	AF284714
(23)	C. *ensifolium* (L.) Sw. 'shaoguanjianlan'	Guangdong; Ye 31	AF284715
(24)	C. *ensifolium* (L.) Sw. 'daqing'	Guangdong; Ye 33	
(25)	C. *ensifolium* (L.) Sw. 'tiegushu'	Yuannan; Sun 98-35	
(26)	C. *sinense*(Jackson ex Andr.) Willd.	Guangdong; Zhang 98-23	AF284716
(27)	C. *sinense*(Jackson ex Andr.) Willd. 'baimo'	Guangdong; Zhang 98-24	AF284717
(28)	C. *sinense*(Jackson ex Andr.)Willd. 'xianyimo'	Yunnan; Sun 98-36	
(29)	C. *sinense*(Jackson ex Andr.) Willd. 'hainanmo'	Hainan; Sun 98-37	
(30)	C. *sinense*(Jackson ex Andr.) Willd. 'heimo'	Guangdong; Sun 98-40	
(31)	C. *sinense*(Jackson ex Andr.) Willd. 'qihei'	Guangdong; Sun 98-41	
(32)	C. *sinense*(Jackson ex Andr.) Willd. 'ruanjianbaimo'	SCBG	
(33)	C. *defoliatum* Y.S. Wu et S.C. Chen.	Yunnan; Sun 98-33	AF284718
(34)	C. *cyperifolium* Wall.ex Lindl.	Yunnan; Sun 98-34	AF284719
(35)	C. *kanran* Makio	Guangdong; Ye 37	AF284720
(36)	C. *kanran* Makio 'qinghanlan'	Guangdong; Sun 98-42	
(37)	C. *kanran* Makio 'zihanlan'	Yunnan; Sun 98-44	
(38)	C. *faberi* Rolfe.	Yunnan; Sun 98-38	AF284721
(39)	C. *goeringii* (Rchb.f.) Rchb.f.	Yunnan; Sun 98-39	AF284722
(40)	C. *goeringii* (Rchb.f.) Rchb.f. 'duoxiang'	Yunnan; Sun 98-45	

Table 1 contd.

Table 1 contd.

(41)	C. *goeringii* (Rchb.f.) Rchb.f. 'doubanlan'	Yunnan; Sun 98-46	
(42)	C. *goeringii* (Rchb.f.) Rchb.f. var.*longibracteatum*	Yunnan; Sun 98-48	
(43)	C. *goeringii* (Rchb.f.) Rchb.f. 'baihuacunjian'	Yunnan; Sun 98-49	
(44)	C. *goeringii* (Rchb.f.) Rchb.f. 'guangdongduoxiang'	Guangdong; Sun 98-50	
(45)	C. *Lianbanlan* Tang et Wang	Yunnan; Sun 98-47	AF284723
(46)	C. *Lianbanlan* Tang et Wang 'huanghualianbanlan'	Yunnan; Sun 98-51	
(47)	C. *lancifolium* Hook.	Yuanan; Sun 98-43	AF284724
(48)	C. *Sinense* x C. *eburneum*	SCBG	
(49)	C. *tracyanum* x C. *eburneum*	SCBG	
(50)	C. x hybridus	SCBG	
	Eulophia graminea Lindl	Yunan; Sun 98-62	AF284726
	Geodorum desiflorum GeLamLa	Yunan; Sun 98-63	AF284727
	Amitostigma pinguiculum (Rchb.f.et S.Moore) Schltr	Yunnan; Sun 98-61	AF284725

The numbers in the bracket are No. of samples for RAPD, and samples for ITS are given the Gene-Bank accession number.

nucleotides on an ABI 377 DNA Sequencer, using at least two of the four primers of N18L18 [24], ITS2, ITS3 and ITS4 [22].

The combined sequences of ITS1, 5.8S and ITS2 were aligned using Clustal X [20], adjusted manually where necessary. Maximum parsimony analyses were performed using PAUP 4.0 [19] with all changes weighted equally, using HEURISTIC searches with TBR branch swapping and 100 random additional sequences. Multiple most parsimonious trees were summarized as strict consensus tree. To assess the relative support for clades found in the parsimony analysis, bootstrap analysis (BS) was conducted using 1,000 replicates and the same tree search procedure as described above, except with a simple taxon addition.

The length of the entire ITS region in the *Cymbidium* species surveyed ranged from 646 bp (base pairs) to 661 bp, with ITS1 ranging from 240 to 245 bp, 5.8 S 162 bp for all species, and ITS2 ranging from 243 to 255 bp. Of the 681 aligned ITS sequence, 132 bp sites were variable with 25 sites being potentially phylogenetically informative. There was one INDEL (insertion/deletion) of an 8-base pair (bp) deletion in the ITS2 region that was observed in subgenus *Cyperorchis* and C. *dayanum* of subgenus *Cymbidium*, but not in other *Cymbidium* species (aligned matrix not shown).

The maximum parsimony analysis, treating gaps as missing data, resulted in 33 most parsimonious trees, with a length of 272 steps, C1=0.66 (excluding uninformative characters), and RI=0.79. These trees essentially differ in the internal composition of the clade comprising subgenus C. *yperochis* and C. *dayanum*, and the clade comprising subgenus *Jensoa* (without C. *lancifolium*). The strict consensus tree is presented in Fig. 2. There is consistent but weak support for a sister-group relationship between C. *lancifolium* and the clade containing all other *Cymbidium* species (BS< 50). The latter consists of two clades, one including three species of subgenus *Cymbidium* section *Floriundum*, with high bootstrap support (BS = 99). The other clade comprises two subclades (both with BS<50); one subclade consists of all species of subgenus *Cyperorchis* and C. *dayanum* of subgenus *Cymbidium* (BS<50), and the other includes two species of subgenus *Cymbidium* (C. *aloifolium* and C. *bicolor*) with 100% bootstrap support, which becomes the sister to the clade containing all species of subgenus *Jensoa* except C. *lancifolium* (BS<50). Poor resolution is presented for species relationships of both subgenus *Cyperochis* and subgenus *Jensoa*.

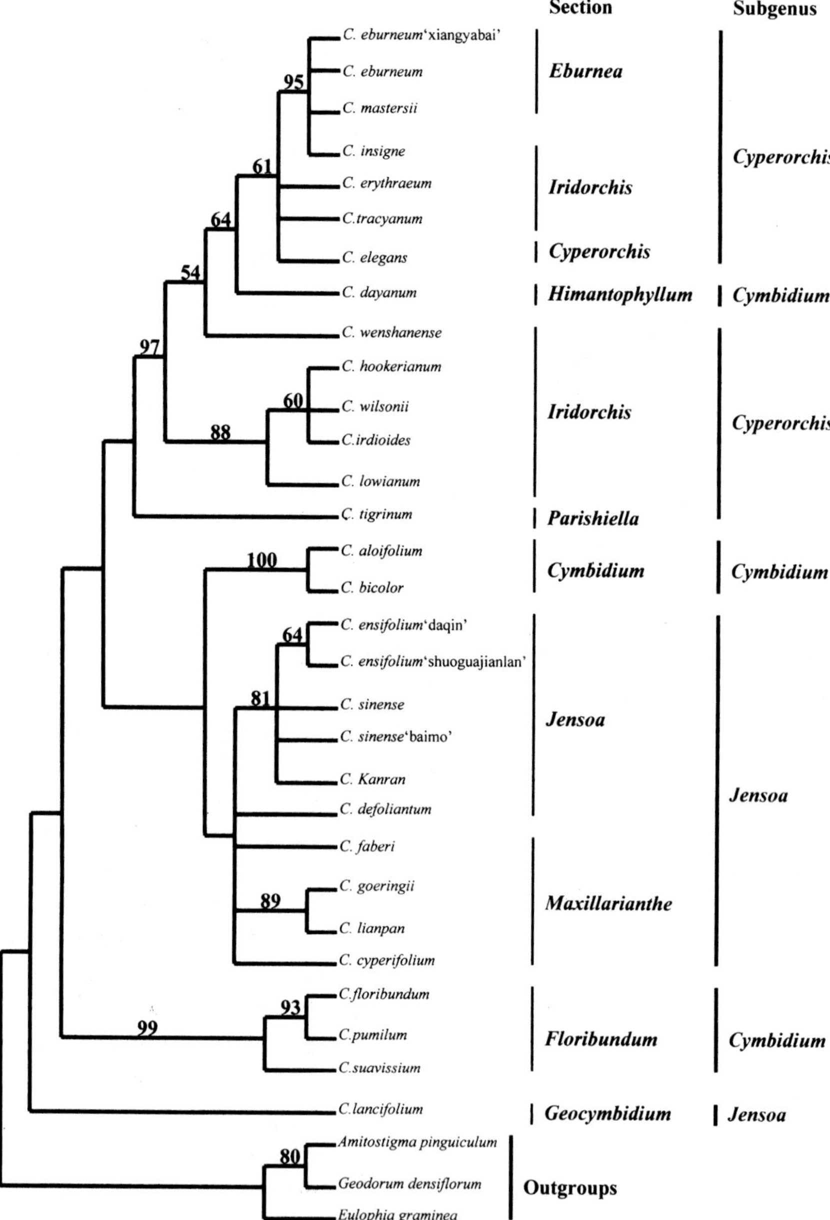

Fig. 2. *Strict consensus of the 33 most parsimonious trees from the entire ITS sequences with gaps coded as missing. Tree length, 272 steps; CI-0.66; RI-0.79. Numbers above lines represent bootstrap values in 1,000 replicates. Supraspecific classification follows Du Puy and Cribb [9]. Figure was from [27].*

Treating gaps as a fifth state, the maximum parsimony analysis generated 33 most parsimonious trees, with a length of 325 steps, a CI of 0.68 (excluding uninformative characters), and RI of 0.83. The strict consensus tree is largely congruent with that of treating gaps as missing data (Fig. 2), except that C. *erythraeum* becomes the sister to the clade comprising C. *eburneum* CV, C. *eburneum*, C. *mastersii*, and C. *insigne* (figure not shown).

Level of variability was low across all accessions of *Cymbidium* species in the ITS region studied, although van den Berg et al. [21] considered the variant of *Cymbidium* as higher than other orchids. The limited number of parsimony-informative characters resulted in relatively weak support for some of the clades identified (especially for the major lineages of *Cymbidium*) and several conclusions discussed below should be tested with additional data.

The ITS tree, however, shows that such subgenus delimitations [9] should be evaluated with more data. Subgenus *Cyperorchis* is not a monophyletic group, with the unexpected nesting of C. *dayanum* (subgenus *Cymbidium*) within it. However, with removal of the C. *dayanum* or considering it as a member of subgenus *Cyperorchis*, it would be a monophyletic group. C. *dayanum*, representing section *Himantophyllum*, is placed in subgenus *Cymbidium* with two cleft pollinia and without any fusion between the lip and base of the column. It is distinctive subgenus *Cymbidium* and superficially resembles some of the species in subgenus *Jensoa*, especially in the vegetative characteristics and the slender, acute, arching leaves [9]. The systematic position of C. *dayanum* clearly merits additional study.

C. *tigrinum*, the single, highly distinctive species representing section *Parishiella*, is sister to the remaining subgenus *Cyperorchis* and C. *dayanum*, although with low bootstrap support (<50%). It is quite different from other *Cymbidium* species, especially vegetative characteristics, but possesses all the diagnostic characters of subgenus *Cyperorchis*. It probably has close affinity with section *Iridorchis*, which the pollinarium shape closely resembles [9].

Subgenus *Jensoa* also appears paraphyletic in the ITS tree (Fig. 2), with C. *lancifolium* becoming the sister-group to the remaining of the genus, although with low bootstrap support (<50%). *Cymbidium lancifolium*, the single species comprising section *Geocymbidium*, is the most widespread species in the genus and highly distinctive in vegetative characteristics. Its habitat (or traits) and flowers indicate a close relationship with the saprophytic species C. *macrorhizon* of section *Pachyrhizanthe*, which lacks leaves and chlorophyll, and is apparent only when in blossom [9]. It is speculated from the early divergence of C. *lancifolium* that ancestors of *Cymbidium* species diversified along two paths, one leading to the main epiphytic or lithophytic species, the other into the saprophytic ones. Apparently, the addition of the saprophytic species of section *Pachyrhizanthe* will be invaluable to verify this hypothesis of evolutionally diversifying paths.

Two species of section *Cymbidium*, C. *aloifolium* and C. *bicolor*, constitute the sister-group to subgenus *Jensoa* (without C. *lancifolium*). The splitting feature of subgenus *Cymbidium* apparent in the ITS tree suggests that subgenus *Cymbidium* is polyphyletic. The number of pollinia is usually considered to be a conservative character and has been employed for the infrageneric classification of *Cymbidium* [9]. Nevertheless, this criterion may not be general: the Borneo species C. *borneense* of section *Borneense* has four pollinia, but has more other diagnostic characters and is more reliably to be placed in subgenus *Cymbidium*. No matter whether the two or four pollinia are advanced, transformation of this character may have occurred more than once in *Cymbidium*.

In comparison to the results of van den Berg et al. [21], phylogenetic relationship of Chinese *Cymbidium* based on ITS sequence is similar: the tree generated from ITS sequences is only partially congruent with the current subgenus classification, although the section groups are better congruent with the traditional classification. Both ITS [21, 27] and *matK* sequences [21] have low variation which could not completely solve the phylogenetic relationship between each species of *Cymbidium*.

RELATIONSHIP BETWEEN SPECIES, CULTIVARS OF *CYMBIDIUM* BASED ON RAPD DATA

Due to insufficiency of informative characters of ITS among most of *Cymbidium* species to solve their phylogenetic relationship, the more effective diversity molecular marker RAPD was further used to evaluate the relationships among *Cymbidium* species and some cultivars. A total of 50 samples (Table 3 comprising 28 original species, 19 cultivars among 8 species, and 3 hybrids) were used in this study, which were taken from Yunnan or Guangdong, and planted in South China Garden of Botany.

The total DNA extraction followed the 2xCTAB protocol [17] from fresh leaves, except dry leaf of C. *insigne*. Amplification of genomic DNA was made on a MJ DNA Cycler using the arbitrary decamers. The 12 oligoprimers from 100 RAPD primer kit (Sangon, Shanghai, China) were selected for conducting polymerase chain reaction (PCR) by pre-screening since the 12 primers could produce diversity bands (Table 4). Amplifications of genomic DNA were performed in 20-μl reaction volumes containing 1.0 units of Taq DNA polymerase (Sino-US, China), 10 mM Tris-HCl (pH 9.0), 50 mM KCl, 2 mM $MgCl_2$, 0.2 mM of each dNTP, 0.3 μM each of random primer and 40 ng of template DNA. The cycle program included an initial 3 minutes denaturation at 94°C, followed by 45 cycles of 1 minute at 94°C, 1 minute at 36°C and 2 minutes at 72°C, with a final extension at 72°C for 10 minutes. RAPD fragments were separated electrophoretically on 1.5% agarose gels in 1X TBE buffer, stained with ethidium bromide, and photographed on a UV transilluminator using a digital camera. DNA from each plant was amplified with the same primer more than once, and the banding patterns were compared.

Table 4. Sequence of the selected random nucleotide primers used in RAPD analysis

Primer	*Sequence*	*Primer*	*Sequence*
S106	ACGCATCGCA	S167	CAGCGACAAG
S121	ACGGATCCTG	S174	TGACGGCGGT
S133	GGCTGCAGAA	S178	TGCCCAGCCT
S140	GGTCTAGAGG	S180	AAAGTGCGGG
S142	GGTGCGGGAA	S198	CTGGCGAACT
S143	CCAGATGCAC	S199	GAGTCAGCAG

Fragment sizes were designated as amplified bands, and bands were shared as diallelic characters (present = 1, absent = 0). Those bands amplified in each instance were scored and included in the analyses. UPGMA [18] clades analysis were computed from Jaccard coefficient using NTSYS-pc V1.80 [14].

A total of 250 amplified bands was obtained from all PCR reactions by 12 selected primers, and their band size ranged from 200 bp to 2,100 bp. Except of the 850 bp band amplified from S143 primer, all other amplified bands were diversity locus, and most of the bands were located between 1,030 bp and 300 bp (Fig. 3).

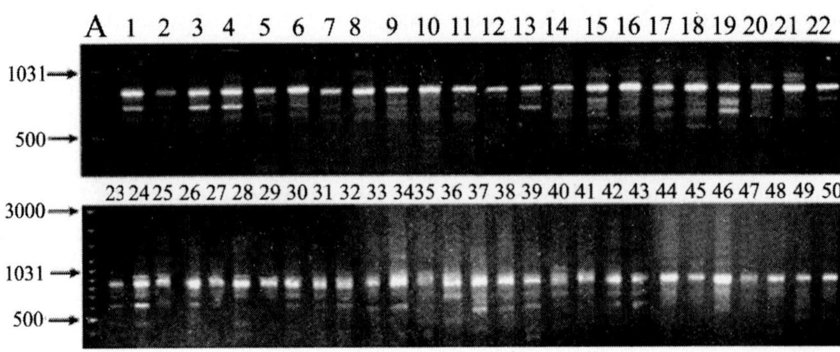

Fig. 3. The RAPD amplified profile using the primer S143. No. of samples was indicated in Table 3.

Jaccard's coefficient (JC) was mainly arranged from 0.100 to 0.250 each other among the 50 samples, The JC of C. *ensifolium* 'daqin' and C. *ensifolium* 'tiegushu' is biggest (0.644) and the JC of C. *goeringii* and C. *lowianum* is smallest (0.050) in all the JCs.

Except C. *eburneum* and C. *goeringii*, the cultivars of C. *aloifolium*, C. *mastersii*, C. *sinense*, C. *ensifolium*, C. *Kanran*, C. *lianpan* and C. *floribundum* were grouped together within their species, respectively, in the UPGMA dendrogram tree generated from the RAPD data (Fig. 4). These results showed that RAPD could be used to evaluate the relationship of species and their cultivars. Six cultivars of C. *goeringii* were divided into three groups and separated, and two cultivars of C. *eburneum* were also separated, which indicated more genetic diversity between these two species than others. From Table 3, two cultivars of C. *eburneum* were obtained from southwest China (Yunnan) and the southernmost island of China (Hainan island), and six cultivars of C. *goeringii* were from Yunnan and Guangdong (south China). These two species were separated by high mountains or strait. This geographic isolation may confer to the greater genetic diversity among the cultivars of C. *eburneum* and C. *goeringii* than other species.

The classic classification [9] put the Chinese *Cymbidium* into 3 subgenus and 11 sections. Subgenus *Cymbidium* was divided into the section of *Cymbidium*, *Himantophyllum* and *Floribundum*, Subgenus *Cyperorchis* into section of *Iridorchis*, *Cyperorchis*, *Eburnea* and *Parishiella*, Subgenus *Jensoa* into section of *Jensoa*, *Maxillarianthe*, *Geocymbidium* and

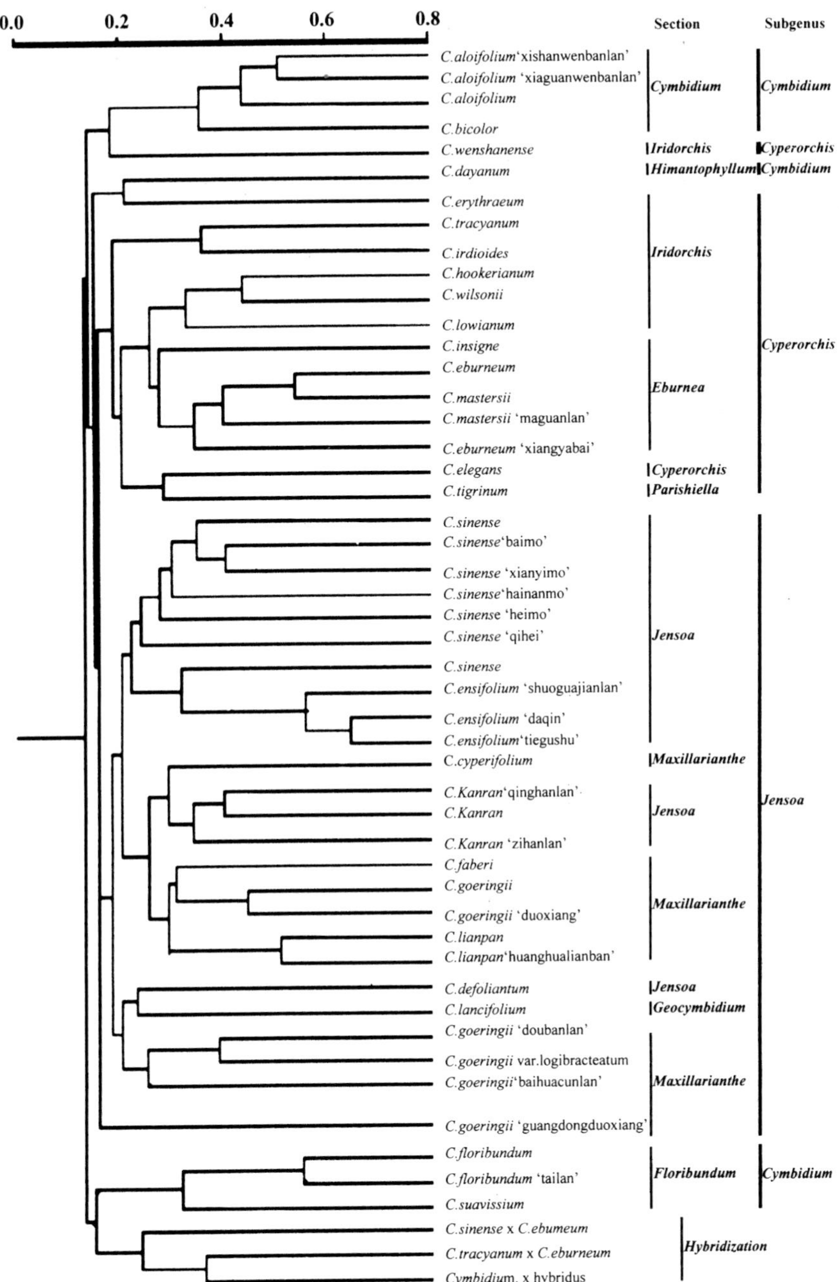

Fig. 4. *UPGMA dendrogram of 50* Cymbidium *species and cultivars in China based on Jaccard's coefficient obtained from 12 RAPD primers.*

Pachyrhizanthe. Comparison among the classifications was based on morphology [9], ITS sequence [27] and the RAPD UPGMA tree (Fig. 4). The phylogenetic trees from ITS and RAPD support the classification of subgenus *Cyperorchis* and *Jensoa*, which appeared monophyletic, only the C. *Wenshanense* is located out of subgenus *Cyperorchis* in RAPD clad tree; three sections of subgenus *Cymbidium* appeared polyphyletic, being split into several clades and intermixed with the main subgenus. However, some apparent incongruence could be noted between the ITS and RAPD results, these could be compared and viewed again to be used in phylogenetic studies.

RAPD results support traditional sectional classification of subgenus *Cymbidium* (section *Cymbidium*, *Himantophyllum* and *Floribundum*), Subgenus *Cyperorchis* (*Iridorchis* excluding C. *tracyanum* and C. *iridioides* C. *cyperorchis, sections Eburnea* and *Parishiella*), Subgenus *Jensoa* (*Jensoa*, Section *Maxillarianthe, Geocymbidium* and *Pachyrhizanthe*).

CONCLUSION

The tree generated from ITS sequences and RAPD data is only partially congruent with the current taxonomic classification of the genus *Cymbidium*, but the section classification is more congruent than the subgenus classification. Phylogenetic relationship of *Cymbidium* based on ITS [21, 27] and plastid *matK* [27] could only partly define some subgenus relationship of the genus, but it was not adequate enough to re-evaluate accurately the traditional taxonomic classification of the genus *Cymbidium* because of the low level of variation of ITS and *matK* genes. More sensitive molecular markers were tried to review the genetic diversity among cultivars and the phylogenetic relationship, which confirmed that geographic isolation could be the original source of genetic diversity in *Cymbidium* and some show phylogenetic relationship of *Cymbidium* from ITS data.

Acknowledgment

This research was supported by operating grants from The National Natural Sciences Foundation of China (research grants 30170061 and 39870086), Department of Science and Technology of Guangdong Province (2003A2010101 and 2004A).

[1] Baldwin BG, Saunderson MJ, Porter JM, et al. The ITS region of nuclear ribosomal DNA: a valuable source of evidence on angiosperm phylogeny. Ann MO Bot Gard 1995; 82: 247-277.

[2] Blume CL. Orchideae. *Rumphia.* 1848; 4: 38-56.

[3] Blume CL. *Cyperorchis. Mus Bot Lugduno-batavia.* 1849; 1:48.

[4] Blume CL. Collection des Orchidees les plus remarquables de l'archipel Indien et du Japon. 1858; 1: 90-93.

[5] Chen SC, Ji ZH, eds. The Orchids of China. Beijing: Forest Science Press, 1998 (in Chinese).

[6] Chen SC. *Cymbidium.* In: Chen SC, ed. Fl Reipubl Popularis Sin. Beijing: Science Press, 1999; 18: 197-227 (in Chinese).

[7] Doyle JJ, Doyle JL. A rapid DNA isolation method for small quantities of fresh tissues. Phytochem Bull 1987; 19: 11-15.

[8] Dressler RL. The Orchids: Natural History and Classification. Cambridge: Harvard University Press, UK 1981: 1-332.

[9] Du Puy D, Cribb PJ. The genus *Cymbidium.* Poregon: Timber Press, 1988: 1-236.

[10] Hooker JD. *Fl British India* 1891; 6: 8-15.

[11] Hunt PF. Notes on Asiatic Orchids 5. *Kew Bull* 1970; 24: 93-94.

[12] Obara-Okey P, Kako S. Genetic diversity and identification of *Cymbidium* cultivars as measured by random amplified polymorphic DNA (RAPD) markers. *Euphytica* 1998; 99: 95-101.

[13] Reichenbach HG. Orchidaceae (*Cymbidium* and *Cyperorchis*). *Walpers Ann* 1852; 3: 547-548.

[14] Rohlf FJ. NTSYS-pc. Numerical taxonomy and multivariate analysis system. Exeter Software, Setauket, N.Y. (Software program). 1993.

[15] Schlechter R. Die Gattungen *Cymbidium* Sw. und *Cyperorchis* Bl. Fedde, Repertorium Species Novarum, 1924; 20: 96-110.

[16] Seth CJ, Cribb PJ. A reassessment of the sectional limits in the genus *Cymbidium* Swarta. Arditti J. Orchid Biology, Reviews and Perspectives. Ithaca, USA: Cornell University Press, 1984, 3: 283-322.

[17] Simon A, Warner J. Genomic DNA isolation and lambda Library construction, plant gene isolation: principles and practice. John Wiley & Sons, 1996: 51-75.

[18] Sneath PHA, Sokal RR. Numerical taxonomy. San Francisco: W. H. Freeman & Co. 1973.

[19] Swofford DL. PAUP 4.0. Phylogenetic Analysis Using Parsimony (and other Methods) UK Sunderland: Sinauer Associates, 1998.

[20] Thomson JD, Gibson TJ, Plewniak F. The Clustal X windows interface: flexible strategies for multiple sequence alignment aided by quality analysis tools. *Nucl Acids Res* 1997; 25: 4876-4882.

[21] van den Berg C, Ryan A, Cribb P, Chase MW. Molecular phylogenetics of *Cymbidium* (Orchidaceae: Maxillarieaeer): sequence data from internal transcribed spacers (ITS) of nuclear ribosomal DNA and plastid. *Matk.* Lindleyana 2002; 17: 102-111.

[22] White TJ, Bruns T, Lee S, Taylor J. Amplification and direct sequencing of fungal ribosomal RNA genes for phylogenetics. In: Innis M, Gelfand D, Sninsky J, White T, eds. PCR Protocols: A Guide to Methods and Application. San Diego, USA: Academic Press, 1990: 315-322.

[23] Wu JX. Chinese *Cymbidium*. Beijing: Forest Science Press, 1991: 10-13 (in Chinese).

[24] Wen J, Zimmer EA. Phylogeny and biogeography of *Panax* L. (the ginseng genus, Araliaceae): inferences from ITS sequences of nuclear ribosomal DNA. *Mol Phylogen Evol*, 1996; 6: 166-177.

[25] Williams JG, Kubelik AR, Livak KJ, et al. DNA polymorphism amplified by arbitrary primer are useful as genetic markers. *Nucl Acids Res* 1990; 18: 6531-6535.

[26] Wong K, Sun M. Population genetic structure and reproductive biology of *Goodyera procera* (Orchidaceae): conservation implications. *Amer J Bot.* 1998; 85:66 [Abstract].

[27] Zhang MY, Sun CY, Hao G, et al. A preliminary analysis of phylogenetic relationships in *Cymbidium* (Orchidaceae) based on nrITS sequence data. *Acta Bot Sin* 2002; 44: 588-592.

CHAPTER

7

Genome Evolution and Population Biology in the Orchidaceae

ELIZABETH ANN VEASEY and GIANCARLO CONDE XAVIER OLIVEIRA

Institute: Escola Superior de Agricultura "Luiz de Queiroz" (ESALQ), University of São Paulo, Department of Genetics, "Paulo Sodero Martins" Orchid, Collection Piracicaba, Brazil

ABSTRACT

Orchidaceae, the largest angiosperm family, is not at the forefront of genomic evolution research since many of the leading approaches in the study of plant genomes has not been applied to this group. The information on chromosome evolution is reviewed in this chapter, with emphasis on the importance of polyploidization in generating new genome sizes and gene pools. The recent advances in molecular phylogenetics of Orchidaceae at several hierarchical levels are gradually providing a framework for testing hypotheses on the evolution at both the morphological and genomic levels. Evolutionary phenomena from the origin of higher taxa down to the microevolutionary processes occurring at the population level have been assessed. The latter area is experiencing fast changes due to the incorporation of all the new molecular techniques developed in the recent decades. Most population biology studies with molecular markers were conducted on terrestrial orchid species, with fewer studies on epiphytic and lithophytic species.

Key Words: Molecular markers, chromosome, phylogeny, population structure, polyploidy, genetic diversity

Address for correspondence: Escola Superior de Agricultura "Luiz de Queiroz", Departamento de Genética, Av. Pádua Dias, 11, Caixa Postal 83, CEP: 13400-970, Piracicaba, SP, Brazil. E-mail: eaveasey@esalq.usp.br; gcxolive@esalq.usp.br.

INTRODUCTION

The main word in the title, "genome", needs a conceptual/historical commentary before the genome evolution of the Orchidaceae Jussieu proper is addressed [sources of historical information in 54, 62, 95, 96], because genomic evolution will be discussed at several levels. The word genome had been associated with the concept of chromosome, and replaced later by the concept of germplasm, created by August Weissman in the 19th century to contrast it against the soma. The germplasm was the substance, and also the part of the body inside the gonads of the animals carrying this substance, responsible for the transmission of hereditary characters to the next generation. In 1869, Johann Friedrich Miescher, discovered a phosphorated, nitrogenated substance in leukocyte nuclei of pus, which he named *nuclein*, and in 1874 discovered that nuclein was composed of two moieties, the first being proteic, and the second viscous and acid, which he called *nucleic acid*.

The existence of the cell nucleus was discovered in an orchid prior to any other organism, by Robert Brown in 1831 [7]. The fact that chromosomes contained nucleic acids was discovered in 1923, when Robert Feulgen applied his coloration method chemically specific for the oxygen atom linked to the 2' carbon of deoxyribose (developed in 1914) to the microscopic observation of tissues.

In 1866 Mendel discovered that heredity was controlled by unitary factors, later called genes. By comparing gene segregation assessed in crossings with chromosome segregation under the microscope, Theodor Boveri and Walter Sutton independently proposed, in 1902, the Chromosome Theory of Heredity.

After the identification of the chromosomes as the carrier of genes, in the beginning of the 20th century, "genome" was introduced with the meaning of the ensemble of all the chromosomes, or of all genetic material. The word "genome" could, and still can, be applied at both the individual and the species level, and later began to be used for organelles, including the nucleus. This word is derived from the contraction of the common Greek word "**gen**os" (clan) with the scientific word of Greek origin "chromosome" and, in the beginning of its history, referred essentially to the chromosome set of an individual, or of a species. See, for instance, the titles of such classical works as [143] and [71], respectively, "Genome-analysis in *Brassica* with special reference..." and "Genomanalyse bei *Triticum* und *Aegilops*". In the DNA era, from 1950 onwards, "genome" was finally associated with DNA.

In the context in which the term was created, "genome" was the set of chromosomes characteristic of one or several related species. The method used for assessing species relatedness consisted primarily of the observation of chromosome pairing in the meiosis of hybrids between diploid species, belonging either to the same genus or to close genera [134]. Pairing degree ranges from null to perfect. A perfect or nearly perfect pairing indicates that species which are crossed belong to the same **genomic group**, which receives an arbitrary capital letter as its symbol (A, B, C, ...) [70]. The first two species to be investigated generally are assigned to genomic group A if their chromosomes turn out to pair perfectly. Otherwise, one species is assigned to genomic group A and the other to genomic group B, and so on. In the hybrid, each chromosome from one species pairs with the corresponding chromosome in the other species. The method assumes that overall genetic differentiation along the chromosomes is reflected in the degree of their inability to pair. Although in most cases the separation into genomic groups is discrete, sometimes intermediate degrees of pairing are found, and in such cases the capital letter is superscripted with a low-case letter in the symbol (A^g, A^l, etc...) [94]. Only a small proportion of the plant genera, however, has been studied at a level of detail sufficient for generating data about hybrid meiosis.

The genome is **compartmentalized** within the cell. If wheat or rice breeders are asked about how many genomes their target taxon is constituted of, they will probably take the question in the genomic group context, which relates only to nuclear DNA, and will answer maybe "two, B and C", or "three, A, B and D", but researchers of other fields, particularly molecular biology of photosynthesis, cell respiration or signal transduction, might think of three genomes for nearly all plants: the nuclear genome, the chloroplast genome (also called *plastome*, or cp DNA) and the mitochondrion genome (the *chondriome*, or mt DNA). It may be practical to refer to different genomes within the same cell in protoctist species with more than one nucleus, or in bacteria, which often have a small, facultative circular DNA molecule (the plasmid, accompanying the main circular DNA molecule).

The third main nuance of the word "genome" appeared with the application of the slab gel electrophoresis technique for DNA sequencing, which was proposed nearly simultaneously in two papers, [85] and [110]. Automation was the technical element necessary for the jump from the relatively small projects involving a few thousand bases, such as the rice chloroplast genome sequencing [65], to genome-wide, high-throughput projects, which came to be known as "**genome projects**". Although the Human Genome Project has been the most widely publicized, the genomes of many animal (www.ri.bbsrc.ac.uk/cgi-bin/arkdb, http://flybase.bio.indiana.edu), plant (http://www.nal.usda.gov/pgdic/Map_proj) and microorganism (http://www.tigr.org/tigr-scripts/CMR2/CMRHomePage.spl) species have been completed or are under way. The abundant information generated by genome projects enabled the rise of the subdiscipline of genomics, which has been gradually incorporating a number of objectives. Presently, the main objectives of genomics are to assemble physical (nucleotide) and genetic (recombination) maps of the genome, with emphasis on expressed genes, often identified partially as ESTs; to annotate the genes found, i.e. compile the information available on their function, expression, natural polymorphism and homology with other genes, both in the same or in other species; to study synteny, or the conservation of gene order across species [19, 49]. A gene is expressed through RNA and protein, and its function is described by its insertion in the metabolic pathways of the organism. Thus, the genome-wide study of gene expression and function entailed the creation of the ancillary disciplines of transcriptomics and proteomics, which analyze the distribution of the RNAs (the transcriptome) and proteins, (the proteome) respectively, produced in the different tissues. The integrated study of the gene functions in a genome-wide context is called metabolomics. Gene structure, expression and functions have been the objective and praxis of geneticists since the beginning of the 20th century, but until the 1990s particular projects were limited to a small number of genes, traits and pathways. The appearance of genomics was the result of a drastic change in the scale of data accumulation capabilities. We have given ourselves the objective of surveying genome evolution in a broad sense. The material ranges from macroevolutionary approaches, such as family-, tribe-, subtribe- and genus-level phylogenetic studies, through chromosome evolution to microevolutionary works addressing basic processes and population structure based on molecular markers. Given the incipience of genomic studies in orchids, macroevolution in this group is still heavily inclined towards straight systematics.

CHROMOSOME EVOLUTION

General Aspects

Chromosome counts in Orchidaceae range from 2n=12 in *Psigmorchis pusilla* [44] to 2n=168 in *Oncidium* Sw. [120, 121]. The commonest numbers are 2n=38, 40 and 42, but all the even numbers from 24 to 60 are well represented [136]. There are very few subtribes of Orchidaceae with constant chromosome numbers (Table 1). Usually, numerical variation is found not only within tribes and subtribes, but also within genera and species. Even the tissues or cells of a single individual may show variation, which is called polysomaty (examples in *Cymbidium* Sw. [47]; *Dendrobium* Sw. [67]; *Phalaenopsis* Blume [86]). The mechanistic cause of polysomaty is endoreduplication, the DNA synthesis without cell division, and its occurrence and rate range across tissues and organs. In *Cymbidium* it is absent from ovaries, rare in leaves and common in roots and flowers. In *Dendrobium* and *Phalaenopsis* it is also common in leaves. Endoreduplication begins after a few weeks in *Papilionanthe hookeriana* (Rchb.f) Schltr. (syn.*Vanda hookeriana)* X *Papilionanthe teres* (Roxb.) Schltr. (syn. *Vanda teres*) embryos [80] and continues throughout life producing 2C to 16C nuclei in adults, distributed unevenly in the tissue types. Root tips are generally endopolyploid in this material and so pose a problem for germ line chromosome counting. Apparently, polysomaty does not have a direct relation to heritable numerical variation because endoreduplication usually occurs in cell lines without a potential for shoot regeneration or gamete formation. The genetic potential for endoreduplication, with the consequent increase in vigor, is probably of adaptive value and must be subjected to selection.

The evolution of the chromosome numbers in a given clade may occur by aneuploidy, euploidy or, more commonly, both. In orchids, the relative weight of each cause in the evolution within the subfamilies is not clear enough for consensus. Five, 6, 7, 10 and 11 have been suggested as the base haploid numbers in various occasions [42]. Part of the problem arises from the possibility that the number of chromosomes may increase through real duplication of the chromosome set (euploidy),

Table 1. Phylogenetic distribution of chromosome numbers (2n) and genome sizes (in pg/2C) over suprageneric taxa of Orchidaceae *(see notes at the end of the table)*

Subfamily[a]	*Tribe*	*Subtribe*	*2n*[b]	*Mean genome size (pg/2C)*
Apostasioideae			**48, 96**, 144	
Cypripedioids [5]			18, 20, *21*, **22**, 24, 26, 27, 28, 29, **30**, 32, 34, 36, *37*, 40, 42, 48, 52, 54, 56 [37]	12.2-69.1 [37]
Vanilloids		Pogoniinae	**18**, 19, **20**, 21, 23, **24**	
		Galeolinae	**28**, 30	
		Vanillinae	**28**, **30**, **32**	**15.19; 14.45**
Orchidoids [74]	Orchideae	Orchidinae [12, 13, 38]	**20, 24, 30, 32, 36, 38, 40, 42, 44, 46, 48**, 63, 64, 72, 73, 80, 84, 85, 100, 120, 126 [15]	
		Habenariinae [13]	14, 16, **28, 30, 32, 34, 36, 38, 40, 42, 44, 46, 48**, 62, 64, 76, 84, 126	
	Diseae	Disinae	**36, 38**	
		Satyriinae	**36**, 41, **42**, 82, 164	

Table 1 contd.

Diurideae (1) [75, 76]	Chloraeinae	**16**
	Pterostylidinae	**42, 50**
Cranichideae [109]	Pachyplectroninae	?
Goodyearinae		**20, 22, 24, 26, 28, 30, 32, 40, 42, 44,** 48, 50, 56, 100, 144
	Prescottiinae	?
	Spiranthinae	**24, 26, 28, 30, 32,** 35, **36, 44, 46,** 56, 60
	Cranichidinae	**46**
Diurideae (2) [76]	Acianthinae	**38,** 40, **44, 54**
	Prasophyllinae	**44**
	Caladeniinae (1)	(for the taxon as a whole) **38, 42, 44, 46, 48**
	Cryptostylidinae	**42, 56**
	Diuridinae	**38, 56**
	Caladeniinae (2)	(for the taxon as a whole) **38, 42, 44, 46, 48**

Table 1 contd.

		Drakaeinae	**40, 44**	
		Thelymitrinae	**24, 26, 32, 56**	
"lower" Epidendroids	Anomalous			
	Nervilieae		**20, 36, 40, 54,** 72, 108, 144	
	Diceratosteleae		?	
	Triphoreae		**44**	
	Tropidieae		**40, 56, 58, 60**	
	Palmorchideae		?	
	Neottieae	Limodorinae	**20**, 24, **32**, **34**, **36**, 38, **40**, 42, 44, 46, **48**, **56**, 60	
		Listerinae	**20, 32, 34,** 35, **36, 38, 40, 42, 46, 56**	
	Epidendreae (1)	Sobraliinae	**54**	
"higher" Epidendroids	Coelogyneae	Coelogyninae [52]	**38, 40, 42, 44,** 76, 80, 120	5.48
		Thuniinae	**38, 40, 42, 44**	
	Arethuseae (1)	Bletiinae (1)	(for the taxon as a whole): 20, **26, 28, 30, 32, 36, 38, 40, 42, 44, 46, 48, 50,** 52, 54, 58, 60, 72	

Table 1 contd.

	Epidendreae (2)	Glomerinae (1)	(for the taxon as a whole) **38, 40, 46**	
	Arethuseae (2)	Arethusinae	**40, 44**	**11.38**
		Bletiinae (2)	(for the taxon as a whole): 20, **26, 28, 30, 32, 36, 38, 40, 42, 44, 46, 48, 50,** 52, 54, 58, 60, 72	
	Epidendreae (3)	Coeliinae	**40**	
		Laeliinae (1)	**42** [for *Dilomilis* only]	
		Pleurothalidinae [103]	**20, 30, 32, 34, 36, 38, 40, 42, 44,** 64	
	Arethuseae (3)	Bletiineae (3)	(for the taxon as a whole): 20, **26, 28, 30, 32, 36, 38, 40, 42, 44, 46, 48, 50,** 52, 54, 58, 60, 72	
	Podochileae	Thelasiinae	**30, 32**	
		Dendrobiinae (1)	(for the taxon as a whole): 18, 20, 30, 32-35, **36, 38, 40,** 41, 57	**1.53-4.23** (for *Dendrobium* alone)
		Podochilinae	**38, 40**	

Table 1 contd.

		Eriinae	**18, 20, 24, 34, 36, 38, 40, 42, 44, 46,** 66	
	Malaxideae		**20, 26, 28, 30, 32, 34, 36, 38, 42, 44, 46,** 72, 80	
	Dendrobieae	Dendrobiinae (2)	(for the taxon as a whole): 18, 20, 30, 32-35, **36, 38, 40,** 41, 57	**1.91;**
		Bulbophyllinae	**36, 38,** 39, **40, 42,** 57, 60, 80	**5.35;**
		Dendrobiinae (3)	(for the taxon as a whole): 18, 20, 30, 32-35, **36, 38, 40,** 41, 57	
	Epidendreae (4)	Glomerinae (2)	(for the taxon as a whole) **38, 40, 46**	
	Cymbidieae (1) [*Govenia*]		?	
	Calypsoeae		**24, 28, 32, 36, 40, 42, 46, 48, 50, 52, 54**	

Table 1 contd.

	Arethuseae (4)	Chrysiinae	?	
		Bletiinae (4)	(for all the clades of the taxon): 20, **26**, **28**, **30**, **32**, **36**, **38**, **<u>40</u>**, **<u>42</u>**, **44**, **46**, **48**, **50**, 52, 54, 58, 60, 72	
	Epidendreae (5)	Meiracylliinae	?	
		Polystachyinae	38, 39, **40**, 80, 81	
		Laeliinae [144, 145]	(for the taxon as a whole) **24**, **38**, 39, **<u>40</u>**, 41, **42**, 44, 48, 54, **56**, 57, 60, 63, 70, 80, 90, 160	**3.29**; **2.12**; **4.98**; **4.99**; **3.29**; **5.97**; **5.31**; **8.55**; **8.13**; **9.29**; **7.30**; **2.87**; **2.45**; **3.51**; **3.89**
		Arpophyllinae	?	
	Vandeae	Aerangidinae	**42**, **44**, **46**, **48**, **<u>50</u>**, **52**, **54**	
		Angraecinae [28]	**34**, **36**, **38**, **40**, **42**, **44**, **46**, **48**, **50**, 63, 76, 95	
		Aeridinae	16, 18, 20, **24**, 28, 30, 32, 34, **36**, **<u>38</u>**, 39, **<u>40</u>**, **56**, 69, 74, 76, 112, 113, 114, 115, 116	**6.40**; **9.25**; **4.73**; **5.53**; **8.65**; **6.02**; **9.65**; **4.19**; **4.10**

Table 1 contd.

	Cymbidieae (2)	Cyrtopodiinae (1)	**32, 38, 40, 42**	**16.5; 3.70; 6.31; 3.44;**
		Eulophiinae	**32, 34, 36, 38, 40,** 41, **42, 44, 46, 48,** 50, **52, 54, 56, 58, 60**, 66, 68, 70, 72, 74, 76, 80, 82, 94, 96, 100, 112	
		Cyrtopodiinae (2)	(for all the clades of the taxon) **32, 38, 40, 42**	
		Catasetinae	**54, 56, 64, 68**, 108, 162	
		Cyrtopodiinae (3)	(for all the clades of the taxon) **32, 38, 40, 42**	
	Maxillarieae [150]	Oncidiinae	**10, 14, 24,** 26, **28, 30,** 33, 34, **36,** 37, **38, 40, 42, 44,** 46, **48, 50,** 52, 54, **56,** 58, 59, **60**, 63, 72, 84, 6, 112, 126, 1338	**3.74; 4.78; 4.74; 3.85; 4.67;**

Table 1 contd.

Table 1 contd.

		Telipogoninae	?	
		Zygopetalinae (1)	(for all the clades of the taxon) **46**, 47, **48**, 50, **52**, 96	
		Cryptarrheninae	?	
		Anomalous	-	
		Zygopetalinae (2)	(for all the clades of the taxon) **46**, 47, **48**, 50, **52**, 96	
		Stanhopeinae	**38**, **40**, 41, **42**, 80	**9.34**
		Lycastinae	**38**, **40**, **44**, **48**, **50**	
		Anomalous	-	
		Maxillariinae	**40**, **42**	

[a]Subfamily, tribe and subtribe names are those given in Dressler [42]. The distribution of subtribes within tribes and of tribes within subfamilies follows the phylogeny hypothesis of Cameron et al. [25]: the first tribe in a subfamily is given in the same line and the other tribes follow downwards until the next subfamily is reached; an analogous arrangement is valid for subtribes within tribes. Some tribes and subtribes are polyphyletic and appear more than once. In this case they are given roman numerals in parentheses, the lowest representing the basalmost clade and the highest representing the most apical in the phylogeny in Cameron et al. [25].

[b]Somatic chromosome numbers (2n) in common roman type refer to Tanaka & Kamemoto [136] and references therein, which are not cited here; numbers in **boldface** refer to Dressler [42]; numbers in *italics* refer to works indicated in between square brackets. Documented natural euploids and aneuploids are included. Modal (most frequent) numbers are underlined, when applicable. Interspecific hybrids, colchiploids and commercial varieties were not included. Genome sizes in **boldface** are from Jones et al. [67].

or by centric fission (aneuploidy). In the latter case, there is some increase in the DNA content, but not proportional to the increase in number. In both cases, in the long run, approximately the same numbers could be achieved. Centric fusions, on the other hand, could reduce the 2n number, causing, for instance, 20 telocentric chromosomes to reduce to 10. Inspite of the great numerical diversity, a large bulk of evidence points to n=10 as the base number for Cypripedioids and n=20 for Epidendroids so that the latter might be tetraploids derived from a common ancestor [42]. The Apostasioids, the basal clade of the family, with 2n numbers equal to exact multiples of 48, is the group least prone to aneuploid variation. It is possibly a tetraploid itself [42], in which case its ancestor (x=n=12) might have an aneuploid relation with the ancestor of the Cypripedioids. Curiously, the chromosomes in Apostasioideae are smaller than in Cypripedioideae, which indicates evolutionary change in DNA contents.

The highest within-genus chromosome number diversity is found in *Angraecum* Bory, *Bulbophyllum* Thouars, *Calanthe* R. Br., *Cattleya* Lindl., *Dactylorchis* (Klinge) Verm., *Epidendrum* L., *Eria* Lindl., *Eulophia* R.Br ex Lindl., *Goodyera* R.Br. in W.T.Aiton, *Gymnadenia* R.Br. in W.T.Aiton, *Habenaria* Willd., *Laelia* Lindl., *Liparis* Rich., *Listera* R.Br. in W.T.Aiton, *Microstylis* (Nutt.) Eaton, *Oncidium*, *Paphiopedilum* Pfizer, *Phalaenopsis*, *Platanthera* Rich., *Spiranthes* Rich., *Vanda* Jones ex R.Br. and *Zeuxine* Lindl. (Table 1). Variation is the rule in large genera and in many smaller genera apparent constancy may be due to incomplete sampling. Besides the loss and gain of ordinary chromosomes by aneuploidy, which is evident from the long list of almost continuous 2n numbers, species vary among themselves in number of supranumerary, or B chromosomes. These chromosomes arise probably by chromosome number reduction and heterochromatinization in formerly normal chromosomes [15], and are not usually essential. They compose the karyotypes of *Coelogyne* Lindl. [136], *Dactylorhiza* Neck. Ex Nevski [15], *Dendrobium*, *Eria*, *Goodyera*, *Listera*, *Mycrostylis*, *Ophrys* L., *Paphiopedilum*, *Pleione* D.Don, *Spiranthes* and *Tainia* Blume. Curiously, many of these genera are also among the most diverse in terms of 2n, perhaps because an increase in the total number of chromosomes facilitates the transformation of a higher number in supernumerary.

Polyploidy often has the effect of enhancing size, color, thickness and durability. As such effects are desirable to some extent in the ornamental market, polyploidy is normally sought by producers, who may just select chance mutants or induce them with colchicine, in which case they are called colchiploids. Much of the cytological diversity in the most popular genera, such as *Cattleya*, *Cymbidium*, *Dendrobium*, *Paphiopedilum*, *Phalenopsis* and *Vanda* is not found in natural populations [136].

Genome Size Evolution

Chromosome number may vary without a proportional change in genome size. Conversely, it may remain constant across species that show considerable genome size differences. In *Dendrobium*, for instance, genome size varies three-fold among species while the 2n number of chromosomes remains almost constant at 38 (Table 1) [67]. On the other hand, *Neofinetia falcata* (Thunb.) Hu and *Oncidium sphacelatum* Lindl. have nearly the same amount of DNA per 2C (4.73 and 4.74 pg, respectively) but very different 2n numbers (38 and 56, respectively) [67]. The nature and evolutionary meaning of the variable DNA is unknown. Ideally, both variables should be studied together, but unfortunately the genome size of very few species is known (Table 1).

Evolution by Allopolyploidy

Polyploidy involves the increase in the numbers of whole haploid chromosome sets, which may or may not be of the same species. In the former case, called autopolyploidy, more than two homologues may pair in the meiotic prophase, and this may cause irregularities in spore formation and sterility, but after selection, meiosis may stabilize. *Serapias lingua* L., for instance, is considered an autotetraploid derived from *S. parviflora* Parl. (2n=36) [15]. In the latter case, called allopolyploidy, the component species generally belong to different genomic groups. Tetraploids, which have two haploid sets of one type and two of another (such as AABB), are called amphiploids. If the chromosomes of the parent species are too similar, they can also form multivalents in prophase I. There are basically two mechanisms leading to allopolyploidy. Unreduced gametes from different species can form a polyploid zygote, which grows into an allopolyploid adult; or an interspecific hybrid may be formed from two normal gametes from different species and become polyploid during its vegetative growth. Allopolyploidy is probably a common speciation mechanism among orchids [128, 130]. *Dactylorhiza insularis* (Sommier ex Martelli) Landwehr, for instance, was shown to be a hybrid of *D. romana* (Sebast. & Mauri) Soó and *D. sambucina* (L.) Soó, which can also form autotriploids by the union of n and 2n gametes [15]. Many species of the Cypripedioids, Orchidoids and Epidendroids with 2n equal to or around 40, 60 or 80 [136] are probably allopolyploids.

In determining the putative parent species several methods should be used. The first method is chromosomal morphological analysis, including a comparison of the karyotype among the three species (hybrid and putative parents). In the orchid genus *Spiranthes*, several allopolyploids with 2n=74 have been found to be the result of hybridization between diploid species with n=15 and n=22 (2 X [15+22]=74) [130]. 2n numbers, fundamental numbers and distribution of chromosome types, as well as banding patterns and fluorescence techniques can be used to identify similarities.

A second method is the comparison of the population genetic structure of parents and hybrids. When the parent species are homozygous for different alleles at the same locus, the allopolyploid will display genotype additivity [10] with "fixed heterozygosity" which appears as two isozyme bands on a gel [130 for *Spiranthes*]. By contrast, the different alleles of the parents will not be fixed in autoploids, but will segregate polysomically, producing genotypes of the types AAAA, AAAa, AAaa, Aaaa and aaaa ('A' and 'a' are alleles). If an allotetraploid shows a number of different alleles per locus higher than four, an additional conclusion can be drawn that more than one hybridization event originated the taxon, because two alleles per locus is the maximum each parent can contribute in one single event. That is the case for *Spiranthes diluvialis* Sheviak [10]. Similar conclusions were reached in relation to the origin of the alloploid orchid *Platanthera huronensis* Lindl., whose probable progenitors are the diploids *Piperia dilatata* (Pursh) Szlach. & Rutk. (syn. *Platanthera dilatata*) and *Platanthera aquilonis* Sheviak [147]. The alloploid shows five different chloroplast RFLP (restriction fragment length polymorphism) haplotypes corresponding to different hybridization events between the diploid parents. The long-distance dispersal by the minute seeds accounts for the wide geographic distribution of the haplotypes in North America.

A third possible method is the comparison of nucleotide sequences of homologous genes retrieved by PCR from the alloploid hybrid and the parents. If the region sequenced is moderately conserved, the diploids are likely to have only one haplotype, which may be different in each species.

An allotetraploid would have then two different haplotypes (considering a nuclear gene), which should be first separated by cloning in bacteria. Phylogenetic analysis is able to indicate which parent is more likely to have contributed each alloploid haplotype (e.g. Ge et al. [48], for rice). If cloning does not separate the alloploid haplotypes, the electrophoregram shows sequence heterogeneity and makes the analysis more difficult and more dependent on chromosome and population genetic analysis [135]. A complementary strategy is reconstructing a cpDNA- or mtDNA-based phylogeny, so that the species closest related to the alloploid can be identified as the maternal parent [97]. The parents of the alloploid American orchid *Spiranthes diluvialis* were identified as the diploids *S. romanzoffiana* Cham. (the mother) and *S. magnicamporum* Sheviak (the father) by a combination of methods [135].

Sometimes the complexity of the data defies an easy explanation and even a combination of methods is unable to shed light on the origin of an alloploid. ITS (internal transcribed spacer) sequencing, AFLPs (amplified fragment length polymorphism) and plastid restriction fragments were analyzed by maximum parsimony methods, in addition to UPGMA (unweighted pair-group methods with arithmetical averages) analysis of the AFLPs, in an attempt to identify the parents of the supposedly allohexaploid orchid *Calopogon oklahomensis* D.H. Goldman, but the results precluded a conclusion about its origin [50]. It is not absolutely clear whether the species is really a hybrid or not. If it is, it may be an ancient, much modified alloploid, which has already accumulated specific mutations.

Genomic groups and their notation have been developed and used in Orchidaceae mostly in the genome breeding literature, to describe the genomic formulae of commercial, artificial hybrids. Curiously, none of the natural alloploids in *Spiranthes* and *Calopogon* R.Br. in Aiton cited above has been described in terms of genomic formulae. Genomic groups have been best studied perhaps in *Dendrobium*, where approximately each section of the genus receives a different capital letter. Thus, section Spathulata is symbolized by C (it was formerly called Ceratobium), Eugenanthe by E, Latourea by L, Formosae by N and Phalaenanthe by P [7]. Species within sections are designated by the section's capital letter superscripted with a species-specific lowercase letter. The genome of *D. gouldii* Rchb.f., for example, is C^g, while that of *D. bigibbum* Lindl. (syn. used in ref. *D. phalenopsis)* is P^p, so that the fertile alloploid hybrid variety obtained from them is symbolized by $P^pP^pC^gC^g$ and has disomic segregation.

MOLECULAR EVOLUTION AND PHYLOGENY

The Orchidaceae Jussieu are the largest angiosperm family, with an estimated number of species ranging from 18,000 to 35,000 [25, 107, 130] and nearly 850 genera, divided into five main clades [25]: Apostasioids, Cypripedioids, Vanilloids, Orchidoids and Epidendroids. Although they do not correspond exactly to the formal subfamilies named by Dressler [42], the differences are not striking and these informal clade names are likely to be referred soon by their Latinized forms. The subfamilies are subdivided into tribes and subtribes (Table 1), and the general structure of the family fits the "hollow curve" [115] displayed by many other families, indicating a strong asymmetry in the distribution of species within genera, genera within subtribes, and subtribes within tribes. In phylogenetic terms, this means that many short branches are concentrated in some apical clades and many long branches in the basal clades.

Descriptive Knowledge of Orchid Genome at the Nucleotide Sequence Level

In spite of its size, the Orchidaceae has not been the focus of the genomic revolution mainly because of its relatively minor economic importance when compared to food crops. A simple comparison between orchids and grasses, shows that molecular research on the former lags way behind. A search for DNA sequences in GenBank (www.ncbi.nlm.nih.gov) retrieved 6,755 records of Orchidaceae, while 6,202,444 records represented Poaceae. There is no orchid genome sequence project or any genetic mapping project, or even a wide-scope EST (expressed sequence tag) sequencing project registered, either completed or under way. As a consequence of this lack of basic information about genes in orchids, there is a marked tendency in evolutionary studies to use regions located between universal primer sites [98]. In Poaceae, by contrast, beside the ITS and chloroplast generic regions, there has been an increasing use of more specific primers, such as those amplifying alcohol dehydrogenase [48] or starch synthase genes [84], which have the advantage of being single-copy genes or being present in few copies in the genome.

Among the main genes already described at the molecular level in orchids are those directly involved with floral biology [156]. Using mainly *Dendrobium*, *Phalaenopsis* and x *Aranda* Auct. systems, several genes were identified, such as (1) those belonging to the *APETALA1/AGL9* subfamily and the *DOMADS* series of the MADS-box; (2) those affecting flower pigmentation, such as dihydroflavonol 4-reductase (DFR), chalcone synthetase, flavanone 3-hydroxylase and Phe ammonia-lyase, whose regulation, though, is not well understood; more recently, the cDNA (complementary DNA) of a flavonoid-3', 5'-hydroxylase gene, which generally produces blue and purple flowers, was isolated from *Phalaenopsis* [129]; (3) O39, which may be involved in the pollination-dependent initiation of ovule development, a process unique to orchids; and (4) the genes involved in ethylene production, and consequently, in ovary maturation, such as ACC (1-aminocyclopropane-1-carboxylic acid) synthase and ACC oxidase. The importance of the molecular dissection and evolutionary studies of these genes can be measured by the centrality of the flower biology in the divergence of Orchidaceae.

The *ndhF* chloroplast gene, which encodes a component of NADH (Nicotinamide Adenine Dinucleotide) dehydrogenase, has been isolated from the orchid *Restrepia* Kunth in F.W.H.von Humboldt, A.J.A.Bonpland & C.S.Kunth (Epidendreae, Pleurothalidinae) and sequenced [88] but no comparison has been done with other orchids, so the utility of the gene for phylogenetic studies within this family remains unclear. Another gene involved in the energy metabolism, *sps1* (sucrose-phosphate synthase), was isolated from the orchid cultivar *Oncidium* Goldiana (Epidendreae) by RT-PCR (Reverse Transcriptase Polymerase Chain Reaction) [78]. The gene is 3,183 bp long, its correspondent polypeptide has 1,061 amino acid residues and a peculiar expression pattern, producing more RNA (ribonucleic acid) in the flowers, than in the leaves, in contrast with *sps* genes found in other families. A second gene affecting the sucrose metabolism, *Msus1*, was isolated from the multigenus orchid hybrid *Mokara* Auct., and is classified as a class I sucrose synthase [79]. This is one of the very few true studies in molecular evolution concerning an orchid. The cDNA, with an open reading frame of 2,447 bp, coding a 816-residue polypeptide, was aligned with similar genes isolated from other species and subjected to a phylogenetic analysis, which produced four clear clades artificially grouped into three classes: (1) the monophyletic Class I, including only monocots, (2) its sister-group Class II, composed of dicots, and (3) the paraphyletic Class II, composed of dicots. The tree contains conspecific paralogues distributed sometimes through two clades, and often has two or more conspecific paralogues in the same clade, indicating that the multigene family evolved by recent (within the species history) and remote (before the species diverged) gene duplication. The conspecific paralogues correspond to isozymes found in electrophoretic studies and sometimes are expressed in different tissues of the same organism, providing metabolic flexibility to the species. The *Mokara sus1* gene is, within the limited monocot sampling available for analysis, closest related to the *Tulipa sus1* and *sus2* genes and is expressed in growing leaves. A more detailed study of gene expression in orchids was performed in *Dendrobium*. *DSCKX1*, a gene coding for a cytokinin oxidase, was sequenced after isolation from *Dendrobium* Sonia [154] and its promoter analyzed by fusing a reporter gene to promoter fragments [155]. The promoter contains regions for the spatial and temporal control of the cytokinin oxidase production and for cytokinin binding, which is necessary for gene transcription.

Phylogenetic Structure of Orchidaceae

A molecular phylogeny generates a hypothesis about the relationships between taxa, which can be used as a frame for devising further research about character evolution, speciation, chromosome and genome size evolution, syntheny, and gene family evolution. Molecular phylogenetic studies in the Orchidaceae have been done at several levels, from family [25, 32] through tribe and subtribe to genus (Table 1). The number of tribes, subtribes and especially genera, which have been the focus of these studies, however, is very limited. Mark Chase of the Royal Botanic Gardens and his collaborators have done most of the family- and tribe-level works. The most comprehensive phylogeny of the family was based on sequences of the plastid gene *rbcL* (RuBisCo large subunit gene) [25], a gene with a remarkably conserved length, thus permitting a straightforward alignment. This study included 171 species and most of the subtribes were represented by 1-6 species. According to the *rbcL* phylogeny, the family is organized in five subfamily-level clades (i.e. monophyletic groups), viz. Apostasioids, Cypripedioids, Vanilloids, Orchidoids and Epidendroids, from the base to the apex. The phylogeny is congruent in many aspects to the most accepted systematic structure of the family [42], but reveals that some taxa are not monophyletic as circumscribed by Dressler. Among the paraphyletic taxa are the genus

Cleistes Rich. ex Lindl., tribes Cranichideae, Diseae, Orchideae, Calypsoeae, and Cymbidieae, and subtribes Habenariinae and Limodorinae. The following taxa are polyphyletic: tribes Diurideae, Arethusae and Epidendreae, and subtribes Bletiinae, Glomerinae, Dendrobinae, Laeliinae, Stanhopeinae, Zygopetalinae and Cyrtopodiinae. While these defects mean further restructuring work for systematists, they pose interesting questions to the evolutionist regarding convergence and heterogeneous rates of change.

The combined phylogenetic analysis of plastid regions *matK* plus *trnL-F* in the subfamily Orchidoideae [76] confirmed the polyphyletic nature of tribe Diurideae. If, however, Chloraeinae and Pterostylidinae are removed, Diurideae is rendered monophyletic and is shown to be composed of three major clades: the first one comprises mostly Drakaeinae, Thelymitrinae, Cryptostylidinae and Diuridinae, the second has only the core Caladeniinae, and the third has Acianthinae and Prasophyllinae. All these subtribes are monophyletic and most have good bootstrap support. The topology based on *matK* and *trnL-F* shows the subfamily Orchidoideae organized in three clades. The basalmost is a combination of Orchideae and Codonorchideae. The second is subdivided in the Diurideae as defined above and a clade, informally called "spiranthid", composed of the Cranichideae and its sequential sisters Pterostylidinae, Megastylidinae (pro parte), and the core Chloraeinae. The genus *Megastylis* (Schltr.) Schltr. is grossly polyphyletic, part of it really belonging to the spiranthids while M. *rara* (Schltr.) Schltr. is placed within Diurideae. The selective forces which caused the morphological convergence lying at the root of this systematic confusion are unknown.

In order to refine the phylogeny of the "spiranthids" and define the limits of Cranichideae, a joint analysis using the plastid regions *rbcL*, *matK* and *trnL-F* plus the nuclear ribosomal ITS, was performed [109]. Pterostylidinae, Megastylidinae, Chloraeinae, Diurideae, Codonorchideae and Orchideae, the outgroups of the analysis, are the sequential sisters to spiranthids, in increasing order of relatedness. The Cranichideae sensu Dressler [42] is actually composed of ten major clades, the five apicalmost of which are redefined as the Spiranthinae. Apart from Spiranthinae, the core spiranthids is composed of Pachyplectroninae, Goodyerinae, Galleottielinae, Manniellinae, Prescottiinae and Cranichidinae. The indels affecting nucleotides normally necessary for proper functioning in the *matK* gene, suggest that this region may have become a pseudogene.

In Cameron et al. [25], the Maxillarieae, one of the largest tribes of the family, was problematic, with Stanhopeinae and Zigopetalinae being polyphyletic. The use of ITS1 and 2, *matK*, the *trnL* intron and *trnL-F* in a broader sample of the tribe considerably clarified the relationships between genera and permitted a better assessment of the monophyly of the subtribes. Six monophyletic subtribes appear on the phylogeny: (1) Eriopsidinae, sister to the rest; (2) Oncidiinae, sister to a clade formed by (3) Stanhopeinae, (4) Coeliopsidinae, (5) Maxillariinae, and (6) Zygopetalinae. *Zygopetalum* Hook., which had been placed in a clade separate from *Dichaea* Lindl., *Koellensteinia* Rchb.f. and *Cryptarrhena* R.Br. in the *rbcL* phylogeny, is joined in a monophyletic Zygopetalinae. Cryptarrheninae is incorporated in Zygopetalinae. All the genera of Stanhopeinae represented by at least two species in the study were monophyletic, and *Lycomormium* Rchb.f., which had been responsible for the polyphyly of the subtribe, is now placed as sister genus to *Coeliopsis* Rchb.f. in Coeliopsidinae.

One of the most spectacular cases of polyphyly in orchids was revealed in *Pleurothallis* R.Br. in W.T.Aiton, in a molecular phylogenetic reconstruction of subtribe Pleurothallidinae [103]. This phylogeny used the ribosomal ITS region and 5.8S gene for the entire taxon sample, complemented by the plastid genes *matK*, *trnL-F* and the *trnL* intron for subsets of genera and produced a structure composed of nine major clades. *Pleurothallis*, as commonly circumscribed, appears in five of these clades, showing extensive polyphyly. Paraphyly is a possible problem in genera *Dracula* Luer, *Myoxanthus* Poepp. & Endl. and *Restrepia* [103], unless nomenclatural changes are proposed. Genus *Dilomilis* Raf., usually included in Laeliinae [42] was found to be a source of problems in the *rbcL* Orchidaceae phylogeny [25] because it appeared as a sister-group of Pleurothallidinae, in a position in the tree that made Laeliinae polyphyletic. *Dilomilis*, together with *Neocogniauxia* Schltr. in I.Urban, appears close enough to Pleurothallidinae to be considered as the basal part of it [103]. This transference is also supported by molecular studies performed in Laeliinae [144, 145], which suggest that *Dilomilis* should be removed from this subtribe.

This sequence of works illustrates how the fine phylogenetic structure of Orchidaceae has been improving from top to bottom with hierarchical increases in taxon sampling towards the different subfamilies, tribes, subtribes and genera, actually encompassing most of the relevant studies at levels higher than genus. In all cases, the phylogeny can be used as an independent historical indicator upon which one can map morphological and ecological traits in order to find evidence for evolutionary phenomena, such as convergences, atrophies, adaptive radiations and other speciation patterns [14 and 68 for *Disa*; 52 for *Coelogyne*; 38 for *Orchys* Tourn. Ex L.; 140 for Oncidiinae; 73 for *Bifrenaria* Lindl. and related genera; 157 for *Kitigorchys* Maek.; 142 for *Phalaenopsis*; 28 for Angraecinae].

MOLECULAR MARKERS AND POPULATION BIOLOGY IN ORCHIDS

To understand the microevolutionary processes such as genetic drift, natural selection, hybridization, mutations, associated with geographical distribution and human activity, molecular markers, such as isozymes, RAPD (random amplified polymorphic DNA), RFLP (restriction fragment length polymorphism), SSR (simple sequence repeats), ISSR (intersimple sequence repeats) and AFLP (amplified fragment length polymorphism), are all important tools utilized in many studies of plant population biology.

The Orchidaceae is considered an excellent group for studying evolution [42]. The sheer size of the family and its numerous taxonomic ambiguities suggest that genetic resources are substantial, evolutionary potential is high, and this diversity is far from static [2]. Population biology studies have been conducted with orchid species using several molecular markers, such as isozymes, RAPD, SSR, AFLP, chloroplast DNA sequencing, chloroplast minisatellite, ISSR, with different objectives: a) the study of levels of genetic variation within and among populations and species, and patterns of population structure [3, 17, 22, 23, 29, 30, 55, 57, 66, 83, 87, 112, 113, 118, 125, 151]; b) effect of genetic drift or small sized populations on genetic variation [20, 30, 39, 66, 118, 125, 131, 139, 146]; c) phylogenetic relationships among species [16, 35, 92, 112, 113, 123, 137]; d) mating systems [126, 131, 132, 133, 151]; e) introgression and hybridization [16, 72, 89, 113, 125]; f) cultivar and species identification [91, 153]; among others.

Most of these studies involve terrestrial orchid species, of the genera *Anacamptis* Rich. [39], *Caladenia* R. Br. [100], *Calopogon* [137], *Calypso* Salisb. [6], *Cephalanthera* Rich. [21, 36, 114], *Cleistes* [123], *Cremastra* Lindl. [36], *Cymbidium* [33, 34, 91, 92, 93], *Cypripedium* L. [1, 18, 22, 29, 30, 72, 149], *Dactylorhiza* [24, 63, 101], *Diuris* Sm. [116], *Epipactis*

Zinn. [23, 43, 61, 66, 112, 126, 127], *Eulophia* [133], *Goodyera* [69, 151]; *Gymnadenia* [55, 56, 57, 111, 124], *Gymnadenia* (syn. in ref.: *Nigritella*) Rich. [64], *Listera* [20], *Ophrys* [26, 125], *Orchis* Tourn. ex L. [8, 9, 108, 113], *Platanthera* [146, 147, 148], *Pseudorchis* Seg. [104], *Pterostylis* R. Br. [117, 118], *Serapias* L. [102], *Spiranthes* [10, 46, 83, 130, 131, 132, 135], *Tipularia* Nutt. [122], *Vanilla* Plum. ex Mill. [16, 89, 90], *Zeuxine* [132, 133], among others.

Although most of the orchid species (about 73%) are epiphytes [11, 42], population biology studies with molecular markers in epiphytic and lithophytic species are modest compared to terrestrial orchids, including the genera: *Catasetum* Rich. ex Kunth [87], *Dendrobium* [153], *Laelia* [105, 138], *Lepanthes* Sw. [139], *Oncidium* (syn. in ref.: *Tolumnia* Rafinesque) [3], *Pleurothallis* [17], *Miltonia* Lindl. and *Prosthechea* Knowles & Westc. [Veasey EA et al. unpublished data].

Population Biology Studies in Terrestrial Orchids

Orchis is the largest European genus, of which 70 terrestrial species are distributed in Europe, the Middle East, north Africa, and temperate regions of Asia from the Himalayas to Japan. The species are extremely diverse in appearance due to wide and varying habitat. To assess the genetic variability among and within species and the phylogenetic relationships within this group, 31 populations of 11 species of the genus *Orchis* were evaluated [113]. Compared with other outbreeding species (H_e=0.086; $\overline{P}$=37.0%), high heterozygosity values were obtained for *Orchis* species (H_e=0.149; $\overline{P}$=43.9%)[51]. A low G_{ST} mean value (0.070) was also observed, practically equal to the one reported for outbreeders (0.071) [81]. Further studies with the species *O. papilionacea* L. and *O. morio* L. [8, 108], have found low interpopulation variation with high levels of gene flow (N_m = 5.9) [8] among populations.

High level of genetic variability was also reported for the outcrossing species *Epipactis helleborine* L. Crantz (H_e=0.233; $\overline{P}$=59.0%) [112], (H_e=0.150; $\overline{P}$=33.2%)[66], (H_e=0.274; $\overline{P}$=73.6%) [43], (H_e=0.294; $\overline{P}$=67.0%) [126], (H_e=0.058–0.164; $\overline{P}$=18.2–40.9%)[23]. Low interpopulation variation was observed for this species as well (G_{ST}=0.033) [112], (F_{ST}=0.134) for rural populations [66], (F_{ST}=0.090) [43], (F_{ST}=0.040-0.130) [23]. Considering the distribution of genetic variability, the majority of the variation was partitioned within populations (96.7%, 91.34%, 72.19%, 76.30%, respectively) [23, 43, 112, 126].

A comparison between populations of *E. helleborine*, introduced in the United States and Europe [126], does not indicate a genetic bottleneck associated with the introduction process. Higher levels of intrapopulational genetic diversity and lower levels of interpopulation differentiation in introduced relative to European native populations were obtained with allozymes and cpDNA RFLPs. Also, an extremely low pollen to seed flow ratio (1.43 : 1) was observed for this species [126], although orchid seeds are considered to play a more important role than pollen in gene flow between populations, with seed dispersal having more influence on the genetic structure in orchid species [99].

E. palustris L. Crantz has also shown high levels of genetic variability (H_e=0.085; $\overline{P}$=29.0%), comparable to the mean value reported for outbreeders [51], whereas *E. microphylla* (Ehrh.) Sw. was monomorphic for all the loci examined in the two populations studied [112], which may be attributed to the fact that the populations originated from the same genetically impoverished ancestral population. The orchid, *E. helleborine* subspecies *helleborine* (syn. in ref.: *E. youngiana* A. J. Richards & A. F. Porter [106]), growing on zinc- and lead-rich sites in Northumberland, England, is supposed to have originated through the stabilization of a hybrid product via autogamy, after hybridization between *E. helleborine* x *E. leptochila* (Godfery) Godfery or *E. helleborine* x *E. phyllanthes* G. E. Sm. A survey of isozyme variation with 17 loci has ruled out the possibility of *E. phyllanthes* being a possible parent of *E. youngiana* [61]. High levels of genetic diversity (H_e=0.11-0.18; $\overline{P}$=40.0–53.0%) and a genetic structure indicative of outcrossing (G_{ST}=0.093) were observed for this species. A hybrid swarm was detected at a Glasgow site [61] for *E. youngiana* with two of its putative parents, *E. helleborine* and *E. leptochila*.

High level of genetic variation was also found in five species of the genus *Calopogon*. There are only about six terrestrial species in this genus distributed in the southeastern United States and one species in eastern Canada. All five species maintained high levels of allozyme variation within their populations (H_e=0.11-0.43; $\overline{P}$=50.0–94.4%). In *C. oklahomensis* D.H.Goldman, *C. pallidus* Chapm. and *C. tuberosus* (L.) Britton, Serns & Poggenb., most of the genetic variation exists within, rather than among populations (G_{ST}=0.037-0.085), and *C. multiflorus*

Lindl., which has the most restricted range and rarest occurrence, had the lowest mean genetic diversity values [137].

Cypripedium is a predominantly insect-pollinated terrestrial genus in the subfamily Cypripedioideae, which contains 30 to 40 species of long-lived, herbaceous perennials ranging from arctic areas to the subtropics of the Northern Hemisphere [82]. Twelve of these species occur in North America [30, 82]. Different genetic variation levels have been reported for *Cypripedium* species. A high level of polymorphism was reported for the American populations of northern *C. parviflorum* var. *pubescens* (Willd.) O. W. Knight and *C. parviflorum* var. *makasin* (Farw.) Sheviak (H_e=0.22-0.29; $\bar{P}$=81.8%) [149], as well as for *C. calceolus* L. 1753 (H_e=0.244; $\bar{P}$=75.0%) [29] and (H_e=0.184; $\bar{P}$=45.5%) [22]. Lower levels of genetic variation were found in *C. reginae* Walter (H_e=0.037; $\bar{P}$=18.0%) [30], *C. acaule* Aiton (H_e=0.0016-0.023; $\bar{P}$=5.3–15.4%) [18], while a total lack of genetic diversity (H_e=0.00; $\bar{P}$=0%) was reported in *C. arietinum* R. Br. in W. T. Aiton populations [18, 30]. The complete lack of genetic variation in this taxon in Michigan at the species level suggests a genetic bottleneck event that has eliminated variation in a population ancestral to the current populations. The same hypothesis is applied to the low genetic diversity observed in *C. reginae* and *C. candidum* Muhl. Ex Willd. populations [30].

Considering the genetic structuring of populations, most of the total species-level variation was found within populations of *C. calceolus* (81%) [29], (98, 4%) [22], *C. acaule* (84%) [30], and *C. candidum* (93%) [30]. These values are in accordance with the results reported previously for *E. helleborine* [23, 43, 112, 126], also an outcrossing orchid species.

Another terrestrial genus, *Spiranthes*, contains about 30 to 50 primarily terrestrial and a few epiphytic or lithophytic species distributed globally. Most species are found in tropical and subtropical regions of the world except Madagascar, tropical America and tropical Africa. *S. spiralis* (L.) Chevall., widely distributed in southern Europe, in the Mediterranean region, and in north Africa, *S. sinensis* (Pers.) Ames and *S. hongkongensis* S.Y.Hu & Barretto, both once widespread in Hong Kong, becoming recently rare, *S. diluvialis* and *S. romanzoffiana*, the latter restricted to the British Isles in Europe, were subjected to population genetic studies [10, 46, 83, 131, 132]. These species have shown contrasting mating systems and genetic diversity characteristics. S.

sinensis, a pollinator dependent outcrosser [131], and *S. spiralis*, where outcrossing is considered possible, being facilitated by protandry and by sequential flowering within inflorescences [83], have shown considerable levels of genetic variability, with low genetic differentiation among populations of *S. spiralis* (F_{ST}=0.022), and *S. sinensis* (G_{ST}=0.174), with 83% of the total gene diversity within populations for the latter [83, 131].

A pollinator-independent selfer allotetraploid, *S. hongkongensis*, exhibited almost complete genetic uniformity both within and among populations [131, 132]. According to the author, this result would not be directly related to the autogamy mating system, but to a population bottleneck hypothesis, where chromosome number and gene duplication at many isozyme loci suggest that this species probably evolved through a single hybridization event between diploid *S. sinensis* and *S. spiralis* [130]. Thus, the population bottleneck associated with its origin would lead to genetic uniformity irrespective of its mating system. Another allotetraploid species, *S. diluvialis*, however, exhibited high intrapopulational genetic variability (A=2.6–3.3; $\overline{P}$=57.1–71.4.0%), and a low interpopulation divergence (F_{ST} = 0.083) among 12 populations from Utah and Colorado, USA [10]. But a subsequent study evaluating 23 populations of *S. diluvialis* representing its entire geographical range [135], revealed no genetic variation within or among populations through PCR-RFLP analysis of the nuclear ribosomal internal transcribed spacer (ITS) and mitochondrial and chloroplast DNA noncoding regions. DNA sequencing revealed that *S. diluvialis* has rDNA of both *S. magnicamporum* Sheviak. and *S. romanzoffiana*, supporting the proposed origin of the allotetraploid.

Different patterns were observed between northern and southern European populations of *S. romanzoffiana*, when investigated with chloroplast microsatellites and AFLP markers [46], which are in agreement with the contrasting published generalizations that orchids show either higher, or lower, levels of population differentiation than other plant families.

Confirming this, Sharma et al. [117, 118] observed unusually high levels of genetic variability in two terrestrial, restricted in distribution, *Pterostylis* species: *P. aff. picta* M.A. Clem. (H_e = 0.284; $\overline{P}$=69.47%) and *P. gibbosa* R.Br. (H_e=0.261; $\overline{P}$=69.0%). *P. gibbosa* is a rare and endangered Australian orchid, distributed on the Central Coast of New

South Wales with disjunct populations near Milbrodale on the North Coast [117]. Despite the potential to extend its geographical range to adjacent similar habitats as seeds are wind blown, high genetic variability and high seed viability (68-90%), *P. gibbosa* is confined to only four relictual sites. Several hypotheses were raised to explain the high genetic variability of these geographicaly isolated populations, such as: the outcrossing nature of this species through specialized pollination system, high fecundity, wind dispersal of seeds and high level of gene flow, and strong summer winds which sometimes occur in October-November at the time of seed capsule dehiscence. An alternative hypothesis is that all these populations were derived from a common ancestral stock and later spread to different areas, but still maintaining the similar genetic structure due to similar evolutionary forces and ecological characteristics in these areas [117]. A population differentiation (G_{ST} = 0.15) comparable to other species with wind dispersal seeds (G_{ST} = 0.14) [81], suggests that the long-distance wind dispersal is an effective means of maintaining gene flow among distant populations. As with other orchids such as *E. helleborine* [112], *Orchis* species [108, 113], *Cypripedium* species [30], *Gymnadenia conopsea* R.Br. in W.T. Aiton [111], *P. aff. picta* [118], most of the variation resides within populations (85%) for *P. sibbosa* [117].

P. aff. picta, another endemic Australian *Pterostylis* species, is a small geophyte restricted to southwestern Australia, growing in deep calcareous sand in shrubby forests. This orchid is considered to be endangered and the total wild population is estimated to be less than 250 plants in nine scattered populations [118]. However, the high genetic diversity found in this species contrasts with the notion put forward by Hamrick and Godt [58] that a narrow geographical range should coincide with low genetic variation. Similar hypotheses to those formulated for *P. gibbosa* [117] were considered by the authors [118] to explain the high genetic diversity in *P. aff. picta.*

A different result was obtained with another threatened species, *Platanthera leucophaea* (Nutt.) Lindl., a long-lived, perennial outcrosssing pollinated by nocturnal hawkmoths. Wallace [146] found very low levels of diversity (H_e = 0.103; $\overline{P}$ =12.0%) and high levels of population differentiation (F_{ST} = 0.75) and high inbreeding coefficients in five of the 10 populations surveyed with allozyme markers, from Michigan and Ohio, USA. In contrast, RAPD markers showed higher levels of polymorphism (P_p = 45.0%) and moderate measures of population differentiation (G_{ST} = 0.26). Genetic and geographic distances were not significantly correlated in this study, suggesting a lack of interpopulation gene flow and/or genetic drift within populations. It is widely believed that *P. leucophaea* speciated from *P. praeclara* Sheviak & M. L. Bowles by adapting to a different suite of pollinators [119]. Furthermore, after the close of the Wisconsinan glaciation, *P. leucophaea* probably rapidly colonized the prairie peninsula around the Great Lakes and invaded an ecologically diverse landscape [146]. Thus, the lack of allelic variation at most loci and fixation of alleles in several populations are consistent with founder events by a small number of individuals and variable source populations. Although it is expected that gene flow was extensive shortly after colonization, recent fragmentation of its habitat and isolation of populations may have increased the allelic differences among populations today [146].

Two other terrestrial orchid species showed low levels of genetic variation with isozyme markers and higher values for RAPD markers. Wong and Sun [151] reported low variation both at the population (H_e = 0.073; $\overline{P}$ =21.78% and species (H_e = 0.15; $\overline{P}$ =33.0%) levels, in comparison with other animal-pollinated outbreeding plant species, and higher levels for both population (H_e = 0.18; $\overline{P}$ =55.13%) and species (H_e = 0.29; $\overline{P}$ =97.03%) levels with RAPD markers in *Goodyera procera* (Ker Gawl.) Hook. However, a restricted interpopulational variation was found for both isozymes (G_{ST} = 0.52) and RAPD (G_{ST} = 0.39) markers, much above the average for outcrossing species, suggesting that gene flow was limited in this species. Another study found a lack of isozyme variation for the species *Zeuxine gracilis* (Breda) Blume, an outcrosser with restricted distribution, *Z. strateumatica* (L.) Schltr., an apomictic colonizer found only in newly available open habitats, and *Eulophia sinensis* Miq., an outcrossing colonizer [133]. Higher genetic variation levels were found at the RAPD loci within populations of *Z. gracilis* (H_e = 0.54-0.076; $\overline{P}$ =15.88–21.65%) and *E. sinensis* (H_e = 0.070-0.084; $\overline{P}$ =17.82–20.97%), but little variation existed within populations of the apomictic *Z. strateumatica* (H_e = 0.011; $\overline{P}$ =2.58–2.84%). Independent of the breeding system, high G_{ST} values were obtained for all three species, indicating that total gene diversity was partitioned primarily between populations [133].

Molecular markers have also been used in orchids to unravel doubts concerning the origin of allotetraploids. The origin of *Phatanthera huronensis* Lindl., considered an allopolyploid derivative of *P. dilatata* (Pursh.) Lindl. ex L.C.Beck. and *P. aquilonis* Sheviak was confirmed from variation at 305 ISSR and RAPD loci and cpDNA patterns generated from amplification and digestion of two noncoding regions, rp116 intron and *trnT-trnF* region [147]. The genetic structure using ISSR markers of these three species revealed that most of the variation occurs within populations for *P. dilatata* (ϕ_{ST} = 0.48) and *P. huronensis* (ϕ_{ST} = 0.36), where ϕ_{ST}, an equivalent measure to F_{ST} used for dominant markers, is the combined percentage of variation occurring among groups and among populations, while most variation occurs among populations for *P. aquilonis* (ϕ_{ST} = 0.69) [148]. According to the author, self-pollination via autogamy is a likely cause of the lower level of variation and greater structure observed in *P. aquilonis*. On the other hand, because *P. dilatata* and *P. huronensis* are thought to be primarily outcrossing, gene flow via pollen may be more extensive in these species, reducing therefore differentiation among populations, while promoting variability within populations.

Allozyme data confirmed the origin of the tetraploid *Gymnigritella runei* Teppner & E. Klein, formed by fusion of an unreduced gamete from *Gymnadenia nigra* (L.) Rchb.f. (syn in ref.: *Nigritella nigra* subspecies *nigra* [L.] Rchb.f.) with a normal, haploid gamete from *Gymnadenia conopsea* (L.) [64]. Also in this study, the multilocus genotype found in *Gymnadenia widderi* (Teppner & Klein) Teppner & E. Klein (syn. in ref.: *Nigritella widderi*) was identical to one of the multilocus genotypes found in *Gymnadenia nigra* (syn. in ref.: *N. miniata* [Crantz] Jach.), indicating a close relationship of these taxa.

Gymnadenia is a genus distributed in the wet grasslands of northern temperate regions, with about 15 terrestrial species, mostly dwarf alpine orchids. Genetic and floral divergence among sympatric populations of G. *conopsea*, a common orchid in central Europe, were conducted by Soliva and Widmer [124], comparing early-flowering populations in Switzerland, recognized as subspecies *conopsea*, and late-flowering populations, recognized as subspecies *densiflora*, which occur in sympatry but with separate flowering periods. Allozyme variation indicated that subspecies *conopsea* was significantly more variable than subspecies *densiflora* and that gene flow between subspecies was low, suggesting that the difference in flowering phenology represented an effective barrier to gene flow and may be associated with genetic divergence and taxonomic diversification. Further studies were conducted with G. *conopsea* [55, 56] using microsatellite markers. High genetic variation within, and low genetic divergence among ten Swedish populations of G. *conopsea* were found, although the correlation between population size and number of alleles was close to significance at the 95% level [55]. This species is pollinated by highly mobile butterflies and moths, and presents wind-dispersed seeds, which according to the authors probably counteracts the effects of fragmentation.

Three species of *Cephalanthera*, a genus of 15 herbs, perennial and achlorophyllous species, distributed in the temperate regions of North Africa, southern Himalayas, Japan, Europe and W. North America [107], presenting different breeding types were evaluated: (1) C. *longifolia* (L.) Fritsch, a normal outbreeder, (2) C. *rubra* (L.) Rich., an outbreeder with facultative vegetative reproduction, and (3) C. *damasonium* (Mill.) Druce, an inbreeding species [114]. The last species showed a total lack of both among and within population genetic variation, probably due to the autogamic breeding type according to the authors, also suggested for the self-pollinating P. *aquilonis* [148], *Epipactis dunensis* (T. Stephenson & T.A. Stephenson) Godfrey, E. *leptochila* and E. *muelleri* Godfery, which were completely homozygous and uniform for the allozyme loci measured [127], in marked contrast to the genetically variable E. *helleborine* [43, 112, 126], as mentioned above, and the outcrossing C. *longifolia* (H_e = 0.168; F_{ST} = 0.104) [114]. These are all examples of how the breeding system in plants can affect their patterns of population genetic variation. The total lack of allozyme variation in these autogamous species may be due to a short time elapsing since one or a few isolated genetically depauperated plants gave origin to a new taxon displaying an autogamous breeding type, where neither mutational events nor natural selection nor drift occurred to alter this homogeneity [114].

Three C. *rubra* populations localized on neighboring mineral islands in the Biebrza National Park, in northeast Poland, were examined with 16 allozyme loci [21], with similar results to those reported for this species [114]. The relatively low levels of genetic variation and clonal diversity in C. *rubra* are mainly a result of the small population sizes, breeding system and type of reproduction, as well as habitat condition. Geographical proximity may be the cause of low genetic differentiation among populations, because it enables gene flow [21]. Two populations of C. *longibracteata* Blume, a self-compatible, mixed-mating species, exhibited low levels of genetic diversity (H_e = 0.036; $\overline{P}$ = 18.0%) and a significant excess of homozygosity (F_{IS} = 0.330), consistent with substantial inbreeding via selfing and/or mating among close relatives in a spatially structured population [35]. Spacial autocorrelation analysis revealed moderate but significant local spatial structure in populations of C. *longibracteata*.

In *Calypso bulbosa* (L.) Oakes, bumblebee-pollinated orchid, self-fertilization and substructuring within sampling units may have contributed to the high inbreeding coefficients observed in many C. *bulbosa* populations (F_{IS} = 0.283), over all loci within populations, and the long-distance seed and pollen dispersal may have accounted for the low to moderate genetic differentiation among populations [6].

Low genetic variation (H_o=0.058; $\overline{P}$ =9.4%; $\overline{A}$ =1.09) associated with small population sizes and genetic drift was observed for *Listera ovata* (L.) R.Br. in W.T. Aiton, localized on mineral islands in the Biebrza National Park of northeast Poland, using 32 allozyme loci [20]. Low allozymic genetic variation within-taxon and within-population in Scandinavian *Pseudorchis albida* (L.) Á.Löve & D.Löve ($\overline{P}$ =6.7%; $\overline{A}$ =2.0) and *P. albida straminea* (Fernald) Soják ($\overline{P}$ =16.7%; $\overline{A}$ =2.0) was attributed to ancient founder events [104]. According to the authors, although the differentiation is small, present-day distributions of taxa suggest that the divergence probably started before the Weichselian glaciation period.

Effective population size (N_e) influences the degree to which random genetic drift changes allele frequencies, increases inbreeding, and decreases genetic diversity, and thus is a parameter of direct relevance to the conservation of rare species [36]. Using isozyme markers and spatial autocorrelation analysis, six populations of the rare *Cremastra appendiculata* (D.Don) Makino were studied in a large (180 ha), undisturbed landscape on Oenaro Island, South Korea [36]. The levels of genetic variation ($\overline{P}$ =34%; $\overline{A}$ =1.40; H_e =0.122) observed were greater than expected, given the estimates of N_e, leading the authors to consider historical factors resulting in N_e being greater in the past than in present-day populations. Spatial autocorrelation analyses indicated significant fine-scale genetic structure, suggesting positive spatial aggregation of clones (clonal structure) and the presence of related genets (family structure) within C. *appendiculata* populations.

A fine-scale population genetic structure was also examined for the terrestrial *Cymbidium goeringii* (Rchb.f) Rchb.f in W.G.Walpers, a small herbaceous perennial, using spatial autocorrelation statistics [34]. All visible individuals (138 and 110, respectively) within 20 x 40 m areas of each of two populations were sampled and their locations mapped. Fourteen allozyme loci were analyzed and Moran's spatial autocorrelation statistics were calculated for a large number of alleles. Results indicated that genetic similarity was shared among individuals within up to a scale of 14 m distance, which was partly due to a combination of limited pollen dispersal and long-distance seed dispersal by wind. The authors recommend, therefore, that sampling within populations should be conducted at 14-16 m intervals [34]. Another study investigated the levels of genetic diversity of 24 populations of *C. goeringii* from Korea and Japan, revealing high levels of genetic variation both at population (H_e = 0.238) and species levels (H_e = 0.260) using 14 allozyme loci [33]. High values of gene flow (N_m=2.06), based on G_{ST} values, were estimated, suggesting that genetic drift is not a major fact in these populations. A significant correlation between geographic distance and genetic distance was found in *C. goeringii.* However, a relatively low interpopulation variation value was observed (G_{ST}=0.029), even though the land connection between the southern Korean peninsula and southern Japanese archipelagos has not existed since the middle Pleistocene [33]. These data are suggestive of the potential for long-distance seed dispersal expected in the Orchidaceae [3], with large numbers of small seeds of *C. goeringii* probably traveling long distances by wind from populations both in Korea and Japan.

A few orchids, such as *Ophrys* species, are pollinated by male bees and wasps through sexual deception. *Ophrys* is a species-rich genus with over 140 species, centered on the coast of the Mediterranean Basin [125]. Sympatric populations representing different species of the *O. sphegodes* Mill. complex, that differ slightly in floral morphology and are pollinated by different solitary bee species, were studied using microsatellites to test whether gene flow across the species boundaries occurs in these sympatric populations, or whether they are reproductively isolated [125]. The authors observed introgression between co-flowering, sympatric populations, therefore contradicting the assumption that species boundaries in *Ophrys* and other sexual deceit pollinated orchids are maintained by strong pollinator specificity [77]. This specificity is achieved through the mimicking of female insects with pollinator-specific odor signals. But in this study gene flow among conspecific *Ophrys* populations is high over short geographic distances and may be the consequence of either pollen flow through pollinator movement, indicating that pollinators occasionally perform heterospecific pollination, or passive seed dispersal [125]. Also observed was an extensive genetic diversity, suggesting that drift plays a minor role in the evolution of this orchid lineage. Another study with *Ophrys* species was conducted with RAPD analysis, evaluating four allopatric populations of the *O. bertolonii* Moretti complex, as well as two populations of *O. bertolonii* s.str. Moretti and *O. sphegodes* subspecies *sphegodes* (syn. in ref.: *O. fuciflora* Curtis), with significant gene diversity observed for all six taxa, suggesting their separation at specific (or subspecific) level [26].

Caladenia tentaculata Schltdl. is also a deceptive orchid species. It possesses both food deceptive and sexually deceptive species. Allozyme analysis was conducted in order to verify if the pollinator behavior is likely to result in outcrossing and long-distance pollen flow in *C. tentactulata* populations [100]. Outcrossing and long-distance pollen flow were confirmed in this study, where the maximum pollen flow distance observed was 58 m. Thus, the authors concluded that deceptive pollination in this species results in long-distance pollen flow. As far as the genetic variability parameters were concerned ($\overline{A}=1.6$; $\overline{P}=24.2\%$; $H_e=0.091$), the mean number of alleles per locus was above average, while $\%P$ and H_e were a little below average for animal pollinated outcrossing species, but close to typical values for mixed-mating plants [58]. Within-population fixation indices (F) close to zero and an unusually low interpopulation differentiation $\theta = 0.034$ (θ=F_{ST}) were observed for *C. tentactulata* [100], in accordance with its outcrossing mating system.

Two different pollination strategies are known for North American *Cleistes* [123], a genus comprising around 56 terrestrial orchid species distributed throughout the Americas, from eastern North America south to Brazil [40]. Its flowers can act as 'food-fraud' mimics [4], where the yellow labellar crest of the nectarless and scentless flower of West Virginia *C. bifaria* (Fernald) Catling & Gregg probably mimics pollen, thus attracting naive bees seeking food, or they can present a reward strategy [53]. For instance, at the Brunswick County savannah in coastal

North Carolina, where a substantial proportion of bumblebee pollinators collect pollen. Flowers of C. *bifaria* emit a strong vanilla scent, whereas those of C. *divaricata* (L.) Ames produce a daffodil-like scent. In this case, floral fragrance is thus associated with pollen reward and may encourage bees to visit the flowers [123]. To test whether the development of different fragrances, as well as presenting peak flowering times one week apart [31], is a possible instance of character displacement and an evidence for selection against hybrid formation, molecular analyses using AFLP, DNA sequencing, and microsatellites, corroborated the absence of gene flow where the two taxa occur sympatrically [123]. In the same study, genetic links occurred between the coastal plain C. *divaricata* and the two mountain populations of C. *bifaria*. This fact raised the question whether gene flow might be occurring among these groups. This possibility was ruled out because of the long distances between the two areas. The most accepted hypothesis was of a recent common ancestor as opposed to contemporary gene flow [123].

A different result was reported for the genus *Vanilla*, composed of 100 terrestrial or hemiepiphytic species, distributed in tropical and subtropical regions of North America, Mexico, West Indies, Central America, South America, Africa, Southeast Asia and West Pacific Islands [107]. The occurrence of natural hybrids between sympatric *V. claviculata* (Sw.) Sw. and *V. barbellata* Rchb.f. species were confirmed using morphological, isoenzymatic and pollination experimental data [89]. A significant surplus of heterozygotes in the deviating individuals corroborates the theory that the individuals were F_1 hybrids, and not backcrosses or F_2 generation individuals, which would result from introgression. Post-pollination barriers that could separate both parental species were ruled out after interspecific artificial crossings were made. However, hybrids were only discovered in the area where the two parentals come into contact, where the two species have almost synchronous flowering times and are likely to share the same pollinator. Also, *V. claviculata* was typically found in moist serpentine forests, while *V. barbellata* seemed to exploit dryer habitats, so hybridization is presumably normally avoided by the prezygotic mechanism of spatial isolation because of minor differences in habitat preference [89].

RAPD markers were used to assess the levels of genetic diversity in cultivated *V. planifolia* Jacks ex Andrews in introduced areas of Reunion

Island (Indian Ocean), the relationships between *V. planifolia* (including the now invalid *V. tahitensis* J.W.Moore) and *V. pompona* Schiede representatives, and the legitimacy of putative *V. planifolia* x *V. tahitensis* hybrids [16]. Low levels of genetic diversity were detected in *V. planifolia*, which is in accordance with the vegetative mode of reproduction of vanilla, and the history of recent introduction in these regions. Based on the RAPD data, the authors suggested that *V. tahitensis* is probably not a species of direct hybrid origin (*V. planifolia* x *V. pompona*) but rather a species related to *V. planifolia*. *V. tahitensis* was eventually included in *V. planifolia* (www.kew.org/monocotChecklist).

Population Biology Studies in Epiphytic and Lithophytic Orchids

The vast majority of epiphytic orchids is animal-pollinated and has tiny wind-dispersed seeds [41]. Epiphytic plants differ from the terrestrial orchids by being distributed in three dimensions, which results in an individual plant surrounded by more individuals than would be possible in two dimensions. This characteristic can affect the fine-scale genetic structure of their populations, that is, the nonrandom distribution of genetically similar individuals within populations, which can, in turn, influence mating patterns and other population phenomena [138]. With the aim of elucidating whether individual clusters in the same tree contain more than one genotype and what is the spatial distribution and fine-scale genetic structure of genotypes within a population, three large populations of *Laelia rubescens* Lindl. were sampled in the Costa Rican seasonal dry forest [138]. Isozyme analysis revealed high levels of genetic diversity at the population level ($\overline{A}$ =2.13; $\overline{P}$ =83.3%; H_e=0.199) and low among-population variation (G_{ST}=0.016). Also, multiple genotypes within a cluster were observed. These multiple genotypes were not the result of somatic mutation within a single individual, but perhaps represented the deposition and preferential establishment of sexually derived progeny in existing clusters [138].

Spatial and temporal genetic structures were also examined across sites on islands and mainlands (continuous forest area) of *Catasetum viridiflavum* Hook. using 17 polymorphic allozyme loci [87]. High levels of allelic diversity were obtained, with a total of 94 alleles detected for 17 polymorphic loci. Low among-population differentiation was observed across the landscape suggesting that the species-specific pollinator and tiny wind-dispersed seeds maintain interconnections among distant patches. This level of differentiation is within the range reported for wind dispersed, long-lived perennial, and endemic species [59]. The tiny orchid seeds are likely to traverse distances of 100 m to > 1 km depending on the height of seed release and local wind speed [87].

Another study made with *Pleurothallis*, a large genus of about 2,500 epiphytic species distributed in the tropics [107], occurring in the Brazilian campo rupestre vegetation in the southeastern and northeastern regions of Brazil, mainly in Minas Gerais and Bahia states, found surprisingly high genetic variation levels ($\overline{A}$ =2.1–3.8; $\overline{P}$ =58–83%; H_e=0.25-0.42) in all five species ((1) *P. johannensis* Barb.Rodri., (2) *P. teres* Lindl., (3) *P. ochreata* Lindl., (4) *P. fabiobarrosii* Borba & Semir, and (5) *P. adamantinensis* Brade), in spite of the fact that the five species are pollinated by small flies whose behavior enables self-pollination [17]. The authors suggest that self-incompatibility, inbreeding depression, and mechanical barriers that prevent self-pollination in these species are responsible for maintaining such high genetic variability. Moderately low values of F_{ST}, interpreted as a low level of genetic structuring, were found in *P. johannensis*, *P. fabiobarrosii* and *P. adamantinensis*. The other two species showed higher interpopulation differentiation (F_{ST}=0.21) and (F_{ST}=0.17) [17].

Seven Brazilian populations of the epiphytic *Prosthechea calamaria* (Lindl.) W.E. Higgins, which originated in the States of São Paulo, Rio de Janeiro and Espírito Santo, were analyzed with six isozyme loci, showing high genetic variation levels ($\overline{A}$ =1.50–2.50; $\overline{P}$ =33.3–100%; H_e=0.19-0.46) (Veasey et al. unpublished data). A moderate interpopulation variation (H_T=0.43; D_{ST}=0.07; G_{ST}=0.15) was found, with most of the genetic variation occurring within populations.

High levels of genetic variation ($\overline{A}$ =1.43; $\overline{P}$ =71%; H_e=0.21) were found in *Oncidium variegatum* Sw (syn. in ref.: *Tolumnia variegata* (Sw.) Braem), a widespread, morphologically variable, twig epiphyte of the Caribbean that frequently occurs in large populations [3]. Nearly all genetic variation occurred within populations (H_T=0.22; D_{ST}=0.03; G_{ST}=0.11), with moderate average gene flow estimated among populations. Comparing mainland and island populations, genetic differentiation was more substantial among islands. A significant negative correlation was observed between geographic distance and either genetic identity or Nm_w (gene flow based on Wright's statistics [152] among populations.

The potential role of genetic drift in orchid populations was investigated by estimating effective population sizes (N_e) in three Puerto Rican species of *Lepanthes*, one of the most species-rich genera in the family [139]. The three species investigated are: (1) *L. rupestris* Stimson, a common lithophyte along riverbeds of the northwestern slopes of the Luquillo Mountains, (2) *L. eltoroensis* Stimson 1970, an epiphyte restricted to mountain ridges along El Toro and Tradewinds trails and Cerro El Cacique in the Caribbean National Forest, and (3) *L. rubripetala* Stimson, a rare epiphyte in Puerto Rico. All estimates of N_e were usually <40% of the standing population size, resulting in values of <20 individuals per population. Restricted gene flow among populations was observed with isozyme analysis, in the range of one or less successful migrant per generation. This result led to another not surprising inference of high genetic differentiation among populations. Therefore, is was concluded that genetic drift is likely to be important for population differentiation in *Lepanthes* as a result of small effective population sizes and restricted gene flow [139].

Molecular markers such as AFLP and RAPD have also been used in epiphyte or lithophytic species to identify species or hybrid genotypes and to estimate genetic relationship among these taxa. Forty-three commercial *Dendrobium* hybrids were subjected to AFLP analysis and each hybrid tested had a distinct AFLP fingerprint profile, except the tissue culture mutants [153]. AFLP fingerprint profiles were uniform in different parts of tested plants, stable among individuals in vegetative propagated populations throughout different growth periods, showing potential to be an integral part of current new plant varieties protection systems. RAPD marker was used to evaluate 24 accessions of subtribe Oncidiinae, of approximately 1,000 species classified into 56-78 genera, including *Oncidium*, *Odontoglossum* Kunth in F.W.H von Humboldt, *Miltonia* Lindl., *Miltoniopsis* God.-Leb., *Brassia* R.Br. in W.T.Aiton, *Ada* Lindl., *Gomesa* R.Br. and *Comparettia* Poepp. & Endl., most commonly used as cross parents [141]. Fourteen primers produced 263 bands, of which 257 revealed polymorphism. Cluster analysis showed six major clusters and one individual not belonging to any of those. The genus *Miltonia* was segregated from other genera among the Oncidiinae in this study.

With the objective of investigating the level of genetic variability among populations of *Miltonia spectabilis* Lindl. and M. *flavescens* (Lindl.) Lindl., patterns of population variability of M. *spectabilis* were compared with the population of M. *spectabilis* var. *moreliana*. Based on these results, the taxonomic status of M. *spectabilis* var. *moreliana* was evaluated and seven isozyme loci were analyzed (Veasey et al. unpublished data). High diversity levels were observed for M. *spectabilis* ($\overline{A}$ =1.43–2.43; $\overline{P}$ =28,6–71,4%; H_e=0.13-0.28) and M. *spectabilis* var. *moreliana* ($\overline{A}$ =2.14; $\overline{P}$ =85.7%; H_e=0.35). Lower values were obtained for M. *flavescens* ($\overline{A}$ =1.14–1.43; $\overline{P}$ =14.3–42.9%; H_e=0.04-0.16). A cluster analysis showed a clear separation between both species, but did not separate M. *spectabilis* var. *moreliana* from the other populations of M. *spectabilis*, which is not in accordance with a morphometric analysis of floral characters showing a clear distinction of the two taxa [27]. Based on morphometric data, the authors proposed an old species name, M. *moreliana*, whose assumption was not supported by the isozyme data. Officially, *Miltonia spectabilis* var. moreliana is now a synonym of M. *moreliana* A.Rich (www.kew.org).

CONCLUSION

There are many research groups in the world tackling the genome evolution of orchids. However, the bewildering size of this family calls for an immense amount of work yet to be done. Among the relatively simple things that should be done is basic cytogenetics. The confusion between the effects of euploidy and Robertsonian aneuploidy may be alleviated by studying the karyotype structure. A trade-off between metacentrics and twice as much telocentrics is a good hint in favor of centric fission/fusion. More complex molecular cytogenetic methods, such as FISH (fluorescence in situ hybridization), and molecular genetic methods, such as the comparison of saturated physical maps, or even whole genome sequencing of critically chosen species, should be used to ascertain homologies among chromosomes of different species and test polyploidization hypotheses. Such an approach would help resolve one of the major problems in orchid genome evolution, viz. the relative importance of euploidy and aneuploidy in the production of current karyotypes.

Considering the pervasiveness of polyploidy, and notedly allopolyploidy [128], it is noteworthy the relative scarcity of studies using any genetic methods whatsoever in the identification of extant closest relatives of the parents of alloploids. It should be very profitable for evolutionary studies involving allopolyploidy to focus on this subject under the point of view of genomic groups. This would involve the systematic observation of meiotic pairing in lattice-like crossings among all species in a genus, but would allow the compartmentalization of the species into gene pool complexes [60] within which gene flow is easy.

Most of the molecular phylogenetic work carried out to date on orchids has been resorted to the ITS and chloroplast regions. Our knowledge about this family would greatly benefit from the extension of the molecular methods to include single- or few-copy nuclear genes and their nontranslated regions for studies at the infrageneric and infraspecific levels, as well as for the use of the coalescent theory in microevolutionary investigations [45] which require a higher level of variability and sampling.

Some deficiencies in the array of techniques employed and some bias in the choice of taxa (chiefly genera or below) included in phylogenetic and genetic studies can be noted. DNA quantification, for instance, is lacking for many species. Most of the genus- and species-level genetic research is done on temperate groups (Orchidinae and Spiranthinae are good examples). This bias is still more pronounced in studies involving ecology. Table 1 may be used as a non-exhaustive survey on the distribution of research over a phylogenetic frame and can aid in enhancing efforts in this area.

The Orchidaceae is also an excellent group for studying microevolutionary processes, such as genetic drift, hybridization, natural selection, gene flow, migration, etc. Most of the population biology studies concerned with such processes using molecular markers reviewed here have been undertaken with terrestrial orchid species (88.7%), while only 11.3% were conducted with epiphytic or lithophytic species, although epiphytic species outnumber the terrestrial species by 73% [11]. Most of these studies have been conducted with temperate species, which indicate that further attention should be given to tropical species, experiencing different environmental conditions. Considering the molecular markers applied in these studies, 60% used isozyme markers, 10% RAPD, 7.5% SSR, 5% IRRS, 3.7% AFLP and 13.8% others (plastid microsatellites, chloroplast minisatellites, cpDNA RFLP, cpDNA single stranded conformation polymorphism–SSCP, DNA sequencing). Therefore, isozymes are still a powerful tool for genetic population and population biology studies in orchids, although other types of molecular markers have gradually been incorporated to the methodology of recent research.

Acknowledgement

The authors wish to thank Dr. Maria Imaculada Zucchi, from Instituto Agronômico (IAC), for reviewing part of the manuscript, and the Brazilian Council for Scientific Research (CNPq) for a grant support for the research conducted by Elizabeth Ann Veasey.

[1] Aagaard JE, Harrod RJ, Shea KL. Genetic variation among populations of the rare clustered lady-slipper orchid (*Cypripedium fasciculatum*) from Washington State, USA. Nat Areas J 1999; 19: 234-238.

[2] Ackerman JD. Evolutionary potential in orchids: patterns and strategies for conservation. Selbyana 1998; 19: 8-14.

[3] Ackerman JD, Ward S. Genetic variation in a widespread, epiphytic orchid: Where is the evolutionary potential? Syst Bot 1999; 24: 282-291.

[4] Ackerman JD, Meléndez-Ackerman EJ, Salguero-Faria J. Variation in pollinator abundance and selection on fragrance phenotypes in an epiphytic orchid. Am J Bot 1997; 84: 1383-1390.

[5] Albert V. Cladistic relationships of the slipper orchids (Cypripedioideae: Orchidaceae) from congruent morphological and molecular data. Lindleyana 1994; 9: 115-132.

[6] Alexandersson R, Agren J. Genetic structure in the nonrewarding, bumblebee-pollinated orchid *Calypso bulbosa*. Heredity 2000; 85: 401-409.

[7] Arditti J. Orchid Biology. New York/Chichester/Brisbane/Toronto/Singapore: John Wiley & Sons, 1992. p. 691.

[8] Arduino P, Cianchi R, Rossi W, et al. Genetic variation in *Orchis papilionacea* (Orchidaceae) from the central Mediterranean region: taxonomic inferences at the intraspecific level. Plant Syst Evol 1995; 194: 9-23.

[9] Arduino P, Verra F, Cianchi R, et al. Genetic variation and natural hybridization between *Orchis laxiflora* and *Orchis palustris* (Orchidaceae). Plant Syst Evol 1996; 202: 87-109.

[10] Arft AM, Ranker TA. Allopolyploid origin and population genetics of the rare orchid *Spiranthes diluvialis*. Am J Bot 1998; 85: 110-122.

[11] Atwood JT Jr. The size of the Orchidaceae and the systematic distribution of epiphytic orchids. Selbyana 1986; 9: 171-186.

[12] Bateman RM, Pridgeon AM, Chase MW. Phylogenetics of subtribe Orchidinae (Orchidoideae, Orchidaceae) based on nuclear ITS sequences. 2. Infrageneric relationships and reclassification to achieve monophyly of *Orchis* sensu stricto. Lindleyana 1997; 12: 113-141.

[13] Bateman RM, Hollingsworth PM, Preston J, et al. Molecular phylogenetics and evolution of Orchidinae and selected Habenariinae (Orchidaceae). Bot J Linn Soc 2003; 142: 1-40.

[14] Bellstedt DU, Linder HP, Harley EH. Phylogenetic relationships in *Disa* based on non-coding *trnL-trnF* chloroplast sequences: evidence of numerous repeat regions. Am J Bot 2001; 88: 2088-2100.

[15] Bernardos S, Tyteca D, Amich F. 2004. Cytotaxonomic study of some taxa of the subtribe Orchidinae (Orchidoideae, Orchidaceae) from the Iberian Peninsula. Israel J Plant Sci 2004; 52: 161-170.

[16] Besse P, Da Silva D, Bory S, et al. RAPD genetic diversity in cultivated vanilla: *Vanilla planifolia*, and relationships with *V. tahitensis* and *V. pompona*. Plant Sci 2004; 167: 379-385.

[17] Borba EL, Felix JM, Solferini VN, Semir J. Fly-pollinated *Pleurothallis* (Orchidaceae) species have high genetic variability: evidence from isozyme markers. Am J Bot 2001; 88: 419-428.

[18] Bornbusch AH, Swender LA, Hoogerwerf DL. Genetic variation in Massachusetts populations of *Cypripedium arietinum* R Brown in Ait and *C. acaule* Ait (Orchidaceae). Rhodora 1994; 96: 354-369.

[19] Brown TA. Genomes. New York: John Wiley & Sons, Oxford: BIOS Scientific Publishers Limited, 2002.

[20] Brzosko E, Wroblewska A. Low allozymic variation in two island populations of *Listera ovata* (Orchidaceae) from NE Poland. Ann Bot Fenn 2003a; 40: 309-315.

[21] Brzosko E, Wroblewska A. Genetic variation and clonal diversity in island *Cephalanthera rubra* populations from the Biebrza National Park, Poland. Bot J Linn Soc 2003b; 143: 99-108.

[22] Brzosko E, Ratkiewicz M, Wroblewska A. Allozyme differentiation and genetic structure of the Lady's slipper (*Cypripedium calceolus*) island populations in north-east Poland. Bot J Linn Soc 2002; 138: 433-440.

[23] Brzosko E, Wroblewska A, Talalaj I. Genetic variation and genotypic diversity in *Epipactis helleborine* populations from NE Poland. Plant Syst Evol 2004; 248: 57-69.

[24] Bullini L, Cianchi R, Arduino P, et al. Molecular evidence for allopolyploid speciation and a single origin of the western Mediterranean orchid *Dactylorhiza insularis* (Orchidaceae). Biol J Linn Soc 2001; 72: 193-201.

[25] Cameron KM, Chase MW, Whitten MW, et al. A phylogenetic analysis of the Orchidaceae: evidence from *rbcL* nucleotide sequences. Am J Bot 1999; 86: 208-224.

[26] Caporali E, Grünanger P, Marziani G, et al. Molecular (RAPD) analysis of some taxa of the *Ophrys bertolonii* aggregate (Orchidaceae). Israel J Plant Sci 2001; 49: 85-89.

[27] Carlini-Garcia LA, Van den Berg C, Martins PS. A morphometric analysis of floral characters in *Miltonia spectabilis* and *Miltonia spectabilis* var. *moreliana* (Maxillarieae: Oncidiinae). Lindleyana 2002; 17: 122-129.

[28] Carlsward BS, Whitten WM, Williams NH. Molecular phylogenetics of Neotropical leafless Angraecinae (Orchidaceae): reevaluation of generic concepts. Int. J. Plant Sci. 2003; 164: 43-51.

[29] Case MA. High levels of allozyme variation within *Cypripedium calceolus* (Orchidaceae) and low levels of divergence among its varieties. Syst Bot 1993; 18: 663-677.

[30] Case MA. Extensive variation in the levels of genetic diversity and degree of relatedness among five species of *Cypripedium* (Orchidaceae). Am J Bot 1994; 81: 175-184.

[31] Catling PM, Gregg KB. Systematics of the genus *Cleistes*, North America. Lindleyana 1992; 7: 57-73.

[32] Chase M, Cameron K, Hills H, Jarrell D. DNA sequences and phylogenetics of the Orchidaceae and other lilioid monocots. In: Pridgeon A, ed. Proceedings of the Fourteenth World Orchid Conference. Glasgow: Her Majesty's Stationary Office, 1994: 61-73.

[33] Chung MY, Chung, MG. Allozyme diversity in populations of *Cymbidium goeringii* (Orchidaceae). Plant Biology 2000; 2: 77-82.

[34] Chung MY, Chung GM, Chung MG, Epperson B. Spatial genetic structure in populations of *Cymbidium goeringii* (Orchidaceae). Genes Genet Syst 1998; 73: 281-285.

[35] Chung MY, Nason JD, Chung MG. Spatial genetic structure in populations of the terrestrial orchid *Cephalanthera longibracteata* (Orchidaceae). Am J Bot 2004a; 91: 52-57.

[36] Chung MY, Nason JD, Chung MG. Implication of clonal structure for effective population size and genetic drift in a rare terrestrial orchid, *Cremastra appendiculata*. Conserv Biol 2004b; 18: 1515-1524.

[37] Cox AV, Abdelnour GJ, Bennet MD, Leitch IJ. Genome size and karyotype evolution in the slipper orchids (Cypripedioideae: Orchidaceae). Am J Bot 1998; 85: 681-687.

[38] Cozzolino S, Aceto S, Caputo P, et al. Speciation processes in Eastern Mediterranean *Orchis* s.l. species: Molecular evidence and the role of pollination biology. Israel J Plant Sci 2001; 49: 91-103.

[39] Cozzolino S, Noce ME, Musacchio A, Widmer A. Variation at a chloroplast minisatellite locus reveals the signature of habitat fragmentation and genetic bottlenecks in the rare orchid *Anacamptis palustris* (Orchidaceae). Am J Bot 2003; 90: 1681-1687.

[40] Dodson CH, Escobar RR, eds. Native Ecuadorian orchids, I. Medellín/Colombia: Companía Litográfica Nacional, 1994.

[41] Dressler RL. The orchids: natural history and classification. Cambridge, MA: Harvard University Press, 1981.

[42] Dressler RL. Phylogeny and classification of the orchid family. Portland, OR: Dioscorides Press, 1993.

[43] Ehlers BK, Pedersen HAE. Genetic variation in three species of *Epipactis* (Orchidaceae): geographic scale and evolutionary inferences. Biol J Linn Soc 2000; 69: 411-430.

[44] Felix LP, Guerra M. 1999. Chromosome analysis in *Psigmorchis pusilla* (L.) Dodson & Dressler: the smallest chromosome number known in Orchidaceae. Caryologia 1999; 52: 165-168.

[45] Felsenstein J. Inferring phylogenies. Sunderland, Massachusetts: Sinauer Associates, Inc., 2004.

[46] Forrest AD, Hollingsworth ML, Hollingsworth PM, Sydes C, Bateman RM. Population genetic structure in European populations of *Spiranthes romanzoffiana* set in the context of other genetic studies on orchids. Heredity 2004; 92: 218-227.

[47] Fukai S, Hasegawa A, Goi M. Polysomaty in *Cymbidium*. Hort Science 2002; 37(7): 1088-1091.

[48] Ge S, Sang T, Lu B, Hong D. Phylogeny of rice genomes with emphasis on origins of allotetraploid species. Proc Natl Acad Sci USA 1999; 96: 14400-14405.

[49] Gibson G, Muse SV. A primer of genome science. Sunderland, Massachusetts: Sinauer Associates, Inc., 2002. 347 p.

[50] Goldman DH, Jansen RK, Van den Berg C, Leitch IJ, Fay MF, Chase MW. Molecular and cytological examination of *Calopogon* (Orchidaceae, Epidendroideae): circumscription, phylogeny, polyploidy, and possible hybrid speciation. Am J Bot 2004; 91: 707-723.

[51] Gottlieb LD. Electrophoretic evidence and plant populations. Prog. Phytochem 1981; 7: 1-46.

[52] Gravendeel B, Chase MW, Vogel EF, et al. Molecular phylogeny of *Coelogyne* (Epidendroideae; Orchidaceae) based on plastid RFLPs, matK, and nuclear ribosomal ITS sequences: evidence for polyphyly. Am J Bot 2001; 88: 1915-1927.

[53] Gregg KB. Defrauding the deceitful orchid: pollen collection by pollinators of *Cleistes divaricata* and *C. bifaria*. Lindleyana 1991; 6: 214-220.

[54] Griffiths AJF, Miller JH, Suzuki DT, et al. An Introduction to Genetic Analysis. New York: W.H. Freeman and Company, 1993.

[55] Gustafsson S. Patterns of genetic variation in *Gymnadenia conopsea*, the fragrant orchid. Mol Ecol 2000; 9: 1863-1872.

[56] Gustafsson S, Thorén PA. Microsatellite loci in *Gymnadenia conopsea*, the fragrant orchid. Mol Ecol Notes 2001; 1: 81-82.

[57] Gustafsson S, Sjogren-Gulve P. Genetic diversity in the rare orchid, *Gymnadenia odoratissima* and a comparison with the more common congener, G *conopsea*. Conserv Genetics 2002; 3: 225-234.

[58] Hamrick JL, Godt MJW. Allozyme diversity in plant species. In: Brown AHD, Clegg MT, Kahler AL, Weir BS, eds. Plant population genetics, breeding and germplasm resources. Sunderland, MA: Sinauer Associates, 1989: 43-63.

[59] Hamrick JL, Godt MJW. Effects of life history traits on genetic diversity in plant species. Philos T Roy Soc B 1996; 351: 1291-1298.

[60] Harlan JR. Crops and man. Madison, Wisconsin: American Society of Agronomy, 1975.

[61] Harris SA, Abbott RJ. Isozyme analysis of the reported origin of a new hybrid orchid species, *Epipactis youngiana* (Young's helleborine), in the British Isles. Heredity 1997; 79: 402-407.

[62] Hausmann R. História da Biologia Molecular. [Translation from Betrachtungen zur Geschichte der Molekularbiologie] Ribeirão Preto, Brazil: Funpec Editora, 2002.

[63] Hedren M. Notes on the esterase variation in Swedish *Dactylorhiza incarnata* S.L. (Orchidaceae). Nord J Bot 1996; 16: 253-256.

[64] Hedren M, Klein E, Teppner H. Evolution of polyploids in the European orchid genus *Nigritella*: Evidence from allozyme data. Phyton-Ann Rei Bot A 2000; 40: 239-275.

[65] Hiratsuka J, Hiroahi S, Whittier R, Ishibashi T, Sakamoto M, Mori M, Kondo Y, Honji Y, Sun C, Meng B, Li Y, Kanno A, Nishizawa Y, Hirai A, Shinozaki K, Sugiura M. The complete sequence of the rice (*Oryza sativa*) chloroplast genome: intermolecular recombination between distinct tRANA genes accounts for a major plasmid DNA inversion during evolution of cereals. Mol Gen Genet 1989; 217: 185-194.

[66] Hollingsworth PM, Dickson JH. Genetic variation in rural and urban populations of *Epipactis helleborine* (L.) Crantz. (Orchidaceae) in Britain. Bot J Linn Soc 1997; 123: 321-331.

[67] Jones WE, Kuehnle AR, Arumuganathan K. Nuclear DNA content of 26 orchid (Orchidaceae) genera with emphasis on *Dendrobium*. Ann Bot 1998; 82: 189-194.

[68] Johnson SD, Linder HP, Steiner HE. Phylogeny and radiation of pollination system in *Disa* (Orchidaceae). Am J Bot 1998; 85: 402-411.

[69] Jung YH, Oh MY. TI Genetic variations in *Goodyera velutina* (Orchidaceae) on Jeju Island, Korea, as determined by single stranded conformation polymorphism analysis. Korean J Genetic 2004; 26: 345-350.

[70] Kihara H. Cytologische und genetische studien bei wichtigen getreidearten mit besondererrücksicht auf das verhalten der chomosomen und die sterilität in den bastarden. Mem Coll Sci Kyoto Imp Uni Ser B 1924; 1: 1-200.

[71] Kihara H. Genomanalyse bei *Triticum* und *Aegilops*. Cytologia 1930; 1: 263-270.

[72] Klier K, Leoschke MI, Wendel JF. Hybridization and introgression in white and yellow ladyslipper orchids (*Cypripedium candidum* and *C. pubescens*). J Hered 1991; 82: 305-318.

[73] Koehler S, Williams NH, Whitten WM, Amaral MCE. Phylogeny of the *Bifrenaria* (Orchidaceae) complex based on morphology and sequence data from nuclear rDANA internal transcribed spacers (ITS) and chloroplast *trnL-trnF* region. Int J Plant Sci 2002; 163: 1055-1066.

[74] Kores PJ, Cameron KM, Molvray M, Chase MW. The phylogenetic relationships of Orchidoideae and Spiranthoideae (Orchidaceae) as inferred from *rbcL* plastid sequences. Lindleyana 1997; 12: 1-11.

[75] Kores PJ, Weston P, Molvray M, Chase MW. Phylogenetic relationships within Diurideae: inferences from plastid *matK* DNA sequences. In: Wilson KL, Morrison DA, eds. Monocots: systematics and evolution. Collingwood, Victoria, Australia: CSIRO Publishing, 2000: 449-456.

[76] Kores PJ, Molvray M, Weston PH, et al. A phylogenetic analysis of Diuridae (Orchidaceae) based on plastid DNA sequence data. Am J Bot 2001; 88: 1903-1914.

[77] Kullenberg B, Bergström G. The pollination of *Ophrys* species. Bot Not 1976; 129: 11-19.

[78] Li CR, Zhang XB, Hew CS. Cloning of a sucrose-phosphate synthase gene highly expressed in flowers from the tropical epiphytic orchid *Oncidium* Goldiana. J Experim Bot 2003; 54(390): 2189-2191.

[79] Li CR, Zhang XB, Huang CH, Hew CS. Cloning, characterization and tissue specific expression of a sucrose synthase gene from tropical epiphytic CAM orchid *Mokara* Yellow. J. Plant Physiol 2004; 161: 87-94.

[80] Lim WL, Loh CS. Endopolyploidy in *Vanda* Miss Joaquim (Orchidaceae). New Phytologist 2003; 159: 279-287.

[81] Loveless MD, Hamrick JL. Ecological determinants of genetic structure in plant populations. Annu Rev Ecol Syst 1984; 15: 65-95.

[82] Luer CA. The native orchids of the United States and Canada excluding Florida. Bronx, NY: The New York Botanical Garden, 1975.

[83] Machon N, Bardin P, Mazer SJ, et al. Relationship between genetic structure and seed and pollen dispersal in the endangered orchid *Spiranthes spiralis*. New Phytol 2003; 157: 677-687.

[84] Mason-Gamer RJ, Weil CF, Kellogg EA. Granule-bound starch synthase: structure, function, and phylogenetic utility. Mol Biol Evol 1998; 15: 1658-1673.

[85] Maxam AM, Gilbert W. A new method for sequencing DNA. Proc Natl Acad Sci USA 1977; 74: 560-564.

[86] Mii M, Mishiba K, Tokuhara K. Polysomaty and ploidy determination in *Phalaenopsis*. Breed Sci 1997; 47: 373.

[87] Murren CJ. Spatial and demographic population genetic structure in *Catasetum viridiflavum* across a human-disturbed habitat. J Evolution Biol 2003; 16: 333-342.

[88] Neyland R, Urbatsch LE. The *ndhF* chloroplast gene detected in all vascular plant divisions. Planta 1996; 200: 273-277.

[89] Nielsen LR. Natural hybridization between *Vanilla claviculata* (W. Wright) Sw. and *V. barbellata* Rchb.f. (Orchidaceae): genetic, morphological, and pollination experimental data. Bot J Linn Soc 2000; 133: 285-302.

[90] Nielsen LR, Siegismund HR. Interspecific differentiation and hybridization in vanilla species (Orchidaceae). Heredity 1999; 83: 560-567.

[91] Obara-Okeyo P, Fujii K, Kako S. Enzyme polymorphism in *Cymbidium* orchid cultivars and inheritance of leucine aminopeptidase. HortScience 1997; 32: 1267-1271.

[92] Obara-Okeyo P, Fujii K, Kako S. Isozyme variation in *Cymbidium* species (Orchidaceae). HortScience 1998; 33: 133-135.

[93] Obara-Okeyo P, Kako S. In vitro and in vivo characterization of *Cymbidium* cultivars by isozyme analysis. J Hortic Sci 1997; 72: 263-270.

[94] Oka HI. Origin of cultivated rice. Tokyo: Japan Scientific Societies Press; Amsterdam: Elsevier Science Publishers, 1988.

[95] Olby R. The path to the double helix. The discovery of DNA. New York: Dover Books, 1994.

[96] Oliveira GCX. O ABC dos genomas: os tradicionais grupos genômicos estão resistindo à nova genômica? Anais do 17°. Encontro sobre temas de genética e melhoramento. "Genômica: Uma abordagem em plantas". 2000; 17: 52-57. Universidade de São Paulo, ESALQ, Departamento de Genética.

[97] Oliveira CXO. A molecular phylogenetic analysis of *Oryza* L. based on chloroplast DNA sequences. Saint Louis, MO: Washington University, 2002. (PhD. Dissertation).

[98] Palumbi SR. Nucleic Acids II: the polymerase chain reaction. In: Hillis DM, Moritz C, Mable BK, eds. Molecular Systematics. Sunderland, Massachussetts: Sinauer Associates, 1996: 205-248.

[99] Peakall R, Beattie AJ. The genetic consequences of worker ant pollination in a self-compatible, clonal orchid. Evolution 1991; 45: 1837-1848.

[100] Peakall R, Beattie AJ. Ecological and genetic consequences of pollination by sexual deception in the orchid *Caladenia tentactulata*. Evolution 1996; 50: 2207-2220.

[101] Pedersen HA. Allozyme variation and genetic integrity of *Dactylorhiza incarnata* (Orchidaceae). Nord J Bot 1998; 18: 15-21.

[102] Pellegrino G, Cafasso D, Widmer A, et al. Isolation and characterization of microsatellite loci from the orchid *Serapias vomeracea* (Orchidaceae) and cross-priming to other *Serapias* species. Mol Ecol Notes 2001; 1: 279-280.

[103] Pridgeon AM, Solano R, Chase MW. Phylogenetic relationships in Pleurothallidinae (Orchidaceae): combined evidence from nuclear and plastid DNA sequences. Am J Bot 2001; 88: 2286-2308.

[104] Reinhammar LG, Hedren M. Allozyme differentiation between lowland and alpine populations of *Pseudorchis albida* s.lat. (Orchidaceae) in Sweden. Nord J Bot 1998; 18: 7-14.

[105] Resende RMS. Aplicação de técnicas de análise multivariada e eletroforese de isoenzimas em estudos de relações fenéticas no gênero *Laelia seção Parviflorae.* Dissertation (Masters Degree). Escola Superior de Agricultura "Luiz de Queiroz"/ University of São Paulo, 1991.

[106] Richards AJ, Porter AF. On the identity of a Northumberland *Epipactis*. Watsonia 1982; 14: 121-128.

[107] Romero-González G, Fernández-Concha GC, Dressler RL, et al. Orchid Family. In: Magnoliophyta: Liliidae: Liliales and Orchidales, v. 26. Flora of North America Editorial Committee, eds. 1993. Flora of North America North of Mexico. 7 vols. New York and Oxford. Available in: http://www.efloras.org/flora_page.aspx?flora_id=1 (12/03/2005).

[108] Rossi W, Corrias B, Arduino P, et al. Gene variation and gene flow in *Orchis morio* (Orchidaceae) from Italy. Plant Syst Evol 1992; 179: 43-58.

[109] Salazar GA, Chase MW, Arenas MAS, Ingrouille M. Phylogenetics of Cranichidae with emphasis on Spiranthinae (Orchidaceae, Orchidoideae): evidence from plastid and nuclear DNA sequences. Am J Bot 2003; 90: 777-795.

[110] Sanger F, Nicklen S, Coulson AR. DNA sequencing with chain-terminating inhibitors. Proc Natl Acad Sci USA 1977; 74: 5463-5467.

[111] Scacchi R, De Angelis G. Isozyme polymorphism in *Gymnadenia* and its inferences for systematics within this species. Biochem Syst Ecol 1989; 17: 25-33.

[112] Scacchi R, Lanzara P, De Angelis G. Study of electrophoretic variability in *Epipactis helleborine* (L.) Crantz, *E. pallustris* (L.) Crantz and *E. microphylla* (Ehrh.) Swartz (fam. Orchidaceae). Genetica 1987; 72: 217-224.

[113] Scacchi R, De Angelis G, Lanzara P. Allozyme variation among and within eleven *Orchis* species (fam. Orchidaceae), with special reference to hybridizing aptitude. Genetica 1990; 81: 143-150.

[114] Scacchi R, De Angelis G, Corbo RM. Effect of the breeding system on the genetic structure in three *Cephalanthera* spp (Orchidaceae). Plant Syst Evol 1991; 176: 53-61.

[115] Scotland RW, Sanderson MJ. The significance of few versus many in the tree of life. Science 2004; 303: 643.

[116] Sharma IK, Jones DL. An electrophoretic study of variability in *Diuris sulphurea* R. Br. (Orchidaceae) in the Canberra region. J Orchid Soc India 1996; 10: 9-24.

[117] Sharma IK, Clements MA, Jones DL. Observations of high genetic variability in the endangered Australian terrestrial orchid *Pterostylis gibbosa* R. Br. (Orchidaceae). Biochem Syst Ecol 2000; 28: 651-663.

[118] Sharma IK, Jones DL, French CJ. Unusually high genetic variability revealed through allozymic polymorphism of an endemic and endangered Australian orchid, *Pterostylis aff. picta* (Orchidaceae). Biochem Syst Ecol 2003; 31(5): 513-526.

[119] Sheviak CJ, Bowles ML. The prairie fringed orchids: A pollinator-isolated species pair. Rhodora 1986; 88: 267-290.

[120] Sinoto Y. Chromosome numbers in *Oncidium* alliance. Cytologia 1962; 27: 306-313.

[121] Sinoto Y. Chromosomes in *Oncidium* and allied genera. I. genus *Oncidium*. La Kromosomo 1969; 76: 2459-2473.

[122] Smith JL, Hunter KL, Hunter RB. Genetic variation in the terrestrial orchid *Tipularia discolor*. Southeastern Nat 2002; 1: 17-26.

[123] Smith SD, Cowan RS, Gregg KB, et al. Genetic discontinuities among populations of *Cleistes* (Orchidaceae, Vanilloideae) in North America. Bot J Linn Soc 2004; 145: 87-95.

[124] Soliva M, Widmer A. Genetic and floral divergence among sympatric populations of *Gymnadenia conopsea* S.L. (Orchidaceae) with different flowering phenology. Int J Plant Sci 1999; 160: 897-905.

[125] Soliva M, Widmer A. Gene flow across species boundaries in sympatric, sexually deceptive *Ophrys* (Orchidaceae) species. Evolution 2003; 57: 2252-2261.

[126] Squirrell J, Hollingsworth PM, Bateman RM, et al. Partitioning and diversity of nuclear and organelle markers in native and introduced populations *of Epipactis helleborine* (Orchidaceae). Am J Bot 2001; 88: 1409-1418.

[127] Squirrell J, Hollingsworth PM, Bateman RM, et al. Taxonomic complexity and breeding system transitions: conservation genetics of the *Epipactis leptochila* complex (Orchidaceae). Mol Ecol 2002; 11: 1957-1964.

[128] Stebbins GL. Variation and evolution in plants. New York, NY: Columbia University Press, 1950.

[129] Su V, Hsu BD. Cloning and expression of a putative cytochrome P450 gene that influences the colour of *Phalenopsis* flowers. Biotechnology Letters 2003; 25: 1933-1939.

[130] Sun M. The allopolyploid origin of *Spiranthes hongkongensis* (Orchidaceae). Am J Bot 1996a; 83: 252-260.

[131] Sun M. Effects of population size, mating system, and evolutionary origin on genetic diversity in *Spiranthes sinensis* and *S. hongkongensis*. Conserv Biol 1996b; 10: 785-795.

[132] Sun, M. Genetic diversity in three colonizing orchids with contrasting mating systems. Am J Bot 1997; 84: 224-232.

[133] Sun M, Wong KC. Genetic structure of three orchid species with contrasting breeding systems using RAPD and allozyme markers. Am J Bot 2001; 88: 2180-2188.

[134] Swanson CP, Merz T, Young WJ. Cytogenetics. The chromosome in division, inheritance, and evolution. New Delhi, India: Prentice-Hall of India, 1982.

[135] Szalanski AL, Steinauer G, Bischof R, Petersen J. Origin and conservation genetics of the threatened Ute ladies'-tresses, *Spiranthes diluvialis* (Orchidaceae). Am J Bot 2001; 88: 177-180.

[136] Tanaka R, Kamemoto H. 1984. Chromosomes in orchids: counting and numbers. In: Arditti J. ed. Orchid biology. Reviews and perspectives, III. Appendix. Ithaca: Cornell University Press, 1984: 323-410.

[137] Trapnell DW, Hamrick JL, Giannasi DE. Genetic variation and species boundaries in *Calopogon* (Orchidaceae). Syst Bot 2004a; 29: 308-315.

[138] Trapnell DW, Hamrick JL, Nason, JD. Three-dimensional fine-scale genetic structure of the neotropical epiphytic orchid, *Laelia rubescens*. Mol Ecol 2004b; 13: 1111-1118.

[139] Tremblay RL, Ackerman JD. Gene flow and effective population size in *Lepanthes* (Orchidaceae): a case for genetic drift. Biol J Linn Soc 2001; 72: 47-62.

[140] Tsai CC, Huang SC. The internal transcribed spacer of ribosomal DNA as a marker for identifying species and hybrids of the Oncidiinae. J Hort Sci Biotech 2001; 76: 674-680.

[141] Tsai CC, Huang SC, Huang PL, et al. Phenetic relationship and identification of subtribe Oncidiinae genotypes by random amplified polymorphic DNA (RAPD) markers. Sci Hortic-Amsterdam 2002; 96: 303-312.

[142] Tsai CC, Huang SC, Huang PL, Chou CH. Phylogeny of the genus *Phalaenopsis* (Orchidaceae) with emphasis on the subgenus *Phalaenopsis* based on the sequences of the internal transcribed spacers 1 and 2 of rDNA. J Hort Sci Biotech 2003; 78: 879-887.

[143] UN. Genome-analysis in Brassica with special reference to the experimental formation of *Brassica napus* and peculiar mode of fertilization. Jpn J Bot 1935; 7: 389-452.

[144] Van den Berg C. Molecular phylogenetics of tribe Epidendreae with emphasis on subtribe Laeliinae (Orchidaceae). Reading, UK: University of Reading, 2000. (PhD. Dissertation).

[145] Van den Berg C, Higgins WE, Dressler RL, et al. A phylogenetic analysis of Laeliinae (Orchidaceae) based on sequence data from nuclear transcribed spacers (ITS) of ribosomal DNA. Lindleyana 2000; 15: 96-114.

[146] Wallace LE. Examining the effects of fragmentation of genetic variation in *Platanthera leucophaea* (Orchidaceae): inferences from allozyme and random amplified polymorphic DNA markers. Plant Species Biol 2002; 17: 37-49.

[147] Wallace LE. Molecular evidence for allopolyploid speciation and recurrent origins in *Platanthera huronensis* (Orchidaceae). Int J Plant Sci 2003; 164: 907-916.

[148] Wallace LE. A comparison of genetic variation and structure in the allopolyploid *Platanthera huronensis* and its diploid progenitors, *Platanthera aquilonis* and *Platanthera dilatata* (Orchidaceae). Can J Bot 2004; 82: 244-252.

[149] Wallace LE, Case MA. Contrasting allozyme diversity between northern and southern populations of *Cypripedium parviflorum* (Orchidaceae): Implications for pleistocene refugia and taxonomic boundaries. Syst Bot 2000; 25: 281-296.

[150] Whitten WM, Williams NH, Chase MW. Subtribal and generic relationships of Maxillarieae (Orchidaceae) with emphasis on Stanhopeinae: combined molecular evidence. Am J Bot 2000; 87: 1842-1856.

[151] Wong KC, Sun M. Reproductive biology and conservation genetics of *Goodyera procera* (Orchidaceae). Am J Bot 1999; 86: 1406-1413.

[152] Wright S. The interpretation of population structure by F-statistics with special regard to systems of mating. Evolution 1965; 19: 395-420.

[153] Xiang N, Hong Y, Lam-Chan LT. Genetic analysis of tropical orchid hybrids (*Dendrobium*) with fluorescence amplified fragment-length polymorphism (AFLP). J Am Soc Hortic Sci 2003; 128: 731-735.

[154] Yang SH, Yu H, Goh CJ. Isolation and characterization of the orchid cytokinin oxidase *DSCKX1* promoter. J Experim Bot 2002; 53: 1899-1907.

[155] Yang SH, Yu H, Goh CJ. Functional characterization of a cytokinin oxidase gene *DSCKXI* in *Dendrobium* orchid. Plant Mol Biol 2003; 51; 237-278.

[156] Yu H Goh CJ. Molecular genetics of reproductive biology in orchids. Plant Physiol 2001; 127: 1390-1393.

[157] Yukawa T, Chung SW, Luo Y, et al. Reappraisal of *Kitigorchis* (Orchidaceae). Bot Bull Acad Sin 2003; 44: 345-351

Karyological Evolution of the Genus *Luzula* DC. (Juncaceae)

M.C. GARCÍA-HERRAN
Institut de Botanique, laboratoire de botanique évolutive, Université de Neuchâtel, Neuchâtel, Switzerland

ABSTRACT

The genus *Luzula* DC. (Juncaceae) possesses the very specific cytological characters of: diffuse centromere, post-reductional meiosis, and variation in the chromosome number by agmatoploidy (chromosome fragmentation), or by symploidy (chromosome fusion), as well as by polyploidy. Agmatoploidy can occur as an exclusive mechanism within a taxon, being total or partial (agmato-dysploidy), as demonstrated by the *L. spicata* complex ($2n$=12 12AL, 24 24BL, 14 10AL+4BL, 16 8AL+8BL, 18 6AL+12BL) and the sub-genus *Pterodes*. However, it can also be present combined with polyploidy (*L. alpina*, $2n$=36 12AL+ 24BL). The best example of karyotype produced by symploidy is $2n$=6A_0L of *L. elegans* (sub-genus *Marlenia*). The sub-genus *Luzula* seems to be the only one representing polyploid taxa in the genus. In spite of the presence of these phenomena, which in the genus *Luzula* can be considered to be evolutionary, the chromosome variation is not generalized, the diploid valence ($2n$=12, 12AL) being the most widespread in the Luzulas. Consequently, x=6 (6AL) can be considered to be the base chromosome number, which forms the starting point from which all chromosomal evolution in the genus *Luzula* can be explained.

Key Words: *Luzula*, agmatoploidy, symploidy, diffuse centromere, agmato-dysploidy

Address for correspondence: Institut de Botanique, laboratoire de botanique évolutive, Université de Neuchâtel. 11, rue Emile-Argand, 2000 Neuchâtel, Switzerland. E-mail: karmegherran@hotmail.com.

INTRODUCTION

The genus *Luzula* DC. belongs to the family Juncaceae (Liliopsida, Monocotyledons). This family comprises seven genera: *Juncus*, *Luzula*, *Oxychloë, Distichia, Patosia, Marsippospermum* and *Rostkovia*, which contain over 440 species. Within Juncaceae, greatest diversity of species is found in temperate and cold areas of the world. Paradoxically, neotropical regions provide a habitat for the largest number of genera. *Luzula*, with 115 species, and *Juncus* L., with more than 315 species, are the largest genera and the most cosmopolitan, although they show a marked preference for the northern hemisphere. The other genera in the family are composed of one to six species and are concentrated in the southern hemisphere, principally in the southern Andes. The genera *Oxychloë* Phil. (five species), *Distichia* Nees and Meyen (three species) and *Patosia* Buchenau (one species) are endemic in South America; the genera *Marsippospermum* Desv. (four species) and *Rostkovia* Desv. (two species) have been observed in Patagonia, in the Falkland Islands and in New Zealand, the genus *Rostkovia* also grows in the Equator and in certain islands in the South Atlantic Ocean.

The genus *Luzula* is formed of herbaceous plants, perennial (with the exception of *Luzula elegans* Lowe), rhizomatous or stoloniferous, and characterized by hairy sided leaves. According to the new classification [93] the genus is subdivided into three sub-genera: (1) *Luzula* (with seven sections: (i) *Alpinae,* (ii) *Anthelaea,* (iii) *Atlanticae,* (iv) *Diprophyllatae,* (v) *Luzula,* (vi) *Nodulosae,* and (vii) *Thyrsanochlamydeae*), (2) *Marlenia*, and (3) *Pterodes*. The Luzulas, almost cosmopolitan, are present in the various stages of vegetation and prefer acidophile or silicicolous environments. They are principally distributed in moderate and even cold climates, while the genus *Luzula* is almost absent in tropical regions. The centres of maximum diversity are the south-west of Europe, the Far East, western North America, South America (the Andes), Australia and New Zealand.

CYTOLOGICAL CHARACTER OF THE GENUS *LUZULA*

Due to its particular cytological character, the genus *Luzula* was frequently studied in the 1940s and 1950s. This body of work contributed to better understanding of its karyological variability and evolution. Malheiros and Castro [62], in their karyological analysis of *Luzula purpurea* Link [in Buch] ex E. Mey (*L. elegans* Lowe), for the first time in 1947, signalled the original nature of the chromosomes in the genus: two centromeres for each chromosome, located in each extremity. In the same year, Malheiros and Gardé [64] showed that the doubling of the number of chromosomes is concomitant with a reduction of about half the chromosome length in Luzulas and concluded that the chromosome fragmentation is the result of an evolutionary process in the genus; whereas, on the other hand, it was estimated that the chromosomes have a non-localized centromere. Malheiros, Castro and Cámara [63] observed a particular process in the chromosomal cleavage of *L. purpurea* during the two divisions occurring in the meiosis. The divisions are reversed: the first is equational (separation of the sister chromatids), and the second is reductional (separation of the homologous chromatids). During metaphase I, the bivalents form two superimposed rings parallel to the equatorial plane (chromosomal auto-orientation [5, 22, 45, 46, 75]). In metaphase II, each ring, formed of two homologous chromatids, is oriented perpendicular to the equatorial plane (co-orientation). After the meiosis, the tetraspores are not separated and they undergo two successive and synchronised mitotic divisions in each tetrad. The divisions linked to the development of the gametophyte occur before the dehiscence of the anther [31]. In their analysis of chromosomal behaviour of *L. purpurea* under the effects of X-rays, Castro, Cámara and Malheiros [10] confirm the lack of a localized centromere in the Luzulas. All of the original chromosomal fragments resulting from irradiation divide independently and are conserved from generation to generation. In 1950, Malheiros-Gardé [65] did the same and showed that the chromosome fragments were inherited from one generation to another in the same way as the polyploid chromosomes, the fragmentation having been caused by an antimitotic agent, morphine. Later, by using colchicine or X-rays, several studies [12, 13, 69] have again demonstrated that the fragmentation of the chromosomes plays a very important role in the evolutionary process of the genus *Luzula* because the fragments experimentally produced are conserved from generation to generation. Braselton [9], in studying the activity of the kinetochores during *Luzula* cell division, concluded that the chromosomes are polycentric, while Godward [36] remarked on the very unusual mechanism of the chromosomes with diffuse centromere: holocentric or holokinetic chromosomes [7].

From the karyological study of various species of the genus *Luzula*, Nordenskiöld [73] observed that not only the number but also the size of the chromosomes could vary from one species to another. In fact, the chromosome size decreases when they increase in number: (a) 4-6μ in length and $\pm 1\mu$ in width for *L. purpurea* ($2n=6$); (b) 1.9μ for *L. sylvatica* (Huds.) Gaudin ($2n=12$), *L. luzuloides* (Lam.) Dandy and E. Willm. ($2n=12$) and *L. nivea* (Nathh.) DC. ($2n=12$); (c) 1.1μ, which is the usual dimension in the genus, for *L. campestris* (L.) DC. ($2n=12$), *L. multiflora* (Ehrh.) Lej. ($2n=36$), *L. frigida s.l.* ($2n=36$), *L. arctica* Blytt ($2n=24$) and *L. parviflora* (Ehrh.) Desv. ($2n=24$); (d) 0.7μ for *L. spicata s.l.* ($2n=24$); (e) 0.3-0.4μ for *L. pilosa* (L.) Willd. ($2n=c.70$) and *L. sudetica* (Willd.) Schult. ($2n=48$). This difference in chromosomal size led to the definition of a specific nomenclature for the karyotypes of the genus *Luzula* by Nordenskiöld as detailed below: in 1951 [74] (1) **AL:** standard chromosome type present in the larger part of species at $2n=12$ and in the euploid series $2n=24$, 36, 48 resulting from polyploidy; (2) **BL**: for chromosomes whose size is less than half that of type AL; (3) **CL:** for chromosomes whose length is less than half that of type BL and a quarter of type AL cells. Malheiros-Gardé and Gardé [67] complete this nomenclature by attributing the type $\mathbf{A_0L}$ to the *L. purpurea* chromosomes, the only species to possess $2n=6$ and more than twice as many AL chromosomes. In quoting the work of Portuguese authors, Nordenskiöld (*l.c.*) admits that the fragmentation may play a very important part in the evolution of the genus, processes linked to the non-localized centromere and to the inverted meiosis. She also demonstrated chromosomal fragmentation in a study of chromosomal behaviour during meiosis in experimental hybrids, which resulted from species possessing different types of chromosomes: *L. campestris* $2n=12$ (12AL) was crossed with *L. sudetica* $2n=48$ (48CL), the hybrid possesses $2n=30$ (6AL+24CL), and in meiosis the AL chromosomes never pair between them, but usually pair with some CL chromosomes.

Several authors discuss the phenomenon of the increasing chromosome number without change in the amount of nuclear DNA: "endo-nuclear polyploidy" Nordenskiöld [73, 74, 75], "pseudo-polyploidy" [6], and "agmato-polyploidy" [37]. Furthermore, the fragmentation cannot occur simultaneously in all the chromosomes. This also explains the intermediate numbers in certain species in the genus

Luzula, which results from the process Nordenskiöld (*l.c.*) named "half-completed endo-nuclear polyploidy". However, the term nowadays accepted is the one proposed by Malheiros-Gardé and Gardé in 1950 [66]: *agmatoploidy*, from the Greek word αγμα, which means rupture, in the multiplication mechanism for the chromosome number linked to the chromosome fragmentation. According to these authors, agmatoploidy is always subordinate to the presence of chromosomes with diffuse centromere. On the other hand, the reverse is not true. In 1964, Ebinger [21] interpreted this phenomenon as follows "...though the chromosome number is doubled, the DNA value remains the same (chromosome ratio 2:1; DNA ratio 1:1). In contrast, in a normal polyploid series, the chromosome number as well as the DNA value is doubled (chromosome ratio 2:1; DNA ratio 2:1)..."

Gardé [34] states that agmatoploidy is the main evolutionary factor in the genus *Luzula*. The results of Nordenskiöld [74] provided Gardé with the proof that the meiosis of chromosomes of different sizes originates from interspecific crossings. However, the cause of agmatoploidy remained unknown to Gardé. Malheiros-Gardé and Gardé [67], who also studied the hybrids and intermediate forms (*L. spicata s.l.*, *L. orestera* Sharsm.), draw the same conclusion. They consider agmatoploidy to be an evolutionary process, but without establishing the causes and the development of the fragmentation. They state that the fragmentation is produced in a median position and that it did not simultaneously intervene strongly in all the chromosomes, which also gives a progressive aspect to the variation of the chromosome number. They call the cytotypes presenting more than one type of chromosome *intermediate forms*. Agmatoploidy has also been defined as "orthoevolution karyotypic", a process implicating all the chromosomes in the karyotype [101]. In 1952, Noronha-Wagner and Castro [77] studied the meiosis of *L. campestris* and *L. nemorosa* Pollich and E. Meyer. They observed that the meiosis is post-reductional, but the mechanism seemed to be different from that observed in *L. elegans* by Malheiros, Castro and Camara [63]: during the two divisions, the bivalents were placed parallel to the spindle fibres and perpendicular to the equatorial plane, undergoing a break in anaphase I, followed by the fusion of the sister chromatid fragments in telophase I and the homologue chromatids in telophase II. All the same, when the authors analyzed the meiosis in several populations of *L. campestris* [32], no signs of chromosome fragmentation during the two divisions were noticed. Through this interpretation of meiosis, Noronha-Wagner and Castro, at least in part, provided an understanding of agmatoploidy. During telophase II the fusion of the homologous chromosomes could fail, resulting in an increase in the chromosome number. The fusion could be partial producing intermediate forms, which shows the progressive character of agmatoploidy.

From the 1960s, several studies were dedicated to quantifying DNA using photometry. Mello-Sampayo [70] and Halkka [39], have also either demonstrated or confirmed the hypothesis, according to which the number of chromosomes increases in agmatoploidy without parallel change in the amount of nuclear DNA. Barlow and Nevin [3] suggested that the amount of DNA and the total volume of the chromosomes in the agmatoploid species at 12AL and 24BL are not constant. Nevertheless, their conclusions do not seem to be well founded since their comparison is based on species, which do not belong to the same section, or at least are not akin. In contrast, the authors admit that the karyological evolution of the genus *Luzula* favours the phenomena of fragmentation, creating agmatoploid series of chromosome numbers. More recently, Sen *et al.* [88] analyzed 6 species of Luzulas produced by in vitro culture, as Inomata [47] had previously done with *L. elegans*. Their conclusions essentially confirm that the increase in chromosomal number is made by fragmentation, without any variability in the amount of DNA and the chromosomes, with diffuse centromeres. Similar conclusions are drawn in recent the work on the genus *Luzula* [20, 53].

In parallel with agmatoploidy and polyploidy, Noronha-Wagner [76] also consider chromosomal fusion to be a possible auxiliary factor in the chromosomal evolution for the genus *Luzula*, which at first sight would amend the opinion of Stebbins [95], which is: "polyploidy is predominantly an irreversible trend from lower to higher levels. This irreversibility is not due to the genetic impossibility of reversal, but to either the lowered overall adaptability or the evolutionary insignificance of this product". Stace [94] indicated that the reduction of ploidy level towards diploidy is an extremely rare phenomenon. Concurring with Raven [84], Stace considers all the evolutionary patterns of vascular plants start at diploid level and that the ample spectrum of polyploids finds its origin in the diploids. However, it has been demonstrated that certain vegetable organisms with localized centromere have sustained a reduction in their degree of ploidy: haploidy [16, 48, 87]. On the other hand, in organisms with diffuse centromere, nothing indicates that fusion is not as probable as fragmentation. From this, Luceño and Guerra [61], implied that chromosomal fusion is another phenomenon permitting chromosomal variability in organisms with holocentric chromosomes. The authors suggest the term *symploidy*, from the συν meaning with or together, expressing the idea of chromosomal union. This could explain the reason for such a low number in *L. elegans* ($2n=6$), which may result from chromosome fusion starting from a karyotype at $2n=12$ [37, 61].

The work on the genus *Luzula* highlights three distinctive cytological characteristics. Firstly, the chromosomes are holocentric because they possess a diffuse or a non-localized centromere. Secondly, the two meiotic divisions are reversed because the equational division, with auto-orientation of the chromosomes, precedes the reductional division (post-reductional meiosis). Thirdly, the variation of chromosome number would result from different processes, that is to say, polyploidy, agmatoploidy (increase in chromosome number by fragmentation or fission of chromosomes, without change in the amount of DNA) and symploidy (decrease in the chromosome number via chromosomal fusion). These characteristics have still not been demonstrated in all the other genera of Juncaceae. According to Bailey (in Stace [94]), the genus *Juncus* L. does not even possess holocentric chromosomes. In fact, these characteristics are rarely widespread in the natural world. Certain works include a list of organisms, which potentially may have holocentric chromosomes [102] without the other two cytological characteristics being present. However, the presence of these chromosomes is reported in very few organisms. In the vegetable kingdom and among unicellular organisms, Geiter [35] was also able to observe reverse meiosis in *Spirogyra* Link (Chlorophyta). Among monocotyledons are the genera *Carex* L. (Cyperaceae) [22, 43, 58] and *Chinographis* (Melanthiaceae) [99], and in dicotyledons *Myristica fragrans* Houtt. (Myristicaceae) [28], the sub-genus *Cuscuta* L. with *C. babylonica* Auche and Choisy (Convolvulaceae) [79]. Only in the genus *Carex* and in the species *Cuscuta* has it been possible to observe reverse meiosis and the phenomena of chromosomal fusion and fission. In the animal kingdom the presence of holocentric chromosomes has been noted in the following:- (1) phylum Nematoda (*Ascaris* subspecie is and *Caenorhabditis elegans*, [1]). (2) In some arthropods: as in Arachnida of the order Scorpions [89]. (3) In insects of the order Lepidoptera [97] and Hemiptera (Heteroptera, [86]). In addition, holocentric chromosomes are also accompanied by a reverse meiosis in the order Odonata [78] and in the Coccidae [44] and Aphidae families (Homoptera, Hemiptera [85]).

Castro [11] estimates that chromosomes with non-localized centromere (chromosomes with diffuse centromere or holocentric chromosomes) are more primitive than those with localized centromere, a hypothesis taken up by Vaarama, Halkka, Hughes-Schrader and Schrader, Sybenga (in Wrensch *et al.* [102]), Battaglia [4] and Hakanson [38]. During the course of the evolution, the first may have experienced mutations leading to the localization of the centromere. On the other hand, in the taxa, which have conserved their primitive character, the chromosomes have experienced the phenomenon of fragmentation or fusion engendering new chromosomes without loss of chromosomal material and accompanied by reversed meiotic divisions. From their study of the sub-genus *Cuscuta*, Pazy and Plitmann [79] deduced that because of the rarity of the chromosomes with diffuse centromere in both vegetable and animal kingdoms, this type of chromosomes must be derived and not primitive. This opinion is shared by Greilhuber [37], who states "the diffuse centromere, combined with inverted meiosis, is the only reliable chromosomal higher-level synapomorphy in monocotyledons".

The larger part of the recent work on the genus *Luzula* does not analyze its karyology closely. Those studies focused on the phylogeny of the genus using a molecular approach relying on sequence analysis of the "internal transcribed spacers" (ITS) of the nuclear ribosomal DNA (nuclear rDNA) (García-Herran, unpublished), or otherwise by analyzing different sequences of chloroplast DNA (cpDNA) [17, 18].

KARYOLOGY OF THE GENUS *LUZULA*

All available karyological data concerning the genus *Luzula* is shown in Table 1, which summarizes the complete (or entire) existing chromosomal counts in the bibliography relating to each and every taxa belonging to the genus (Species Plantarum [93], site http://www.unine.ch/caryo/luzula).

It seems that the chromosome variability in the Luzulas is not very wide because the larger part of the species present the diploid valence $2n=12$ (12AL) [32]. This valence is almost omnipresent in the *Anthelaea*

Table 1. Variability of chromosome numbers in the genus *Luzula* DC[×]

Taxon	*Diploid chromosome number*	*Origin of chromosome number*
Subg. *Marlenia*		
L. elegans	$2n=6$ ($6A_0L$)	Symploidy
Subg. *Luzula*		
Sect. *Anthelaea*		
L. seubertii	$2n=12$	
L. lactea var. *lactea*	$2n=12$ (12AL)	
L. canariensis	$2n=12$	
L. nivea	$2n=12$ (12AL)	
L. lutea	$2n=12$ (12AL)	
L. luzuloides s.l.	$2n=12$ (12AL)	
L. sylvatica s.l.	$2n=12$ (12AL)	
Sect. *Atlanticae*		
L. atlantica	$2n=12$ (12AL)	
Sect. *Nodulosae*		
L. nodulosa	$2n=$?	
Sect. *Diprophyllatae*	$2n=12$ (12AL)	
L. wahlenbergii	$2n=24$ (24BL)	Agmatoploidy
L. piperi	$2n=24$	
L. hitchcockii	$2n=24$	
L. glabrata	$2n=12$	
L. desvauxii	$2n=12$ (12AL)	
L. parviflora subsp. *parviflora*	$2n=22$(2AL+20BL)	Agmatoploidy+Symploidy
L. parviflora subsp. *fastigiata*	$2n=24$ (24BL)	Agmatoploidy
L. parviflora subsp. *melanocarpa*	$2n=24$ (24BL)	Agmatoploidy
L. alpinopilosa subsp. *alpinopilosa*	$2n=12$ (12AL)	
L. alpinopilosa subsp. *obscura*	$2n=12$ (12AL)	
Sect. *Alpinae*		
L. ulophylla	$2n=48$ (48CL)	Agmatoploidy
L. traversii var. traversii	$2n=46$ (46CL)	Agmatoploidy
L. traversii var. tenuis	$2n=46$ (46CL)	Agmatoploidy
L. celata	$2n=12$ (12AL)	
L. pindica	$2n=24$ (24BL)	Agmatoploidy
L. spicata (see the table 2)	$2n=12$, 14, 16, 18, 24	Agmatoploidy

Table 1 contd.

Table 1 contd.

L. pediformis	2n=12 (12AL)	
L. caespitose	2n=12 (12AL)	
L. alopecurus	2n=24	
L. chilensis	2n=24	
L. racemosa	2n=24	
Sect. *Thyrsanochlamydeae*		
L. subcongesta	2n=24 (24BL)	Agmatoploidy
L. kjellmaniana	2n=36	
L. arcuata subsp. *arcuata*	2n=36, 42, 48	
L. arcuata subsp. *unalaschkensis*	2n=36	
L. confusa	2n=24, 36, 42, 48, 44-48	
L. nivalis	2n=24 (24BL)	Agmatoploidy
Sect. *Luzula*		
L. pallescens	2n=12 (12AL)	
L. alpina	2n=36(12AL+24BL)	Agmatoploidy+ Polyploidy
L. calabra	2n=24 (24BL)	Agmatoploidy
L. fallax	2n=24 (24BL)	Agmatoploidy
L. taurica	2n=12 (12AL)	
L. divulgata	2n=24 (24BL), 13 (12AL+1B)	Agmatoploidy
L. campestres s.l.	2n=12 (12AL), 13 (12AL+1B)	
L. sudetica	2n=48 (48CL)	Agmatoploidy
L. congesta	2n=36 (36AL), 48 (48AL)	Polyploidy
L. multiflora	2n=24 (24AL), 36 (36AL)	Polyploidy
L. multiflora subsp. *monticola*	2n=24 (24BL)	Agmatoploidy
L. mannii subsp. *mannii*	2n=42	
L. manníi subsp. *gracilis*	2n=24	
L. abyssinica	2n=24	
L. stenophylla	2n=24 (24BL)	Agmatoploidy
L. capitata	2n=12	
L. oligantha	2n=36	
L. lutescens	2n=12 (12AL)	
L. nipponica	2n=12 (12AL)	
L. leptophylla	2n=12	
L. crinita s.l.	2n=12 (12AL)	

Table 1 contd.

Table 1 contd.

L. crenulata	$2n=12$ (12AL)	
L. rufa s.l.	$2n=12$ (12AL)	
L. pumila	$2n=12$ (12AL)	
L. colensoi	$2n=12$ (12AL)	
L. decipiens	$2n=12$	
L. picta s.l.	$2n=12$ (12AL)	
L. banksiana s.l..	$2n=12$ (12AL)	
L. meridionalis	$2n=12$ (12AL)	
L. flaccida	$2n=12$ (12AL)	
L. densiflora	$2n=12$ (12AL)	
L. australasica s.l.	$2n=12$ (12AL)	
L. novae-cambriae	$2n=12$ (12AL)	
L. modesta	$2n=12$ (12AL)	
L. acutifolia	$2n=12$	
L. groenlandica	$2n=24$ (24AL)	Polyploidy
L. orestera	$2n=20$ (4AL+16BL), 22 (2AL+20BL)	Agmatoploidy+ Symploidy
L. subsessilis	$2n=12$	
L. comosa var. *comosa*	$2n=24$	
L. comosa var. *laxa*	$2n=12$ (12AL)	
L. echinata	$2n=12$ (12AL)	
L. bulbosa	$2n=12$ (12AL)	
L. hawaiiensis var. *glabrata*	$2n=28$	
Subg. *Pterodes*		
L. johnstonii	$2n=42$ (6BL+36CL)	Agmato-dysploidy
L. forsteri subsp. *forsteri*	$2n=24$ (24BL)	Agmatoploidy
L. luzulina	$2n=24$ (24BL)	Agmatoploidy
L. pilosa	$2n=62$, 66 (66CL), *ca.*70, 72	Agmato-dysploidy
L. acuminata	$2n=48$ 48CL	Agmatoploidy
L. plumosa subsp. *plumosa*	$2n=46$	
L. jimboi	$2n=24$	
L. rufescens var. *rufescens*	$2n=24$, 26, 48, 52 (52CL)	Agmato-dysploidy

Data from Species Plantarum [93] and from the site http://www.unine.ch/caryo/luzula

[x]Taxa having known chromosome numbers are exclusively indicated.

and *Atlantica* sections. Taxa with a diploid valence can be observed in the other sections, as well as taxa with a variable chromosome number, resulting from the processes of polyploidy and agmatoploidy in the *Luzula* section, and agmatoploidy in the *Diprophyllatae, Alpinae* and *Thyrsanochlamydeae* sections as well as for the sub-genus *Pterodes*. On the other hand, the two processes of polyploidy and agmatoploidy are not exclusive to the same species, as in the case of *L. alpina* Hoppe with $2n=36$ (12AL + 24BL), in which the chromosome number results from the combination of the two processes. The monospecific sub-genus *Marlenia* is an exception: *L. elegans* $2n=6$ ($6A_0L$), which presents an original karyotype formed by symploidy (descending agmatoploidy [32]), as does the taxa *L. orestera* (sect. *Luzula*) or *L. parviflora* (Ehrh.) Desv. subspecie *parviflora* (sect. *Diprophyllatae*). With regard to the *Nodulosae* section, at the time of writing no chromosome number has been associated with its species *L. nodulosa* [Bory and Chaub.] E. Meyer.

The phenomenon of agmatoploidy is always associated with the genus *Luzula*, and yet in reality it is not so frequent in the genus, as the bibliographical data shows (Table 1). However, there is a collective species, *L. spicata*, in which agmatoploidy is very common. It presents a wide spectrum of numbers, all originating from the diploid valence ($2n=12$, 12AL) by ascending agmatoploidy which has led to improved knowledge of the phenomenon of agmatoploidy.

AGMATOPLOIDY

The work done on duplicated *L. spicata* (García-Herran, in progress) reveals that different chromosome numbers are associated with this complex species, with each demonstrating a very specific distribution (Table 2). The diploid valence $2n=12$ (12AL) occupies a very disjointed area in Eurasia: southern Jura, several regions of the Alps, Italian Peninsula, Tatry in Slovakia, Chinese Daban Shan, Altai and Tien Shan Mountains. On the other hand, the number $2n=12$ found in the Upper Atlas Mountains published by Quézel [82], is a particular case since it was not confirmed by Favarger, Galland and Küpfer [25]. Populations offering $2n=24$ (24BL) are the most widespread, occupying vast regions. They thus extend beyond the Eurasian domain, reaching Greenland and North America. In Eurasia, they occupy the northern territories, Iceland and Scandinavia, all the domain in central and southwestern Europe, Greece, and also North Africa, passing through Corsica. On the other

Table 2. Chromosome numbers of *L. spicata* s.l.

2n	*Karyotype*	*Origin*	*Author*
2n=12	**12AL**	Jura Mountains	García-Herran [30]
		Upper Alps	García-Herran [31]
		Central Alps	Favarger [24]
			García-Herran [31]
		Eastern Alps (Carinzia)	Nordenskiöld [74]
		Eastern Alps (Slovenia)	Druskovic [19]
		Abruzzi	Favarger (unpublished)
			García-Herran (unpublished)
		South Italy	García-Herran (unpublished)
		Tatry	Michalska [71]
			Murín & Paclova [72]
		Daban Shan (China)	García-Herran [33]
		Altai (China)	García-Herran [33]
		Tien Shan (China)	García-Herran [33]
	Agmatoploidy		
2n=24	**24BL**	Greenland	Böcher [8]
		North America	Nordenskiöld [74]
		Iceland	Löve & Löve [55, 56]
		Sweden	Nordenskiöld [74]
		Central Europe	Nordenskiöld [74]
			Favarger [24]
			García-Herran [30]
		Northern Apennines	García-Herran (unpublished)
		Massif Central	Chassagne [14]
			García-Herran [31]
		Iberian Peninsula	Küpfer [52]
			García-Herran [33]
		Greece	Strid & Anderson [96]
		North Africa	Favarger Galland & Küpfer [25]
		Corsica	García-Herran [30]

Table 2 contd.

Table 2 contd.

Agmato-dysploidy			
$2n=14$	**10AL+4BL**	Eastern Alps (Styria)	Nordenskiöld [74]
		Balkans	García-Herran [30]
$2n=16$	**8AL+8BL**	Minor Caucasus	García-Herran [33]
$2n=18$	**6AL+12BL**	Great Caucasus	Sokolovskaja & Strelkova [91]
		Upper Alps	García-Herran [33]

hand, intermediate numbers are more isolated and less numerous: $2n=14$ (10AL+4BL) in the Styrian Alps and in the Balkans, $2n=16$ (8AL+8BL) in the Minor Caucasus, and $2n=18$ (6AL+12BL) in the Upper Alps and in the central Great Caucasus.

Until proven otherwise, it can be affirmed that the increase in the chromosome number in *L. spicata* is always attributable to a mechanism of chromosome fragmentation (agmatoploidy). While in certain works the phenomenon is not mentioned, each time authors look into the phenomenon, or when they provide precise comparative images of metaphases, the variation in the chromosome number is confirmed as having its origin in agmatoploidy.

The agmatoploidy is most frequently total (total agmatoploidy [57]) and concerns the two genomes. Consequently, all the chromosomes of a zygotic complement belong to the same size class: in *L. spicata* $2n=24$BL derives from $2n=12$AL by ascending agmatoploidy and is defined as *agmato-tetraploid*. However, karyotypes with chromosomes in intermediate number have also been observed, between 12 and 24, and including two different sizes. Firstly, the intermediate numbers could be explained by the fragmentation, which affects only one part of the chromosomes (*partial agmatoploidy* or *agmato-dysploidy*). As a result, it is better to distinguish the phenomenon of agmato-dysploidy, in which there are structural arrangements, from the phenomenon of *agmato-aneuploidy* with accidental numbers due to the heterozygotic fragmentations, and even from some relevant cases of polysomy (aneuploidy sensu stricto or quantitative aneuploidy). In agmato-dysploidy, the chromosomes, in pair number, all possess a homolog of the same length (Fig. 1). The chromosome number in the population will be fixed and stable, which leads to a variation of the base chromosome number.

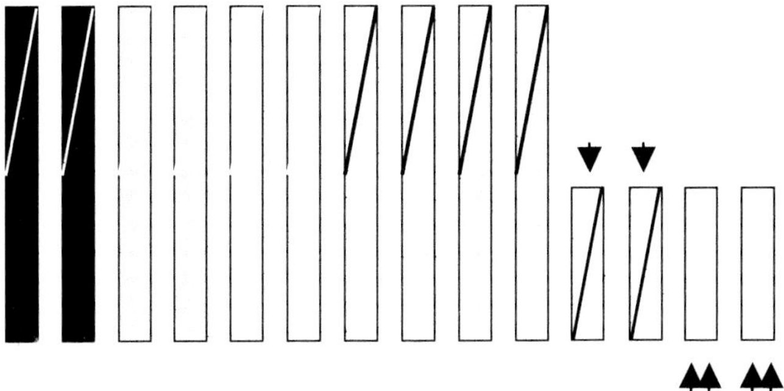

Fig. 1. *Partial agmatoploidy or agmato-dysploidy, e.g. at $2n=14$. The homologies concern chromosomes of the same size. The meiosis will be regular and the cytotype will be the same for all individuals in the population. The genetic equilibrium is respected.*

In agmato-aneuploidy, all the chromosomes do not possess a homolog of the same length because the fragmentation affects one of the two chromosomes in a pair of homologues (Fig. 2). Also, the genetic equilibrium in an agmato-aneuploidy is not strongly maintained and the chromosome number will be unstable within a population.

Secondly by the intermediate numbers may well be of hybrid origin. This may be the case with *L. spicata* at several stations analyzed in the Upper Alps, where the individuals at $2n=18$ (6AL +12BL) were found in some diploid populations ($2n=12$, 12AL). The fact that individuals at $2n=24$ have not been found is evidence in favour of a new fragmentation. However, it must be added that the populations at $2n=24$ (24BL) are not very elongated (Alps in Upper Province), making the hypothesis of hybridization feasible. Analyzing the pairings to study the meiosis could in this case concretely define the nature of the intermediate numbers.

The analysis of *L. spicata* provides the foundation for the formulation of a hypothesis regarding the age of the process of agmatoploidy. *L. spicata* is an Alpine species, which is found only locally in the sub-Alpine stage, and rarely in the upper mountain stage. In Mediterranean mountains, it is integrated in formations whose floral composition includes a high proportion of species of central European origin. The establishment of this contingent has been subject to differing evaluations. Quézel [81] states "les grandes glaciations ont apporté à Sierra Nevada (...) un contingent assez important de types orophiles eurasiatiques". The

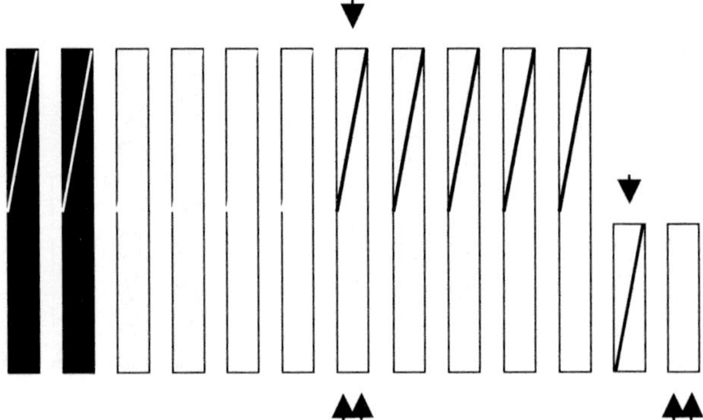

Fig. 2. *Agmato-aneuploidy, e.g. at 2n=13. The homologies concern chromosomes of different sizes. The cytotype will be isolated or more or less in a minority in the population.*

migratory movement could even have been prolonged as far as in the Atlas Mountains [83]. In view of the criteria adopted, there is no doubt that Quézel includes *L. spicata* among the glacial immigrants. On the other hand, for the Atlas domain, Galland [29] takes up the conclusions of Contandriopoulos and Gamisans [15], according to whom "les espèces arctico-alpines de Corse sont arrivées dans l'île dès le Tertiaire". Only the latter point of view can be supported not only because of the paleogeographic but also the biogeographical and karyological facts. Thus, the *L. spicata* complex is represented by the agmato-tetraploidy valence in Corsica, in the Sierra Nevada, and in the Atlas Mountains. Furthermore, in boreal-arctic areas of both the Eurasiatic continent and in North America, *L. spicata* was everywhere subject to karyological control at $2n=24$ (24BL); and yet, Kulczynski [51] places *L. spicata* precisely in the contingent of the species occupying a large area of America, of southern Europe (Corsica, Sierra Nevada) as well as North Africa since the Tertiary. Evidently, a polytypic origin of the agmatoploidy can be postulated. However, it should be pointed out that in all the aforementioned domains the original populations at $2n=12$ may have disappeared since it was deduced that the agmato-tetraploid numbers $2n=24$ (24BL) derive from fragmentation of $2n=12$ (12AL). $2n=24$ (24BL) is equally found in *L. abyssinica* Parl. of the central-east African mountains. Even though the affinities between *L. abyssinica* and *L. spicata s.l.* are not discussed by Hedberg [42], they seem to be close. A common origin for the agmato-tetraploids of *L. spicata* and *L. abyssinica* from the same stock at $2n=24$ cannot be excluded. This hypothesis would still support the idea of an ancient initial phase of agmatoploidy prior to the extension of the complex *L. spicata* to North America and Africa. The source of agmatoploidy can hardly be determined as there are many uncertainties.

Agmatoploidy and Polyploidy

In spite of the fundamentally different character of agmatoploidy and polyploidy, on a more or less large scale their biogeographical patterns of distribution do not seem to be very different. In *L. spicata*, all the northern populations linked with the much glaciated regions in both America and Eurasia during the later glaciations possess $2n=24$ (24BL). It is, therefore, probable that *L. spicata* spread into its northern area from southern regions where diploid populations at $2n=12$ (12AL) and agmatoploid populations at $2n=24$ (24BL) coexisted. *L. spicata* represents a complex orophilous Arctic-Alpine agmatoploid from the northern hemisphere [51]. In the same biogeographical domain, the numerous authors inspired by the classic research of Hagerup and Tischler [23, 26, 27, 52, 54, 68, 92,] have presented evidence of different polyploid complexes, being particularly well studied in the Alps and Pyrenees. The principal steps in the history of the orophilous flora are outlined, considering essentially the Quaternary glaciations. In fact, in several complexes of Alpine flora, the diploid race or races occupy the "refuge" territories, only partially affected or totally unaffected by the ice, while the polyploid race or races were able to easily colonize the territories vacated by the glaciers. With regard to polyploid taxon, there is complete lack of knowledge about factors favouring the postglacial extension of *L. spicata* populations at $2n=24$ (24BL) compared to those at $2n=12$ (12AL). However, the populations at $2n=12$ (12AL) in the Jura, Alps, Tatry, and Altai Mountains in any case testify to the existence of potential sources of northward immigration from the lowest valence in *L. spicata*.

In contrast to the polyploidy, it cannot be excluded that agmatoploidy does not confer a selective advantage. The inverse situation is also just as plausible. The fragmentation of the chromosomes does not affect the number of genes (or it is only partially affected). The expression of characters depending on genes with cumulative effect without dominance is, therefore, not modified. If it is difficult to find a selective advantage for agmatoploidy, it is also difficult to find a negative value. The phenomenon could be totally neutral, in which case, from the orographic and ecological points of view, the colonization of regions left free of ice during the glacial retreat would have initially been achieved by the nearest or the best placed peripheral populations, whatever their chromosome number might be. Perhaps the complexity of the distribution of *L. spicata* populations at $2n=12$ (12AL) and $2n=24$ (24BL) in central Europe can be attributed to such a process.

SYMPLOIDY

All attempts to demonstrate that polyploidy was reversible and that polyhaploidization played an important role in the evolutionary processes have been very frequently denied. The major reason is the fact that during the course of their evolution, the polyploids suffered mechanisms of genome reorganization resulting in the initial homologies progressively disappearing. In contrast, in the genus *Luzula* symploidy or chromosome fusion (descending agmatoploidy) seems to intervene as an evolutionary mechanism, because it seems that, due to the presence of chromosomes with diffuse centromere, there is no genetic material loss.

L. elegans is the only species in the genus which presents the type of annual biology and the lowest chromosome number $2n=6$. Its karyological analysis has shown that its chromosomes are of a great size, being more than three times larger than that of diploid taxa. All this leads to the belief that the *L. elegans* chromosomes originate from chromosome fusion. The karyotype $2n=6$ ($6A_0L$) also derives from fusion of all the chromosomes of an endowment at $2n=12$ (12AL) (total symploidy). Other Luzulas exist in which the chromosomal fusion would have been able to intervene: *L. orestera*. Two chromosome numbers are associated with this species of western North America: $2n=20$ (16BL+4AL) and $2n=22$ (20BL+2AL) [74]. The presence of type AL chromosomes could also be explained just as easily by chromosomal fusion as by chromosomal fission. According to Luceño and Guerra [61], the most plausible hypothesis is the one in which these numbers derive from a karyotype $2n=24$ (24BL) by one or two fusions (partial symploidy): 2BL → 1AL. In the same way it could explain the karyotype of *L. parviflora* var. *parviflora* $2n=22$ (2AL+20BL) [40]. In order to clarify the origin of this type of karyotype, it would be helpful to understand the mechanisms which "control" the phenomena of chromosomal fusion and fission.

Several authors agree to the great importance of chromosomal fusion in the genus *Carex*. In the opinion of Hartvig [41] *Carex divulsa* Stokes subspecie *leersii* (Kmeuker) W. Koch $2n=57$ derives from $2n=58$ by fusion of two chromosomes. According to Luceño [59] the same applies to *C. caryophyllea* Latourr. $2n=67$ as it derives from $2n=68$. Fusion and fission can be present simultaneously as in *C. laevigata* [58], showing that in addition, both can be considered to be evolutionary phenomena. According to Sheikh *et al.* [90] on *D. dichrosepala* Turz, it seems that this could be the case in *Drosera* L. (Droseraceae). However, in nature the phenomenon of chromosomal fusion is not very widespread. Up to the time of writing, it has not been found in other vegetal organisms, and it remains very rare in the animal kingdom [100]. In the light of available data in the literature, chromosomal fusion has only been recognized in certain orders of arthropods: Scorpions [80] and Lepidoptera [98].

ORIGIN OF THE CHROMOSOME VARIABILITY IN THE GENUS *LUZULA*

It is principally due to the presence of holocentric chromosomes that chromosomal variability can be observed in the genus *Luzula*. This chromosomal variability results from very specific phenomenon such as agmatoploidy, symploidy or polyploidy. This variability implies chromosomal evolution within the genus which might be explained from the diploid valence and primitive valence $2n=12$ (12AL), possessing the base chromosome number $x=6$ (6AL). Furthermore, Fig. 3 attempts to show the polarity in the evolution of the chromosome number in the genus *Luzula* DC.

The existence of *L. elegans* at $2n=6$ ($6A_0L$) could support the belief that the chromosomal fragmentation is effected from $2n=6 \rightarrow 12 \rightarrow 24 \rightarrow$... and also considering $x=3$ as being the primitive base number. Some data would support this hypothesis. On the one hand, *L. elegans* belongs to a larger centre of biodiversity and to ancient species conservation centre, the Canary Islands and Madeira. Besides, the inflorescence structure in loose anthela with unifloral glomerules, represents a primitive type if one follows the evolutionary scheme proposed by Balslev [2]. On the other hand, *L. elegans* is an annual species introduced into Portugal where it behaves like a "rudérale" species. Nevertheless, these characteristics are generally found in the derived taxa and are rarer in the pale-endemics. In addition, the chromosome number $2n=6$ is only present in one species. Even though the karyological inventory of Luzulas

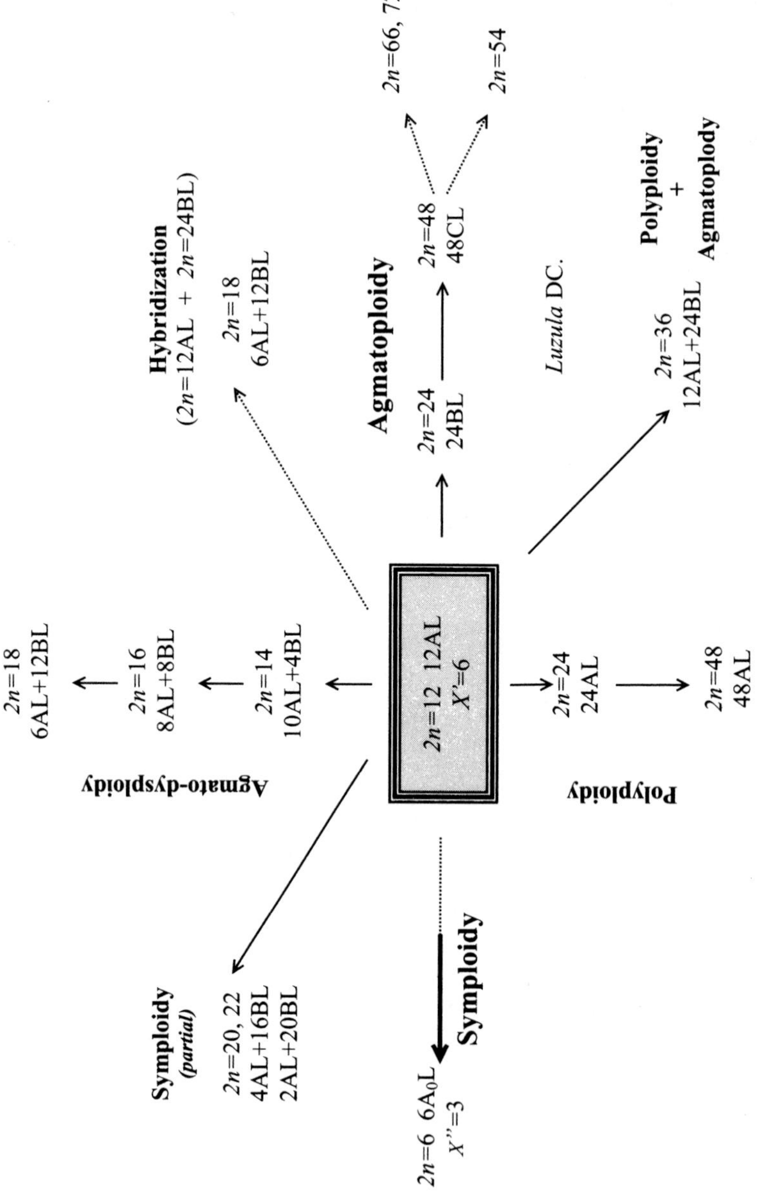

Fig. 3. *Origin of the chromosome variability in the genus efface: Luzula DC.*

has not yet been completed at the time of writing, the data is sufficient to make discovery of other species at $n=3$ unlikely. If need be, it is still necessary to confirm that the same base number $x=3$ in the two different species does have symplesiomorphy rather than homoplasy. In angiosperms, to have a base chromosome number lower than $x=5$ [95, 84] or higher than 11 or 13 [94] is not usual. The fact that chromosome numbers between $n=3$ and $n=6$ have not been observed supports the hypothesis of the authors according to which the somatic number 6 would be derived from 12, without doubt from chromosomal fusion $x=6$ (6AL) $\rightarrow$ $x=3$ ($3A_0L$) or from symploidy. Therefore, the base number $x'=3$ would have been derived.

Furthermore, it no longer seems particularly conjectural to state that $x=6$ represents the primitive base number in the genus *Luzula* in view of the widespread distribution, on all continents, of the numbers $2n=12$ (12AL) which are considered to be the diploid valence. The fact that no species at $2n=12$ (12AL) in the sub-genus *Pterodes* have been found would validate this hypothesis. All the species are agmatoploid and the number $2n=24$ (24BL) with x"=12 (12BL) (*L. forsteri* (Sm.) DC. or *L. luzulina* (Vill.) Racib.) appears to be the most ancient of the series. Some species such as *L. pilosa* (L.) Willd. $2n=62$, 66 (66CL), ca. 70, 72 have undergone several phases of fragmentation. In the opinion of the authors, these data are more a proof of the age of the process of fragmentation rather than the demonstration of the plesiomorphic character of $2n=24$. The sub-genus *Pterodes* would be some kind of a paleo-agmatoploid. As indicated above, other examples in the genus lead to the consideration that the agmatoploidy must have intervened very early: the agmato-tetraploid populations of *L. spicata* are disseminated in an almost continuous manner from Morocco to Scandinavia, and from California to the central Alps. Or yet again, the presence of *L. racemosa* Desv. $2n=24$ (24BL) in America and of *L. abyssinica* $2n=24$ (24BL) in the mountains of central-eastern Africa are agmato-tetraploid taxa both presenting incontestable affinities with *L. spicata*. The presence of agmato-tetraploids in the Atlas Mountains, in the Sierra Nevada, in Corsica and in North America favors a pre-glacial event inasmuch as the agmatoploidy in the complex *L. spicata* may be monophyletic. In the present state of knowledge, there are still no indications suggesting the contrary.

As a consequence of the diffuse character of the centromere, the rupture of a chromosome does not carry the loss of chromosomal fragments, or its corollary, the loss of information. Therefore, following what happens in the genus *Carex* [22], the variability of the chromosome number could have been particularly important in the genus *Luzula*. However, karyological analysis demonstrates that the genus is relatively stable, conserving the diploid valence $2n=12$ (12AL) in a good number of species. This leads to the belief that the Luzulas are particularly sensitive to the genetic equilibrium, either on the gametophytic level or on the sporophytic level. A strong selective constraint would favour the structural homozygotes and would eliminate a certain number of genotypes presenting fragmentations in their initial phase, heterozygote. The results presented in this paper tend to support such a hypothesis, the intermediate numbers being less frequent.

The polyploidy seems to be limited to the sub-genus *Luzula*, in which it can present parallel, or combined with the agmatoploidy, as is the case in the *Luzula* section: *L. sudetica* ($2n=48$ (48CL), [75]), *L. alpina* ($2n=36$ (12AL+24BL), [49]). The species belonging to this section are the only ones in which the taxonomic treatment is strongly modelled on the karyological data. *L. campestris* is always diploid, whereas polyploid taxa are found in the *L. multiflora* complex. There is an exception in *L. multiflora* subspecies *monticola* $2n=24$ (24BL), agmato-tetraploid, described by Kirschner [50], who considers them to be the only known European diploid belonging to the *L. multiflora* complex.

Otherwise, agmato-aneuploidy seems to be non-existent in the Luzulas, or at least very rare. The supernumerary chromosomes have always been interpreted as B chromosomes [30, 32] from the fact that they presented a different size from that of the AL chromosomes and that they were heterochromatic. The situation seems to be fundamentally different in the genus *Carex* in which, among others, Luceño and Castroviejo [58] and Luceño [60] highlighted aneuploid series (that the authors call agmato-aneuploids) in several species. In particular, the work of the two authors shows that variation was not anarchic: *C. laevigata* presents an agmato-aneuploid gradient with a progressive increase in the chromosome number in the Iberian Peninsula, from the Basque country to the south of Andalusia. Majority of the populations present several different chromosome numbers, the phenomena of fission, incidentally of fusion, still being in a phase, not fully stabilized. Although the genus *Carex* has many similarities with the genus *Luzula*, other differences must be mentioned. Whereas the total agmatoploidy in the genus *Luzula* is very frequent, in *Carex* the variation in the chromosome number can result from two processes: on the one hand, by addition or by subtraction of one or more chromosomes by chromosome fission or fusion; and on the other hand, by polyploidy [57]. Another distinct characteristic corresponds to the size of the chromosomes. The chromosomal fissions in the *Carex* are most often asymmetric, generating fragments of a generally unequal size. This can be deduced from the explicative scheme for the fragmentation of bivalents in the genus *Carex* [60]. On the other hand, the fragmentation in the Luzulas seems to be median. The karyotypes of the agmatoploid are, therefore, relatively symmetrical (García-Herran, in progress), only certain sizes of chromosomes can consequently be observed and associated with the nomenclature of Nordeskiöld [74]. Consequently, the chromosomal variability in the genus *Carex* due to chromosomal fission or fusion can be defined as being agmato-aneuploidy, so as to distinguish it from the agmatoploidy (total or partial) observed in *L. spicata*.

If the intermediate numbers in *L. spicata* are closely analyzed, the agmato-dysploidy could have played an important role in affecting the *Luzula* karyotype. In fact, the widespread distribution of number $2n=14$ (10AL+4BL) in the Balkans supports the belief that the phenomenon may have been monophyletic, old and stabilized, with a variation in the base number. In addition, this number forms part of a progressive gradient of agmatoploid cytotypes towards the east: $2n=12$ (12AL) (Hohe Tauern, Carinzian Alps), $2n=14$ (10AL+4BL) (Styrian Alps and Balkans), $2n=16$ (8AL+8BL) (Minor Caucasus), $2n=18$ (6AL+12BL) (central Great Caucasus). If this series of numbers is not a fortuitous occurrence, it could result from a partial agmatoploidy, progressive but stabilized. To this can be added the opinions of Malheiros-Gardé and Gardé [67] and Noronha-Wagner and Castro [77], who in their studies of the development of meiosis, explain the progressive character of the agmatoploidy by starting with the presence of intermediate numbers. On the other hand, it is also possible to state that these numbers result from a more recent agmatoploidy, stopping in a still inchoative phase, before perhaps leading to a new taxon at $2n=24$ (24BL)? Such an interpretation postulates particular affinities between populations of these geographical areas. However, numerical analysis of the *L. spicata* complex in its Eurasiatic area of distribution (García-Herran, in progress) shows no morphological similarity between the individuals of different regions from the eastern Alps to the Caucasus. It seems that the phenomenon of chromosome fragmentation in the Luzulas is significantly different from that presented by C. *laevigata*. Perhaps *Luzula* and *Carex* are found in another phase of their evolution, inchoative for *Carex* and completed in *Luzula*.

Concerning polarity, some facts, however, also speak in favour of the evolution in the genus *Luzula* by chromosomal fragmentation $2n=12AL \rightarrow 2n=24BL$ rather than by fusion $2n=24BL \rightarrow 2n=12AL$. Several species (*L. sylvatica*, *L. lutea*, etc.) present $2n=12$ (12AL) as a unique chromosome number. Moreover, most of the agmato-tetraploid taxa belong to complexes in which the zygotic number $2n=12$ still exists. An exception is the sub-genus *Pterodes* and the *L. multiflora* complex. In spite of this, it seems the evolution follows the direction of an increase in the chromosome number by fragmentation from species at $2n=12$ (12AL) to species at $2n=24$ (24BL) (ascending agmatoploidy). The inverse polarity, due to symploidy (descending agmatoploidy), is much less frequent. Until now this has been associated with *L. elegans* ($2n=6 \leftarrow 2n=12$) and perhaps with *L. orestera* and *L. parviflora*. If the cases above are abstracted, the base chromosome number recognized by Nordenskiöld [73] can be agreed with, corresponding to the number of link groups, tending to increase in the Luzulas, both by agmato-dysploidy ($x=6 \rightarrow 7 \rightarrow 8 \rightarrow ?\ 9$) and by agmatoploidy ($x=6AL \rightarrow x=12BL$). It is worth noting that in a different evolutionary stage, the greater polarity (increasing), of the gametic number gradient is essentially the same as that observed in the genus *Carex*. Is the phenomenon fortuitous? Would such pressure of selection interact on the genotype to favour the zygotic numbers above $2n=12$? Questions which for the moment remain open.

Acknowledgement

This work was supported by Basque Government Grant No. BFI89.056-AK.

[1] Albertson DG, Thomson JN. Segregation of holocentric chromosomes at meiosis in the nematode *Caenorhabditis elegans*. Chromosoma, 1993; 1: 15-26.

[2] Balslev H. Juncaceae. Fl Neotrop Monogr 1996; 68: 1-168.

[3] Barlow PW, Nevin D. Quantitative karyology of some species of *Luzula*. Pl Syst Evol 1976; 125(2): 77-86.

[4] Battaglia E. Assenza di centromere localizzato in *Heliocharis uniglumis* (Link) Schult. Caryologia, 1954; 6: 319-332.

[5] Battaglia E, Boyes JW. Post-reductional meiosis: its mechanism and causes. Caryologia 1955; 8: 87-134.

[6] Battaglia E. The concept of pseudopolyploidy. Caryologia 1956; 8: 214-220.

[7] Bauer H. Die Chromosomen in Soma der Metazoen. Zool Anz (suppl Bd) 1952; 17: 252.

[8] Böcher TW. Zur Zytologie einiger arktischen und borealen Blütenpflanzen. Sven Bot Tidskr, 1938; 32: 346-361.

[9] Braselton JP. The ultrastructure of the non-localised kinetochores of *Luzula* and *Cyperus*. Chromosoma 1971; 36: 89-99.

[10] Castro D, Cámara A, Malheiros N. X-Rays in the centromere problem of *Luzula purpurea* Link (1). Genét Ibér, 1949; 1(1): 49-54.

[11] Castro D. Notes on two cytological problems of the genus *Luzula* DC. Genét Ibér, 1950; II(2-3): 201-209.

[12] Castro D, Mello-Sampayo T. Poliploidia induzida pontes e fragmentaçao em *L. purpurea*. Genét Ibér 1953; V(1-2): 1-16.

[13] Castro D, Noronha-Wagner M, Camara A. Two X-ray induced translocations in *Luzula purpurea*. Genét Ibér 1954; VI (1-2): 3-9.

[14] Chassagne M. Inventaire analytique de la flore d'Auvergne I. 1956: 166-171.

[15] Contandriopoulos J, Gamisans J. A propos de l'élément arctico-alpin de la Corse. Bull Soc Bot Fr 1974; 121: 174-204.

[16] De Wet JMJ. Polyploidy and evolution in plants. Taxon 1971; 20: 29-35.

[17] Drábkobá L, et al. Phylogeny of the Juncaceae based on *rbc*L sequences with special emphasis on *Luzula* DC. and *Juncus* L. Pl Syst Evol 2003; 240: 133-147.

[18] Drábkobá L, et al. *Trn*L-*trn*F intergeneric spacer and *trn*L intron define major clades within *Luzula* and *Juncus* (Juncaceae): importance of structural mutations. J Mol Evol 2004; 59: 1-10.

[19] Druskovic B. IOPB chromosome, Data 9. Taxon 1995; 24: 11-14.

[20] Dubas E, et al. Karyology and amounts of nuclear DNA in *Luzula* L- A genus with holokinetic chromosome. Cell Mol Biol Lett 2002; 7: 44.

[21] Ebinger JE. Taxonomy of the subgenus *Pterodes* genus *Luzula*. Mem NY Bot Gard 1964; 10(5): 279-304.

[22] Faulkner JS. Chromosome studies on *Carex* section *Acutae* in north-west Europe. Bot J Linn Soc 1972; 65: 271-301.

[23] Favarger C. Contribution de la biosystématique à l'étude des flores alpine et jurassienne. Rev Cytol Biol Veg 1962; 25: 398-410.

[24] Favarger C. Notes de caryologie alpine IV. Bull Soc Neuchâtel, Sci Nat 1965; 88: 5-60.

[25] Favarger C, Galland N, Küpfer P. Recherches cytotaxonomiques sur la flore orophile du Maroc. Nat Monspeliensia Ser Bot 1979; 29: 1-64.

[26] Favarger C. Liens génétiques entre la flore orophile des Tatras et celle des Alpes à la lumière de quelques complexes polyploïdes. Polish Bot Stud 1991; 2: 23-38.

[27] Favarger C. Flore et végétation des Alpes. Delachaux et Niestlé, éd. 3 1-2, 1995.

[28] Flach M. Diffuse centromere in a dicotyledonous plant. Nature 1966; 209: 1369-1370.

[29] Galland N. Recherche sur l'origine de la flore orophile du Maroc. Travaux de l'Institut Scientifique, Ser botanique Rabat 1988; 35(22): 106-121.

[30] García-Herran MC. In: Kamari G, et al. Mediterranean chromosome number, reports-4. Flora Mediterranea 1994; 4: 280-284.

[31] García-Herran MC. Polymorphisme et affinités des espèces ibériques du genre *Luzula* DC. Thèse, Université de Neuchâtel, Switzerland, 1998: 179.

[32] García-Herran MC. Polymorphisme caryologique du genre *Luzula* DC (Juncaceae) dans le sud-ouest et le centre de l'Europe. Bull Soc Neuchâtel, Sci Nat 2001; 124: 59-72.

[33] García-Herran MC. In: Kirschner et al. Juncaceae 1: *Rostkovia* to *Luzula.* Species Plantarum: Flora of the World part 6, 2002.

[34] Gardé A. A importância da fragmentaçao cromosómica na evoluçao. Broteria, 1951; 20: 159-176.

[35] Geiter L. 1930. In: Bolkhovskiktt, Z et al. Chromosome numbers of flowering plants. Russie 1969.

[36] Godward MBE. The kinetochore. Int Rev Cytol 1985; 94: 77-105.

[37] Greilhuber J. Chromosome of the Monocotyledons (General aspects). In: Rudall PJ et al., *eds.* Monocotyledons: Systematics and Evolution. Royal Botanic Gardens, Kew 1995: 379-414.

[38] Hakansson A. Holocentric chromosome in *Eleocharis*. Hereditas 1958; 44: 531-540.

[39] Halkka O. A photometric study: the *Luzula* problem. Hereditas 1964; 52: 81-88.

[40] Hämet-Ahti L, Virrankoski V. Cytotaxonomic notes on some monocotyledons of Alaska and northern British Columbia. Ann Bot Fenn 1971; 8: 156-159.

[41] Hartvig P. Chromosome numbers in Nordic populations of the *Carex muricata* group (Cyperaceae). Acta Univ Ups Symb Bot Ups 1986; 27: 127-138.

[42] Hedberg O. Afroalpine vascular plants. A taxonomic revision. Uppsala 1957.

[43] Hoshino T. Karyomorphological and cytogenetical studies on aneuploidy in *Carex*. J Sci Hiroshima Univ, Ser Bot, Div 2 Bot 1981; 17: 155-238.

[44] Hughes-Schrader S. A study of the chromosome cycle and the meiotic division-figure in *Llaveia bouvari-* a primitive coccid-. Z Zellforsch Mikrosk Anat 1931; 13(4): 742-769.

[45] Hughes-Schrader S. The chromosomes of the giant scale *Aspidoprotus maximus* Louns (Coccoidea-Margarodidae) with special reference to asynapsis and sperm formation. Chromosoma 1955; 7: 420-438.

[46] Hughes-Schrader S, Schrader F. The kinetochore of the Hemiptera. Chromosoma 1961; 12: 327-350.

[47] Inomata N. Chromosome variation in the callus cells of *Luzula elegans* with nonlocalized kinetochores. Jap J Gen 1982; 57(1): 59-64.

[48] Kimber G, Riely R. Haploid angiosperms. Bot Rev 1963; 29: 480-531.

[49] Kirschner J, Engelskjon T, Knaben GS. *Luzula alpina* Hoppe a neglected Alpina species. Preslia 1988; 60: 97-108.

[50] Kirschner J. *Luzula* sect *Luzula* (Juncaceae) in Spain. Pl Syst Evol, 1996; 200: 1-11.

[51] Kulczynski S. Das boreale und arktisch-alpine Element in der mittel-europäischen Flora. Bull Int Acad Pol Sci, Lett Cl, Sci Math Nat Ser B 1923; 127-214.

[52] Küpfer P. Recherches sur les liens de parenté entre la flore orophile des Alpes et celle des Pyrénées. Boissiera 1974; 23: 1-322.

[53] Kuta E, et al. Chromosome and nuclear DNA study on *Luzula* – a genus with holokinetic chromosome. Genome 2004; 47 (2): 246-256.

[54] Landolt E. Die Artengruppe des *Ranunculus montanus* Willd. in den Alpen und im Jura. Ber Schweiz Bot Ges 1954; 64: 9-83.

[55] Löve A, Löve D. Chromosome numbers of Northern plant species. Reykjavík 1948.

[56] Löve A, Löve D. Cytotaxonomical conspectus of the Icelandic flora. Acta Horti Gotob 1956; 20(4): 65-291.

[57] Löve A, Löve D, Raymond M. Cytotaxonomy of *Carex* section *Capillares*. Can J Bot 1957; 35: 715-761.

[58] Luceño M, Castroviejo S. Agmatoploidy in *Carex laevigata* (Cyperaceae). Fusion and fission of chromosomes as the mechanism of cytogenetic evolution in Iberian populations. Pl Syst Evol 1991; 177: 149-159.

[59] Luceño M. Chromosome studies on *Carex* L. section *Mitratae* Kükenth (Cyperaceae) in the Iberian Peninsula. Cytologia 1993; 58: 321-330.

[60] Luceño M. Cytotaxonomic studies in Iberian Balearic North African and Macaronesian sepcies of *Carex* (Cyperaceae) II. Can J Bot 1994; 72: 587-596.

[61] Luceño M, Guerra M. Numerical variations in species exhibiting holocentric chromosomes: a nomenclatural proposal. Caryologia 1996; 49(3-4): 301-309.

[62] Malheiros N, Castro D. Chromosome numbers and behaviour in *Luzula purpurea* Link. Nature 1947; 160: 156.

[63] Malheiros N, Castro D, Cámara A. Cromosomas sem centrómero localizado. O caso de *Luzula purpurea* Link. Agron Lusit, 1947; 9(1): 51-71.

[64] Malheiros N, Gardé A. Contribuiçoes para o estudo citologico do genero *Luzula* Link. Agron Lusit 1947; 9(1): 75-79.

[65] Malheiros-Gardé N. Algunos efectos de la morfina en la mitosis de *Luzula purpurea* Link. Genét Ibér 1950; II(1): 29-38.

[66] Malheiros-Gardé N, Gardé A. Fragmentation as a possible evolutionary process in the genus *Luzula* DC. Genét Ibér 1950; II(4): 257-262.

[67] Malheiros-Gardé N, Gardé A. Agmatoploidia no genero *Luzula* DC. Genét Ibér 1951; III: 1-22.

[68] Manton I. The problem of *Biscutella laevigata* L. Ann Bot N S, 1937; 1: 439-462.

[69] Mello-Sampayo T, Castro D, Malheiros-Gardé N. Observaçoes sobre a autotetraploidia induzida pela colchicina em *Luzula purpurea* Link. Agron Lusit 1951; 13(1): 1-10.

[70] Mello-Sampayo T. Differential polyteny and karyotype evolution in *Luzula*: A critical interpretation of morphological and cytophotometric data. Genét Ibér, 1961; XIII (1-2): 1-22.

[71] Michalska A. Badania cytologiczne nad rodzajem *Luzula*. Acta Soc Bot Poloniae 1953; 22(1): 169-186.

[72] Murín A, Paclová L. IOPB chromosome number reports. Taxon 1979: 28.

[73] Nordenskiöl H. The somatic chromosomes of some *Luzula* species. Bot Not 1949; 1: 81-92.

[74] Nordenskiöl H. Cyto-taxonomical studies in the genus *Luzula* I. Somatic chromosome and chromosome numbers. Hereditas 1951; 37: 325-355.

[75] Nordenskiöl H. Studies on meiosis in *Luzula purpurea*. Hereditas 1962; 48: 503-519.

[76] Noronha-Wagner M. Subsidio para o estudo citologico do genero *Luzula* DC (1). Genét Ibér 1949; 1(1): 59-67.

[77] Noronha-Wagner M, Castro D. Interpretaçao dum comportamento meiótico observado em *Luzula*. Sci Genet 1952; 4(3): 154-161.

[78] Oksala T. Zytologische Studien an Odonaten I. Ann Acad Sci Fenn A 1943; 4: 1-65.

[79] Pazy B, Plitmann U. Holocentric chromosome behavoir in *Cuscuta*. Pl Syst Evol 1994; 191: 105-109.

[80] Piza S de T. Observações cromossómicas em escorpiões brasileiros. Cienc Cult 1950; 2: 202-206.

[81] Quezel P. Contribution à l'étude phytosociologique et géobotanique de la Sierra Nevada. Memórias da Sociedade Broteriana 1953: 9.

[82] Quezel P. Peuplement végétal des hautes montagnes de l'Afrique du Nord. Encyclopédie biogéographique et écologique 1957: 390-426.

[83] Quezel P. Les hautes montagnes du Maghreb et du Proche-Orient: Essai de mise en parallèle des caractères phytogéographiques. Actas III, Congr Optima, Anales Jar Bot Madrid 1981; 37: 353-372.

[84] Raven PI. The bases of angiosperm phylogeny: cytology. Ann Missouri Bot Gard 1975; 62: 724-764.

[85] Ris H. A cytological and experimental analysis of the meiotic behaviour of the univalent X-chromosome in the bearberry aphid Tamalia (Phyllaphis) coweni (Chll). J Exp Zool 1942; 90: 267-330.

[86] Schrader F. The chromosome cycle of *Protortonia primitiva* (Coccidae) and a consideration of the meiotic division apparatus in the male. Ztschr Wissensch Zool 1931; 138 (3): 385-408.

[87] Schulz-Schaeffer J. Cytogenetics. Plants, animals, humans. New York: Springer-Verlag, 1980: 466.

[88] Sen J, Mukherjee S, Sharma AK. Study of chromosomes, DNA amount and in vitro growth in different species of *Luzula*. Genome 1990; 33: 143-147.

[89] Shanahan C. Cytogenetics of Australian scorpions I. Interchange polymorphism in the family Buthidae. Genome 1989; 32: 882-889.

[90] Sheikh SA, Kondo K, Hoshi Y. Study on diffused centromeric nature of *Drosera* chromosomes. Tokyo, Cytologia 1995; 60: 43-47.

[91] Sokolovskaja AP, Strelkova OS. Geograficeskoje raspredelenije poliploidov. III. Issledovanije flory alpijskoj oblasti Centralnovo Kavkazskovo chrebta. Ucenye Zapiski Pedagogicesky Inst Imeni Gercena [Sec Federov AA Ic], 1948; 66: 195-216.

[92] Söllner R. Recherches cytotaxonomiques sur le genre *Cerastium*. Ber Schweiz Bot Ges 1954; 64: 221-354.

[93] Species Plantarum. Kirschner, et al. Juncaceae 1: Rostkovia to *Luzula*. Species Plantarum: Flora of the World part 6, 2002.

[94] Stace CA. Cytology and cytogenetics as a fundamental taxonomic resource for the 20th and 21st centuries. Taxon 2000; 49: 451-477.

[95] Stebbins GL. Chromosomal evolution in higher plants 216 pp. Edward Arnold Ltd., London, 1971.

[96] Strid A, Anderson IA. In Index to plant chromosome numbers 1984-1985. Missouri Bot Gard Bull 1985.

[97] Suomalainen E. The kinetochore and the bivalent structure in the Lepidoptera. Hereditas 1953; 39: 88-96.

[98] Suomalainen E. Chromosome evolution in the Lepidoptera. Chromosoma 1966; 2: 132-138.

[99] Tanaka N, Tanaka N. Chromosome studies in *Chionographis* (Liliaceae) III. The mode of meiosis. Cytologia 1980; 45: 809-817.

[100] Ueshima N. Hemiptera II: Heteroptera. In: John B, ed. Animal Cytogenetics. Vol 3: Insecta 6. Berlin, Stuttgart Gëbrüder Borntraeger 979: 117.

[101] White MJD. Animal cytology and evolution. Ed 3, London: Cambridge University Press, 1973.

[102] Wrensch DL, et al. Cytogenetics of holokinetic chromosomes and inverted meiosis: keys to the evolutionary success of Mites with generalizations on eukaryotes. 283-343. In: Houck MA, ed. Mites Ecological and evolutionary analyses of the life-history patterns. New York, London, 1994.

CHAPTER

9

Phylogenetic Relationships and Systematics in Genus *Bromus* (Poaceae)

TATJANA OJA
Institute of Botany and Ecology, University of Tartu, Estonia

ABSTRACT

Bromus is a taxonomically complex genus with about 130 species of annual and perennial, diploid and polyploid brome grasses of wide geographic distribution. The morphological diversity in the genus *Bromus* and great variation within its species are well-known and have caused substantial difficulties in taxonomic delimitations for various species. Recently many authors have studied genetic relationships and variation of many of the species in the genus *Bromus,* but despite this information, phylogenetic relationships among the subdivisions of *Bromus* are still far from clear. The results obtained from different molecular markers (allozymes, cp DNA, rDNA) highlight the problematic issues for future investigation.

Key Words: Bromus, taxonomy, phylogenetic relationships

Abbreviations: AFLP = Amplified Fragment Length Polymorphism, cpDNA = chloroplast DNA, ITS = Internal Transcribed Spacer, MLAL = Multilocus Allozyme Lineage, MLIL = Multilocus Isozyme Lineage, nrDNA = nuclear DNA, RAPD = Random Amplification of Polymorphic DNA, rDNA = ribosomal DNA, SSR = Simple Sequence Repeats or microsatellites, UPGMA = Unweighted Pair Group Method with Artithmetic Mean.

Address for correspondence: Institute of Botany and Ecology, University of Tartu, 40 Lai Str., 51005, Tartu, Estonia. E-mail: tatjana.oja@ut.ee.

INTRODUCTION

The grass genus *Bromus* L. belongs to the subfamily *Pooideae* A. Br. tribe *Bromeae* Dum. It contains herbaceous self- and cross-pollinating species of different ploidy levels with the basic chromosome number x = 7, and has long been known for interspecific hybridization, variability, and taxonomic complexity [100]. Polyploid species with chromosome numbers as multiples of 14 occur in all of sections of the genus [15]. The genus *Bromus* probably originated in western Eurasia [83, 100] and is now extremely widespread throughout Eurasia, extending to North and South America, Africa and Australia. *Bromus* is a taxonomically complex genus with about 130 species of annual and perennial, diploid and polyploid brome grasses of wide geographic distribution. Phylogenetically, it has been shown [29, 93] to be basally sister to the tribe *Triticeae* Dum., that contains economically important crops – barley, rye and wheat. Brome-grasses may give suitable genes for the improvement of different agronomic properties of the cereal crops, e.g. disease resistance and tolerance to various stressful conditions [64].

The genus has been divided in various ways either into seven sections [84], seven subgenera [100], or even seven different genera [102] as summarized in Table 1. The number and rank of the divisions depend on the characters used as a basis of their delimitation (serological, cytological and morphological, respectively), and phylogenetic relationships among them are still inconsistent. Sales [74] doubted the reality of the section *Genea* as an independent taxonomic unit. She noted that there is a continuous range of variation between sections *Bromus* and section *Genea via* the *B. pectinatus* complex of

Table 1. Genus *Bromus* taxonomic delimitation by different authors

Tzvelev (1976)	*Smith (1972, 1985)*	*Stebbins (1981)*
- 7 genera	- 7 sections	- 7 subgenera
Anisantha	Genea	Stenobromus
Bromus	Bromus	Bromus
Bromopsis	Pnigma	Festucaria
Ceratochloa	Ceratochloa	Ceratochloa
–	*Neobromus*	*Neobromus*
Boissiera	Boissiera	Boissiera
Nevskiella	Nevskiella	Nevskiella
Littledalea		

section *Bromus*. This complex was suggested to link the *Genea* species with the section *Bromus* through the diploid *B. japonicus* [74, 80, 84].

Some aspects of genetic relationships in genus *Bromus* have been investigated using molecular methods with cpDNA and rDNA markers [4, 65, 68]. Pillay and Hilu [68], on the basis of the cpDNA data, found that subgenera *Stenobromus* and *Bromus* are not distinct entities and probably originated from similar ancestors related to the subgenus *Festucaria*. They examined cpDNA restriction site variation in 32 species, representing five subgenera of *Bromus*. The analyses indicated two major clades within the genus. One clade includes subgenus *Neobromus* and subgenus *Ceratochloa*. The other is composed of subgenera *Festucaria*, *Stenobromus* and *Bromus*. Species of subgenus *Festucaria* appeared in three separate lineages with very little resolution of relationships between them. Pillay [65] also concluded that it was unlikely that the subgenera vs. sections of *Bromus* had independent origins. However, Ainouche and Bayer [4] showed sections *Pnigma*, *Genea* and *Bromus* in separate monophyletic clades on the rDNA ITS sequence cladogram. Thus, despite this molecular information, complete detailed clarification is still required regarding phylogenetic relationships between the subdivisions of *Bromus*.

In addition to inconsistent intrageneric classification, there are a lot of problems about the species delimitation within the sections, due to a great and frequently continuous morphological variability and less known genetic relationships between them. Among the diploids, some species are easily recognized by unique morphological characters, like *B. pumilio* and *B. danthoniae*, but other species are hard to identify because their diagnostic characters are often overlapping and difficult to define [84]. The morphological diversity in the genus *Bromus* and great variation within its species are well-known and have caused substantial difficulties in taxonomic delimitations for various species. The high degree of phenotypic plasticity in morphology depending on the environmental conditions, complicate the use of morphometric characters for species identification.

Section *Genea*

Section *Genea* Dum [86], synonyms: subgenera *Stenobromus*, genera *Anisantha*, of the genus *Bromus* L. comprises weedy, annual brome grasses widely distributed in Mediterranean countries, Southwest Asia, extending to northern Europe. The centres of diversity of section Genea are in the Mediterranean region.

Section *Genea* has been regarded as an evolutionarily advanced section in the genus *Bromus* that is still in the process of further specialization [76]. Species of the section often display remarkable inter- and intraspecific variation in morphology. The section includes the following species: diploids (2n=14) *B. sterilis* L., *B. fasciculatus* Presl and *B. tectorum* L., tetraploids (2n=28) *B. rubens* L. and *B. madritensis* L., hexa- and octoploid *Bromus diandrus* Roth (2n=42, 56) and *B. rigidus* Roth (2n=42, 56) with two intraspecific ploidy cytotypes [18, 19, 34, 36, 63, 86, 100, 102]. Naganowska [51] studied karyotypes of five *Bromus* species of *Genea* section. She revealed significant differences in the chromosome features between diploids *B. fasciculatus, B. sterilis* and *B. tectorum* and similarities between tetraploids *B. madritensis* and *B. rubens*. Genome affinity between diploid *B. tectorum* and tetraploids *B. madritensis* and *B. rubens* was also found.

It has been supposed [74, 80] that sections *Genea* and *Bromus* may be less distinct than recognized before. Analysis of flower microstructural variation within the genus *Bromus* showed [39] that sections *Bromus* and *Genea* are evolutionarily close and they distinctly overlap each other in the minimum spanning tree space. Cladistic analysis of isozyme variation data among the bromus species nested the two *Genea* diploids *B. tectorum* and *B. sterilis* among the diploids of section *Bromus* [60]. Sales [74] questioned the reality of the section *Genea* as an independent taxonomic unit. Following the previous Scholz's study [80], she also suggested that sections *Genea* and *Bromus* may be linked through *Bromus pectinatus* of the section *Bromus*. Sales [74] supposed that *B. pectinatus* of section *Bromus* is closely related to section *Genea*. The placement of *B. pectinatus* in section *Bromus* is also questionable because the characters of its spikelets (wedge-shaped, broader to the top) clearly match with those of the section *Genea*. Scholz [81] in his study about the *B. pectinatus* complex suggested the hybrid origin of this complex. The morphological characteristics of *B. pectinatus* are something between *B. japonicus* of section *Bromus* and *B. tectorum* of section *Genea*. This species resembles *B. tectorum* in general morphology and may have taken part in the species formation in *Genea*. The descriptions of *B. pectinatus* [69, 74, 81] often mentioned the robustness of this species, which may be due to its tetraploid nature [49, 84]. Kosina [38], on the basis of 14 parameters of the embryo morphology, also confirmed the intersectional status of *B.*

pectinatus. Isozyme data confirm the allotetraploid nature of *B. pectinatus* and its intermediate position between section *Genea* and section *Bromus*. Several heterozymes are consistent with the view that *B. pectinatus* may be the result of the hybridization between *B. japonicus* of section *Bromus* and *B. tectorum* of the section *Genea* (Oja, in preparation).

Bromus tectorum L. and *B. sterilis* L. are annual diploids that are predominantly cleistogamous, colonizing weeds. They often share the same habitats and grow intermixed and sympatrically, but are easily distinguishable based on morphology [74, 76, 84]. The two species are clearly differentiated karyotypically [40], indicating their reproductive isolation and specific distinctness. All three diploid species of section *Genea* have diverged from each other at a number of isozyme loci [42, 57]. Genetic diversity among 50 accessions of *Bromus tectorum* and 43 of *B. sterilis* from different sites of their Eurasian ranges has been studied by electrophoretic analysis of ten enzymes encoded by 18 loci by Oja [55]. The two species proved clearly differentiated by alternate allozymes of seven isozymes. Populations of both taxa showed differentiation into eleven (*B. tectorum*) and six (*B. sterilis*) multilocus allozyme lineages (MLALs). The extent of interspecific allozyme divergence estimated by Manhattan distance exceeded more than three times intraspecific differentiation between the multilocus lineages. Only two MLALs in each species have wide geographical distribution from Near East to Europe. Other MLALs were found each for only one or two populations and were region-specific. Most geographically marginal European populations had widespread MLALs. Nuclear microsatellite analysis of *Bromus sterilis* from three English farms [27] revealed that *B. sterilis* exists as numerous separate and genetically different lines, which are maintained by inbreeding with occasional outcrossing.

Bromus tectorum was introduced into North America, South America, Australia and New Zealand, where it rapidly developed into a noxious weed. Novak et al. [53] detected only low allozyme variation within *B. tectorum* populations in North America. Bartlett et al. [16] found that the invasion of North America by *B. tectorum* occurred through multiple introductions on both coasts. Eurasian populations of *B. tectorum*, however, exhibited somewhat greater allozyme diversity than those in North America, and most differentiated were populations from Southwest Asia, the presumed ancestral native region of this species [52]. This is consistent with the view that the majority of allozyme diversity in autogamous species occurs between populations and that within-population diversity is, as a rule, relatively low [28]. Ramakrishnan et al. [71] using SSR and AFLP data, showed the presence of significant genetic variation among *B. tectorum* populations. They described 40 self-pollinating lines from four populations of *B. tectorum* in the US and suggested that the present genetic diversity is the result of selection in original founder populations rather than the result of mutations.

Pillay and Hilu [68], on the basis of the cpDNA restriction site data, suggested that *B. sterilis* and *B. tectorum* are recently derived species that share a common maternal ancestry.

Among diploids, *B. fasciculatus* is sporadically distributed in the Mediterranean region and has a more restricted range, being confined to East Mediterranean and western parts of Southwest Asia. In contrast with other species of section *Genea, B. fasciculatus* has quite a uniform morphology. The general habit of *B. fasciculatus* shows resemblance to tetraploid *B. rubens* and to depauperate specimens of tetraploid *B. madritensis*, but the diploid can be easily distinguished from both tetraploids by very narrow, needlelike glumes and lemmas, and by strongly outcurved, often twisted, narrow grains. *Bromus fasciculatus* is most distinct among diploids with nine species-specific allozymes [56]. A characteristic feature of isozyme variability in three diploid bromes is almost total absence of two-or three-banded heterozygous allozyme phenotypes at polymorphic isozymes. All variations were observed as single-banded electrophoretic variants attributed to allozymes, indicating that selfing is characteristic of all diploids examined [56, 60]. This is in accordance with other data on autogamy and self-fertilization of *Genea* bromes [43, 44, 53, 84].

The polyploids of the section *Genea* are mainly or exclusively Mediterranean taxa. *Bromus madritensis* L. sensu lato, including *B. madritensis* L. subsp. *kunkelii* H. Scholz together with *B. haussknechtii* Boiss., *B. flabellatus* Boiss., and *B. rubens* L., form a complex of morphologically similar taxa with small lemmas and erect, more or less contracted panicles at flowering time. These characters separate the *Bromus madritensis* complex from the remaining species in section *Genea*. All members of this complex are annual, predominantly self-fertilizing tetraploid grasses with 2n=28 [51, 63, 75]. *Bromus rubens* showed no outcrossing [30, 31].

Species belonging to the complex exhibit continuous morphological variation ranging from typical *B. madritensis* (with longer panicle branches and looser panicles) to typical *B. rubens* (with brush-like condensed panicles), with *B. flabellatus* Boiss, *B. haussknechtii* and newly described by H. Scholz [80] *B. madritensis* subsp. *kunkelii* showing intermediate characters. *Bromus madritensis* and *B. rubens* have traditionally been recognized as separate species on the basis of the panicle and spikelet characters, whereas *B. haussknechtii* and *B. flabellatus* have mostly been considered only as subspecies of *B. madritensis* [62, 73, 79]. The great morphological diversity within the group has also been explained by phenotypic plasticity depending on the environmental growth conditions [19, 20, 22, 76, 106].

Isozyme electrophoresis was used to study genetic diversity and divergence among the brome grass species of the *B. madritensis* complex, comprising *B. madritensis* s.l. and *B. rubens* [57]. In total, 40 multilocus isozyme lineages (MLILs) were determined for 152 accessions of the complex from different sites throughout their geographic distribution. Two most common lineages were found for *B. madritensis* and one for *B. rubens*. At the isozyme level, a clear division corresponding to the two species was evident by species-specific phenotypes of four isozymes. Cluster analysis based on isozyme phenotypes also supported the recognition of two species in the complex. The UPGMA dendrogram showed that *B. madritensis* MLILs could be divided into two major groups or putative subspecies, but they did not correspond to differentiation by morphological characters. Morphologically defined *B. madritensis* subsp. *kunkelii* was clubbed together with *B. rubens* in the same cluster, supporting the recognition of this subspecies within *B. rubens*, not under *B. madritensis*.

Isozymes have been frequently employed to establish allo- versus autoploid nature of plant polyploids, and to identify their putative diploid progenitors [17, 26, 92]. Alloploids are characterized and distinguished by fixed enzyme heterozygosities derived from their diploid progenitors and show disomic inheritance [32, 72]. Kahler et al. [32] found that electrophoretic phenotypes of enzymes could help to distinguish the *Genea* species. The two tetraploids, *B. madritensis* and *B. rubens* exhibited different fixed heterozygosities of several heterozymes, indicating their alloploid nature and independent origins [59]. Oja [56] has analyzed and compared allozyme variation of three diploid species with that for polyploid species of section *Genea*. Allozymeso characteristic for diploids *B. tectorum* and *B. fasciculatus* are combined in fixed heterozygous phenotypes of tetraploid *B. rubens*, suggesting that tetraploid *B. rubens* might have been derived from a hybrid of *B. fasciculatus* with *B. tectorum*. Fixed heterozygous phenotypes of tetraploid *B. madritensis* combine one allozyme of *B. fasciculatus* with another of diploid *B. sterilis* at each of the loci studied. *Bromus fasciculatus* thus appears to be a third diploid ancestor for the two tetraploids of section *Genea*, *B. rubens* and *B. madritensis*.

B. diandrus and *B. rigidus* are indigenous to the Mediterranean region, *B. rigidus* is restricted more to coastal regions than *B. diandrus*. Kon and Blacklow [37] showed that outcrossing in *B. diandrus* was less then 1%. *Bromus diandrus* and *B. rigidus* have been reported as polyploids with different intraspecific ploidy levels: *B. diandrus* (2n = 28, 42, 56) *B. rigidus* (2n = 28, 42, 56, 70) [21, 35, 77]. *Bromus diandrus* and *B. rigidus* are morphologically quite variable [23, 74]. Although traditionally *B. diandrus* and *B. rigidus* are usually treated as separate species, some authors do not agree. Ovadiahu-Yavin [63] proposed their recognition as two subspecies of *B. rigidus*. Esnault and Huon [24] supposed to consider them as members of a polyploid complex with hexa- and octoploid cytotypes. Sales [74] had treated these species as two different varieties of *B. diandrus* because of the occurrence of a lot of individuals with intermediate characters and unclear geographical and ecological separation. The existence of intermediate types makes identification difficult, and morphological characters may not alone permit resolution of these species. Isozyme data [59] also showed that *B. diandrus* and *B. rigidus* are genetically closely related and do not deserve the rank of separate species. Almost all brome grass investigators have had complications in distinguishing the taxa solely on the basis of overall morphology in the *B. diandrus-rigidus* polyploid complex. Chromosome numbers, morphological characters and isozymes of seven enzymes were studied by Oja and Laarmann [62] to assess relationships between species of the *Bromus diandrus-rigidus* polyploid complex and *B. sterilis* that is a closely related diploid (2n = 14) species [20, 40, 78]. It has been shown to be a putative genome donor for the *B. diandrus-rigidus* polyploid complex by the isozyme evidence [59]. Some authors give two chromosome numbers for *B. sterilis*, 2n=14 and 2n=28 [20, 48]. *Bromus sterilis*, *B. diandrus* and *B. rigidus* are quite similar in overall morphology, the last two being more robust plants [74, 76, 83, 86]. The four different cytotypes detected, 2n=14, 28, 42, and 56, could be divided into two species: *B. sterilis* (2n=14, 28) and *B. diandrus* (2n=42, 56) by morphological features and isozymes [62]. The shape of the scar of rachilla segments in the floret proved to be a suitable character for distinguishing the two species, but not for distinguishing the chromosomal races within species. The tetraploid shared homozygous isozyme phenotypes with diploid *B. sterilis* at all loci except one, suggesting that it is autopolyploid. No diagnostic isozymes could be found to distinguish between hexa- and octoploid cytotypes in the *B. diandrus-rigidus* complex.

Isozyme electrophoresis was also used to study the origin of the *B. diandrus-rigidus* polyploid complex. *Bromus diandrus* and *B. rigidus* revealed identical zymograms, providing new support for their conspecific recognition. Different fixed heterozygosities of several heterozymes suggest their alloploid nature and independent origin from different diploid progenitors [59]. The pattern of isozyme variation evidence against the participation of diploids *B. fasciculatus* and *B. tectorum* and tetraploids *B. rubens* and *B. madritensis* in the hybrid origin of the hexa-octoploid *B. diandrus-rigidus* complex [57]. Of the three diploids studied, only *B. sterilis* fits well for a role of a genome donor for the polyploid *B. diandrus-rigidus* complex, while the diploid progenitors of other genomes yet remain to be established.

Section *Bromus*

Section *Bromus* of the genus *Bromus* L. [87], synonyms: subgenus *Bromus*, genus *Bromus*, together with section *Genea* is considered to be the most advanced section in the genus that probably arose during the Pleistocene in southwestern Asia and the Mediterranean area [100]. It includes annual or biennial, predominantly self-fertilizing species that are widespread in Eurasia. Many of them have been distributed into the New World, Africa and Australia [73, 83, 100]. Southwest Asia and Eastern Mediterranean area are considered as the centre of diversity for this group [84, 74]. The number of species included in the section varies remarkably (from 30 to 40), depending on the author of the classification and on the accepted synonymy [79, 84, 102]. Polyploidy is common in this section and is restricted to the tetraploid level [100]. The species of the section *Bromus* have the largest genome size in the genus [15]. Most of species are well-known weedy grasses with wide geographic distribution. Several features, including colonizing success, morphological variability, phenotypic plasticity, hybridization, and polyploidy, make this group of considerable evolutionary interest [100]. Section *Bromus* contains several species complexes with taxonomically problematic species which are closely related and difficult to delimit, e.g. the *B. mollis* s.l. complex [82] or *B. commutatus* and *B. racemosus* [85]. Already Linnaeus was in trouble with species delimitation in section *Bromus*. He changed his concept of some brome species several times [82]. Later, it was one of the reasons for a number of misdescriptions. This section has been regarded as a taxonomically difficult group for a long time and even nominated as "taxonomic nightmare" [88]. All authors realize the difficulty of assessing interspecific relationships and phylogeny in section *Bromus* by using only morphological characters, as is typical for many grasses [33].

Ainouche and Bayer [3] found a weak divergence among the diploid species within the section Bromus by the ITS sequences of the nrDNA. Phylogenetic relationships and genetic differentiation between some diploid annual brome species were evaluated by cladistic and phenetic analysis of allozyme diversity performed by Oja and Jaaska [59]. The diploids of the section *Genea* and section *Bromus* were distinguished into separate subclusters on both cladistic and phenetic allozyme trees. Diploids *Bromus pumilio*, *B. danthoniae*, *B. scoparius* and *B. alopecuros*, despite intraspecific allozyme polymorphism of several heterozymes, lacked heterozygous allozyme phenotypes, indicating prevalent self-fertilization. The important consequence of autogamous breeding system in these diploid bromes is their intraspecific differentiation into distinct multilocus isozyme lineages. Two more distinctly differentiated species among diploids of section *Bromus* seem to be *B. pumilio* and *B. danthoniae*, which are placed basally paraphyletic to the other species of the section *Bromus*. Smith [90] also supposed that *B. pumilio* is most closely related to *B. danthoniae* and *B. alopecuros*. The clear allozyme differentiation of *B. danthoniae* is congruent with its divergence by a unique morphological character (this species has three-awned lemmas).

Bromus arvensis, *B. japonicus* and *B. squarrosus* are morphologically similar diploids of section *Bromus* with largely overlapping specie descriptions and inconsistent taxonomy. In the first description of *B. japonicus*, Thunberg [103] emphasized its close resemblance to *B. arvensis*, whereby having oblong spikelets and divaricated awns differentiates it. Taxonomists have always recognised that *B. japonicus* is a greatly variable taxon with at least two subspecies: typical subsp. *japonicus* Thunb. and subsp. *anatolicus* (Boiss & Heldr.) Penzes. The descriptions of subsp. *anatolicus* and of *B. squarrosus* are obviously overlapping [86, 89]. Acedo and Llamas [2] in their monograph about genus *Bromus* in the Iberian Peninsula, suggest that *B. japonicus* is nothing more than *B. arvensis* with cleistogamous florets and small anthers and, therefore, with selfing breeding behaviour, respectively. Krechetovich and Vvedensky [41] placed *B. arvensis* in a new series *Macrantherae*, and positioned *B. japonicus*, *B. squarrosus*, *B. anatolicus* Boiss. et Heldr. and *B. briziformis* Fisch. et Mey. in another new series *Squarrosae*. This system was later followed by Scholz [79] and supported by Smith [84], Tzvelev [102] and Stace [97]. Most Floras still recognize three different species *B. japonicus*, *B. squarrosus* and *B. arvensis* and may underestimate the presence of remarkable variability with intermediate morphological forms between them. *Bromus arvensis* has a wide geographic range from Southwest Asia, throughout the Mediterranean area to Northern Europe [86, 89]. *Bromus japonicus* and *B. squarrosus* are also widely distributed throughout the Mediterranean region, but extend to central Asia and have been introduced to America and Australia [89].

Serological study [84] showed that *B. squarrosus* and *B. japonicus* are closely related, whereas *B. arvensis* is distinct from them. The RAPD [5] and ITS data [4] also showed that *B. squarrosus* and *B. japonicus* are sister species but *B. arvensis* is somewhat different. Isozyme results [61] revealed that *B. japonicus* and *B. squarrosus* do not have species-specific isozymes and do not form separate clusters on the dendrograms based on the isozyme data. *Bromus arvensis* has two species-specific allozymes and is distinguishable from *B. japonicus* and *B. squarrosus*.

Estimates of outcrossing rate ($t = 1$) in *B. arvensis*, indicating essentially complete allogamy with random mating, whereas *B. japonicus* and *B. squarrosus* ($t = 0.00$) are extreme selfers. It is suggested that *B. japonicus* and *B. squarrosus* are autogamous relatives of the outcrosser *B. arvensis*, but not its immediate progenitors [61]. Twenty-four multilocus isozyme lineages (MLILs) were detected among the 42 accessions of *B. japonicus* and *B. squarrosus*, with some MLILs containing accessions of both taxa. No geographic pattern was found among the accessions or MLILs of *B. japonicus* and *B. squarrosus*.

According to classificatory discriminant, canonical discriminant, principal component and cluster analyses of the morphological characters, the accessions of *B. arvensis*, *B. japonicus* and *B. squarrosus* were separated into three moderately distinct groups that corresponded to the three traditional species. The results showed that qualitative characters were the best for the delimitation of the taxa by statistical analyses.

Bromus intermedius Guss. is an annual diploid (2n=14) that has mostly Mediterranean distribution: Southern Europe, Northern Africa and Southwest Asia. In the Iberian Peninsula it is found only in the southern part of Spain [2]. The main difference between *B. japonicus* and *B. intermedius*, according to a recent very accurate key to Bromeae in the Mediterraneaen climatic zones [96], is the dry lemma texture which is papery, usually with protruding veins in *B. intermedius* and leathery, usually without protruding veins in *B. japonicus*. Inflorescence structure, panicle branch length and spikelet measurement are greatly influenced by environmental conditions during the vegetation period [89, 91] and are of limited value to distinguish the two species. Unfortunately, pubescence can hide protruding veins and frequently in *B. japonicus* leathery lemmas can be quite thin and thus with protruding veins. Spalton emphasized this fact in the notes of his paper [96]. *Bromus japonicus*, *B. squarrosus* and *B. arvensis* form a similar cluster of closely related species [61], and *B. intermedius* also belongs to this complex. Genetic diversity and differentiation among the *B. intermedius* accessions of different geographic origins have been recently studied by me, using isozyme analysis. The mating system was evaluated on the basis of allozyme polymorphism. Outcrossing rate (t) in *B. intermedius* was mostly 0, except one population with t = 0.16, indicating nearly complete autogamy in this species. Given that *B. arvensis* and *B. intermedius* had common allozymes of all isozymes studied, it is suggested that *B. intermedius* may be a direct autogamous derivative of the outcrosser *B. arvensis*. Contrary to expectations, the allozyme diversity in selfing *B. intermedius* was higher than in outcrosser *B. arvensis*, comprising 23 and 16 allozymes, respectively [58].

An artificial group of 12 brome grasses with small spikelets was rewieved by Smith and Sales [91]. They supposed that four small-spikelet bromes (*B. brachystachys*, *B. pseudobrachystachys*, *B. lepidus* and *B. scoparius*) would be the most highly evolved species in the section *Bromus*.

Bromus racemosus, *B. commutatus* and *B. secalinus* are related tetraploids (2n=28) of the section *Bromus*. Commonly they are recognized as different species, however some authors [2, 7, 45, 50] treated *B. commutatus* as a subspecies of *B. racemosus*. Smith [85] admitted that they are not identical but overlap ecologically. Lloret [46] regarded *B. commutatus* as a subspecies of *B. secalinus*. Morphologically, *B. commutatus* appears to be in a central position between *B. racemosus* and *B. secalinus*. Spalton [95] supposed that "these three taxa may have originated from an unidentified common diploid ancestor which may no longer be existing". He also analyzed the morphological characters traditionally used in identification of these taxa and proposed a new detailed key for them. Allozyme data [54] show that tetraploids of section *Bromus* exhibited different fixed heterozygosities of several heterozymes, suggesting their alloploid nature and independent origins. *Bromus secalinus* and *B. commutatus* displayed homologous variation with shared morphs at several heterozymes, indicating their strong genetic affinity. The isoenzyme results on the close phylogenetic affinity of *B. secalinus* and *B. commutatus* are consistent with the morphological similarity of these species, their serological affinity [84] and hybridization data [105]. At the same time, a marked difference between them was revealed by one isozyme and it supports their recognition as separate species.

Ainouche et al. [6] studied allozymic genetic diversity among the four species belonging to section *Bromus* in the Mediterranean region. *Bromus intermedius*, *B. squarrosus*, *B. lanceolatus* and *B. hordeaceus* displayed substantial genetic similarity. At the same time, the species were clearly differentiated with tetraploid *B. hordeaceus* being more closely related to diploid *B. squarrosus*, and tetraploid *B. lanceolatus* to diploid *B. intermedius*, respectively. They found that self-fertilizing diploids *B. intermedius* and *B. squarrosus* may have substantial levels of allogamy and are genetically less diverse than widespread tetraploids *B. lanceolatus* and *B. hordeaceus*.

The diploid *B. pseudosecalinus* is morphologically very similar to the tetraploid *B. secalinus*. The two species were found to be serologically very different [84], and allozyme results [54] also showed that *B. pseudosecalinus* could not be one of the diploid progenitors for *B. secalinus*.

Smith [84] placed *B. secalinus* and *B. alopecuros* in separate groups on the basis of morphological and serological differences. On the other hand, Pillay and Hilu [68] suggested close relationships between them on the basis of the cpDNA. Allozyme results [54] suggest that *B. alopecuros* does not suite as a diploid parent for tetraploid *B. secalinus*. Smith [84] concluded that *B. danthoniae* is likely to be one of the diploid ancestors of *B. lanceolatus* because of the serological similarity. This is, however, not supported by isozymes. Instead, allozyme data [54] indicate that *B. alopecuros* could be one of the genome donors for the tetraploid *B. lanceolatus*.

Bromus hordeaceus is an allotetraploid predominantly self-pollinated invasive weed with a very wide distribution [100]. Traditionally, four subspecies were recognized in the morphologically considerably varying *B. hordeaceus* complex. Three of them are habitat specific "ecotype-subspecies": subsp. *ferronii*, subsp. *thominii* and subsp. *molliformis* [87]. *Bromus hordeaceus* sp. *hordeaceus* (syn. *B. mollis* L.) is a type subspecies that grows in very different ecological and geographical conditions. Outcrossing rate in *B. hordeaceus* differed from 1-18% [30, 31] in different geographical regions. Lönn [47] reported that *B. hordeaceus* is a largely self-fertilizing species with restricted gene flow, because only one heterozygote out of 239 electrophoretically screened individuals was found in Öland, Sweden.

Ainouche et al. [3] have shown that the interpopulational differentiation within *B. hordeaceus* is based on geographic, rather than on subspecific taxonomic divergence. Similarly, intraspecific allozyme variation observed in *B. hordeaceus* [54] was independent from the morphological differentiation that recognized three ecotype-subspecies in *B. hordeaceus*. Ainouche et al. [5] studied genetic diversity at enzyme loci and ITS sequences from the nrDNA in 15 Mediterranean and Atlantic populations of *B. hordeaceus* and found no genetic differentiation among the four subspecies. All the populations studied were homozygous, suggesting selfing. The tetraploids *B. hordeaceus* and *B. interruptus* are very close morphologically. In addition to this external resemblance, they showed a 75-80% similarity in seed proteins [84] and shared identical isozyme phenotypes [54]. Thus, isozyme data support the suggestion of Smith [87] that *B. interruptus* may be interpreted as an ecotype-subspecies of *B. hordeaceus*.

Spalton [94] has described a new subspecies *B. hordeaceus* subsp. *longipedicellatus* that resembles *B. commutatus* in general appearance. He found that *B. hordeaceus* subsp. *longipedicellatus* is most closely related to *B. hordeaceus* subsp. *hordeaceus* and could be distinguished from the latter by having some pedicels and branches longer than the spikelets that they bear. Spalton supposed that subsp. *longipedicellatus* may be the result of a gene transfer from *B. hordeaceus* into *B. racemosus* or *B. arvensis* with *hordeaceus* genes becoming dominant. Before arriving at a definite conclusion about a gene transfer between *B. hordeaceus*, *B. racemosus* or *B. arvensis,* an analysis of the extent to which these species are self-fertilizing or cross pollinated, should be made. Wilson [105] produced fertile interspecific hybrids of *B. racemosus* and *B. commutatus*, which were morphologically intermediate between the parents and proposed to group them indiscriminately as one species. Smith [85] also suggested that *B. commutatus* and *B. racemosus* hybridize in Britain but argued against lumping them. Ainouche et al. [6] reported substantial amount of allogamy for *B. hordeaceus*. Oja et al. [61] estimated outcrossing rate of *B. arvensis* as approximately $t = 1$, indicating that *B. arvensis* is a complete outcrosser with random mating. Thus, the high level of gene flow in these taxa is real and Spalton's hypothesis could be true, but needs more evidence, perhaps more effectively from molecular markers.

Acedo and Llamas [1] described two new annual brome grasses from the Iberian Peninsula: *Bromus cabrerensis* and *B. nervosus*, belonging to section *Bromus*. *Bromus cabrerensis* is a tetraploid and closely related to *B. hordeaceus*, from which it clearly differs by large panicles with numerous spikelets. *B. nervosus* has a rather isolated position within section *Bromus* and its chromosome number and other features are not studied yet, as it is only known from two herbarium specimens.

Section *Pnigma*

Section *Pnigma* [83], synonyms: subgenus *Festucaria* and genus *Bromopsis*, is the largest section in the genus *Bromus*. It contains about 60 perennial, predominantly outcrossing species with two major centres of distribution in Eurasia and North America. Eurasian species of the section *Pnigma* are mainly polyploids (11 species), only 3 diploid species are known there. High level of polyploidy, predominantly hexa- or octoploid, is characteristic for *Pnigma* species in Eurasia [98, 100]. On the other hand, polyploidy is less frequent in the North American species of the section *Pnigma* and is mostly limited at the tetraploid level. The majority of North American species are diploids. Two endemic species from South America are hexaploid (*B. auleticus* and *B. uraguayensis*). Relationships of these species to the other species of the section *Pnigma* are unknown [15]. Pillay and Hilu [67], on the basis of the cpDNA data, confirmed the monophyly of section *Pnigma* that has been proposed by Stebbins

[99] and Armstrong [11]. Cytotaxonomic studies of hybrids between *Pnigma* species by Armstrong [8, 9, 100] revealed that *B. pumpellianus* is an autoalloploid with an AAAABBBB genome formula, *B. erectus* is an autotetraploid with an AAAA, genome formula and *B. inermis* is auto-allo-octoploid with AAAABBBB genome composition. Armstrong [12], on the basis of the cytological observations, showed that Eurasian species *B. benekenii* (Lange) Trimen and *B. ramosus* Huds. are more closely related to the American species *B. ciliatus* L., *B. latiglumis* Hitchc., *B. pacificus* Shear and *B. richardsonii* Link than to other Eurasian species, and suggested that section *Pnigma* may contain two distinct groups of species with different chromosome size. He also suggested [13] that the large (predominantly Eurasian) and small (predominantly American) chromosome species have followed different evolutionary pathways. Chromosome pairing suggested [13] that the *B.variegatus* Bieb. genome was differentiated from the A and B genomes of octoploid *B. inermis*. All polyploid species of the section, except *B. auleticus,* have identical cpDNA, whereas diploid species of the section showed various degrees of cpDNA divergence [67].

Some species of section *Pnigma* have economic significance. Among them, *B. inermis* is most important as a pasture and forage plant widely cultivated in Northern Europe, Russia, the USA and Asia. *Bromus inermis* (octoploid, $8x=2n=56$) and *B. riparius* (decaploid, $10x=2n=70$) are the most commonly cultivated perennial brome grasses in North America [104]. They are cross-pollinated species, but self-pollination has been reported for *B. riparius* [34]. Hybrids between the two species have been discovered through RAPD and AFLP markers by Ferdinandez and Coulman [25]. It has been supposed that diploid *B. riparius* found in Kazakhstan could be a progenitor of the *B. inermis* complex [14].

Pillay and Armstrong [66] examined the inheritance of cpDNA in F1 progeny of interspecific crosses of *B. arvensis* with *B. inermis* and *B. erectus.* No intraspecific cpDNA variability was detected. All the F1 progeny examined exhibited the maternal inheritance of cpDNA.

Section *Ceratochloa*

Section *Ceratochloa* [84], synonyms: subgenera *Ceratochloa* and genera *Ceratochloa*, is a small section in the genus *Bromus*, consisting of 16 polyploid perennial and annual species [15, 67, 100]. No diploid or tetraploid species were found in this section. The section *Ceratochloa* has the smallest (most primitive) genome size in the genus [15]. The species of section *Ceratochloa* break up into two morphologically distinct groups: the *B. catharticus* hexaploid complex, which is endemic to South America, and the *B. carinatus* octoploid complex, found mainly in North America [98, 67]. All species of subgenus *Ceratochloa* display identical cpDNA sequences. The *B. carinatus* complex appears to be phylogenetically closely related to the diploid *B. anomalus* of section *Pnigma* [67]. The six octoploid species of the *B. carinatus* complex are considered to be intersectional amphidiploids between diploids of sect. *Bromopsis* and hexaploid species of sect. *Ceratochloa* [99, 15]. The hexaploid species of section *Ceratochloa* are all found to be strict allopolyploids with genomic formula AABBCC [101].

Bromus catharticus is a native of South America and is a predominantly cross-pollinated hexaploid (2n=42) species. Puecher et al. [70], using RAPD and AFLP markers, demonstrated great genetic similarity between the morphologically contrasting populations of the *B. catharticus* in Argentina.

In addition, genus *Bromus* also includes three small monospecific sections: (1) section *Nevskiella* contains a single diploid species *B. gracillimus* distributed in Central Asia, Iran and Afghanistan, (2) section *Neobromus* contains a single polyploid species *B. trinii* distributed along the Pacific coast of North, Central and South America, and (3) section *Boissiera* with the only diploid species *B. pumilio*. *Bromus pumilio* is native in Central and Southwest Asia and eastern Mediterranean area. *Bromus pumilio* of the monotypic genus *Boissiera* was firstly considered [83] to belong to section *Bromus*, then was placed in its own section *Boissiera* [90]. The placement of *B. pumilio* in its own section *Boissiera* was supported by allozyme study [60]: this taxon has a basal position in a separate clade in both cladistic and phenetic analyses. This is in accordance with its divergence by morphology and dispersal mechanism [90].

The phylogenetic relationships between these small sections of the genus *Bromus* are obscure and still waiting for investigation. Already in 1981, Stebbins [100] had admitted that genus *Bromus* "is a favourable object for learning more about the evolution of grasses" and it still offers this opportunity. Recently many authors have studied genetic relationships and variation of many of the species of the genus *Bromus*. The results obtained from different molecular markers highlighted the problematic issues for future investigation.

Acknowledgement

I am grateful to Vello Jaaska for his valuable comments on the manuscript. This work was supported by grant ESF 5739 from the Estonian Science Foundation.

[1] Acedo C, Llamas F. *Bromus cabrerensis* and *Bromus nervosus* two new species from the Iberian Peninsula. Willdenowia 1997; 27: 47-55.

[2] Acedo C, Llamas F. The genus *Bromus* L. (Poaceae) in the Iberian Peninsula. J. Cramer in der Gebrüder Borntraeger Verlagsbuchhandlung, Berlin, Stuttgart. 1999.

[3] Ainouche M, Bayer RJ. On the origins of two Mediterranean allotetraploid *Bromus* species: *Bromus hordeaceus* L. and *B. lanceolatus* Roth. (Poaceae). Am J Bot. 1996; 83: 135.

[4] Ainouche M, Bayer RJ. On the origin of the tetraploid *Bromus* species (section *Bromus*, Poaceae) - insights from internal transcribed spacer sequences of nuclear ribosomal DNA. Genome 1997; 40: 730-743.

[5] Ainouche M, Bayer RJ, Gourret J, et al. The allotetraploid invasive weed *Bromus hordeaceus* L. (Poaceae): genetic diversity, origin and molecular evolution. Folia geobotanica 1999; 34: 405-419.

[6] Ainouche M, Misset MT, Huon A. Genetic diversity in Mediterranean diploid and tetraploid *Bromus* L. (section *Bromus* Sm.) populations. Genome 1995; 38: 879-888.

[7] Ammann K. Bestimmungsschwierigkeiten bei europäischen Bromus-Arten. Bot Jahrb Syst 1981; 102: 459-469.

[8] Armstrong KC. Chromosome pairing in hexaploid hybrids from *Bromus erectus*, (2n = 28) X *B. inermis* (2n = 56). Can J. Genetics and Cytology 1973; 3: 427-436.

[9] Armstrong KC. Genome relationships in *Bromus erectus*, *B. pumpellianus* ssp. *dicksonii* and *B. pumpellianus*. Can J. Genetics and Cytology 1975; 3: 391-394.

[10] Armstrong KC. Hybrids of the annual *Bromus arvensis* with perennial *B. inermis* and *B. erectus*. Z Pflanzenzücht 1977; 79: 6-13.

[11] Armstrong KC. The evolution of *Bromus inermis* and related species of *Bromus* sect. *Pnigma*. Bot Jahrb Syst 1981; 102: 427-443.

[12] Armstrong KC. The relationship between some Eurasian and American species of *Bromus* section *Pnigma* as determined by the karyotypes of some F1 hybrids. Can J Bot 1983; 61: 700-707.

[13] Armstrong KC. The genomic relationships of the diploid *Bromus variegatus* to *B. inermis*. Can J Genetics and Cytology 1984; 4: 469-474.

[14] Armstrong KC. Chromosome numbers of perennial *Bromus* species collected in the USSR. Can J Bot 1987; 67: 267-269.

[15] Armstrong KC. Chromosome evolution of *Bromus*. In: Armstrong KC, Tsuchiya T, Gupta P, eds. Chromosome engineering in plants: genetics, breeding, evolution. Part B. Amsterdam, Netherlands; Elsevier Sci Publ. 1991: 366-377

[16] Bartlett E, Novak SJ, Mack RN. Genetic variation in *Bromus tectorum* (Poaceae): differentiation in the eastern United States. Am J Bot 2002; 89: 602-612.

[17] Crawford DJ. Electrophoretic data and plant speciation. Syst. Bot. 1985; 10: 405-416.

[18] Cugnac A. Sur quelques Bromes et leurs hybrides. III. Donnes biometriques comparatives sur quelques caracteres distinctifs de *Bromus rigidus* Roth, *B. gussonii* Parl. et *B. sterilis* L. Bull Soc Bot Fr 1934; 81: 318-323.

[19] Cugnac A. Sur quelques'Bromes et leurs hybrides. VII. B. sterilis var. velutinus Volkart obtenu par synthese experimentale a partir du croisement de *B. madritensis* L. par *B. sterilis* L. Bull Soc Bot Fr 1937; 84: 711-713.

[20] Devesa JA, Lugue T, Gomez P. Numeros cromosomaticos de plantas occidentales, 591-601. Anales Jard Bot Madrid 1990a; 47: 411-417.

[21] Devesa JA, Ruiz T, Ortega A, et al. Contribucion al conocimiento cariologico de las Poaceae en Extremadura (Espana). I. Bol Soc Brot 1990b; 63: 29-66.

[22] Esnault MA. Etudes sur la variabilite morphologique de *Bromus madritensis*. Phytomorphology 1984; 34: 91-99.

[23] Esnault MA, Huon A. Application des methodes numeriques a la systematique du genre *Bromus* L. sect. *Genea* Dumort. Bull Soc linn Provence 1985; 37: 69-77.

[24] Esnault MA, Huon A. Etudes morphologiques et caryologiques de *Bromu rigidus* et *Bromus diandrus* Roth: relations taxonomiques. Bull Soc bot Fr 1987; 134 Lettres bot 3: 299-304.

[25] Ferdinandez Y, Coulman B. Evaluating genetic variation and relationships among two bromegrass species and their hybrid using RAPD and AFLP markers. Euphytica 2002; 125: 281-291.

[26] Gottlieb LD. Electrophoretic evidence and plant populations. Progr Phytochem 1981; 7: 1-46.

[27] Green JM, Barker JHA, Marshall EJP, et al. Microsatellite analysis of the inbreeding grass weed Barren Brome (*Anisantha sterilis*) reveals genetic diversity at the within- and between- farm scales. Mol Ecol 2001; 10: 1035-1045.

[28] Hamrick J, Godt M. Allozyme diversity in plant species. In: Brown A, Clegg M, Kahler A, Weir B, eds. Plant Population: Genetics, Breeding, and Genetic Resources. Sunderland, UK: Sinauer, 1989: 43-63.

[29] Hsiao C, Chatterton NJ, Asay KH, Jensen KB. Phylogenetic relationships of the monogenomic species of the wheat tribe, Triticaceae (Poaceae) inferred from nuclear rDNA (internal transcribed spacer) sequences. Genome 1995; 38: 211-223.

[30] Jain SK. Inheritance of phenotypic plasticity insoft chess, *Bromus mollis* L. (Gramineae). Experientia 1978; 34: 385-386.

[31] Jain SK, Marschall DR, Wu K. Genetic variability in natural populations of softchess (*Bromus mollis* L.). Evolution 1970; 24: 649-659.

[32] Kahler AL, Krzakowa M, Allard RW. Isosyme phenotypes in five species of *Bromus* sect. *Genea*. Bot Jahrb Syst 1981; 102: 401-409.

[33] Kellogg EA, Watson L. Phylogenetic studies of a large data set. I - *Bambusoideae, Andropogoneae,* and *Pooideae* (Gramineae). Bot Rev 1993; 59: 273-343.

[34] Knowles PF. Interspecific hybridizations of *Bromus*. Genetics 1944; 29: 128-140.

[35] Kon KF, Blacklow WM. Identification and population variability of great brome (*Bromus diandrus* Roth) and rigid brome (*Bromus rigidus* Roth). Australian J. Agri Resources 1988; 39: 1039-1050.

[36] Kon KF, Blacklow WM. The biology of Australian Weeds *Bromus diandrus* Roth and *B. rigidus* Roth. Plant Protection Quarterly 1989; 4: 51-60.

[37] Kon KF, Blacklow WM. Polymorphism, outcrossing and polyploidy in *Bromus diandrus* and *B. rigidus*. Aust J Bot 1990; 38: 609-618.

[38] Kosina R. Embryo morphology in the genus *Bromus* (Poaceae). Fragm Flor Geobot 1996; 41: 563-576.

[39] Kosina R. Patterns of flower microstructural variation within the genus Bromus. Acta Soc Bot Poloniae 1999; 68: 221-226.

[40] Kozuharov S, Petrova A, Ehrendorfer F. Evolutionary patterns in some brome grass species (*Bromus*, Gramineae) of the Balkan Peninsula. Bot Jahrb Syst 1981; 102: 381-391.

[41] Krechetovich VI, Vvedensky AJ. Subgenus *Zeobromus* Griseb. In: Komarov VL, Roshevitz RJ, Schischkin BK, eds. Flora URSSR, vol. 2. Leningrad: Acad Sci URSS, 1934: 574-583.

[42] Krzakowa M, Kraupe A. Isozyme investigation of natural populations of the cheatgrass (*Bromus tectorum* L.). Bot Jahrb Syst 1981; 102: 393-399.

[43] McKone MK. Reproductive biology of several bromegrasses (*Bromus*): breeding system, pattern of fruit maturation, and seed set. Am J Bot 1985; 72: 1334-1339.

[44] McKone MK. Sex allocation and outcrossing rate: test of theoretical predictions using bromegrasses (*Bromus*). Evolution 1987; 41: 591-598.

[45] Lauber K, Wagner G. Flora Helvetica. Bern: Verlag Paul Haupt, 1996.

[46] Lloret FJ. Sobre Nomenclatura del genere *Bromus* L. Poaceae. Collectanea Botanica (Barcelona) 1993; 22: 151.

[47] Lönn M. Genetic structure and allozyme-microhabitat associations in *Bromus hordeaceus*. Oikos 1993; 68: 99-106.

[48] Luque T, Diaz Lifante Z. Chromosome numbers of plants collected during Iter Mediterraneum I in the SE of Spain. Bocconea 1991; 1: 303-364.

[49] Mehra PN, Khosla PK, Kohli BL, Koonar JS. Cytological studies in the North Indian grasses (Part I). Res Bull Punjab Univ New Ser Sci 1968; 19: 157-230.

[50] Meijden R. Heukels´ Flora van Nederland. Croningen, 1996.

[51] Naganowska B. Karyotypes of five *Bromus* species of *Genea* section. Genetica Polonica 1993; 34: 197-213.

[52] Novak SJ, Mack RN. Genetic variation in *Bromus tectorum* (Poaceae)- comparison between native and introduced populations. Heredity 1993; 71: 167-176.

[53] Novak SJ, Mack RN, Soltis DE. Genetic variation in *Bromus tectorum* (Poaceae): population differentiation in its North American range. Am J Bot 1991; 71: 1150-1161.

[54] Oja T. Isoenzyme diversity and phylogenetic affinities in the section Bromus of the grass genus *Bromus* (Poaceae). Biochem Syst Ecol 1998; 26: 403-413.

[55] Oja T. Allozyme diversity and interspecific differentiation of the two diploid bromegrass species, *Bromus tectorum* L. and *B. sterilis* L. (Poaceae). Plant Biology 1999; 1: 679-686.

[56] Oja T. *Bromus fasciculatus* Presl – a third diploid progenitor of *Bromus* section *Genea* allopolyploids (Poaceae). Hereditas 2002; 137: 113-118.

[57] Oja, T. Genetic divergence and interspecific differentiation in the *Bromus madritensis* complex (Poaceae) based on isozyme data. Biochem Syst Ecol 2002; 30: 433-449.

[58] Oja, T. Isozyme evidence on the genetic diversity mating system and evolution of *Bromus intermedius* (Poaceae) Pl Syst Evol 2005; 254: 199-208.

[59] Oja T, Jaaska V. Isoenzyme data on the genetic divergence and allopolyploidy in the section *Genea* of the grass genus *Bromus* (Poaceae). Hereditas 1996; 125: 249-255.

[60] Oja T, Jaaska V. Allozyme diversity and phylogenetic relationships among diploid annual bromes (*Bromus*, Poaceae). Annales Botanici Fennici 1998; 35: 123-130.

[61] Oja T, Jaaska V, Vislap V. Breeding system, evolution and taxonomy of *Bromus arvensis*, *B. japonicus* and *B. squarrosus* (Poaceae). Plant Syst Evol 2003; 242: 101-117.

[62] Oja T, Laarmann H. Comparative study of the ploidy series *Bromus sterilis*, *B. diandrus* and *B. rigidus* (Poaceae) based on chromosome numbers, morphology and isozymes. Plant biology 2002; 4: 484-491.

[63] Ovadiahu-Yavin Z. Cytotaxonomy of genus *Bromus* of Palestine. Isr J Bot 1969; 18: 195-216.

[64] Pillay M. Chloroplast DNA similarity of smooth bromegrass with other Pooid cereals: implications for plant breeding. Crop Science 1995; 35: 869-875.

[65] Pillay M. Genomic organization of ribosomal RNA genes in *Bromus* (Poaceae). Genome 1996; 39: 198-205.

[66] Pillay M, Armstrong KC. Maternal inheritance of chloroplast DNA in interspecific crosses of *Bromus*. Biologia plantarum 2001; 44: 47-51.

[67] Pillay M, Hilu KW. Chloroplast DNA variation in diploid and polyploid species of *Bromus* (Poaceae) subgenera *Festucaria* and *Ceratochloa*. Theor Appl Genet 1990; 80: 326-332.

[68] Pillay M, Hilu KW. Chloroplast DNA restriction site analysis in the Genus *Bromus* (Poaceae). Am J Bot 1995; 82: 239-249.

[69] Phillips S. *Bromus* L. In: Hedberg I, Edwards S. eds. Flora of Ethiopia and Eritrea. 1995; 7: 54-55.

[70] Puecher DI, Robredo CG, Rios RD, Rimieri P. Genetic variability measures among *Bromus catharticus* Vahl. populations and cultivars with RAPD and AFLP markers. Euphytica 2001; 121: 229-236.

[71] Ramakrishnan AP, Meyer SE, Waters J, Stevens MR, Coleman CE, Fairbanks DJ. Correlation between molecular markers and adaptively significant genetic variation in *Bromus tectorum* (Poaceae), an inbreeding annual grass. Am J Bot 2004; 91: 797-803.

[72] Ramsey J, Schemske DW. Pathways, mechanisms, and rates of polyploid formation in flowering plants. Annu Rev Ecol Syst 1998; 29: 467-501.

[73] Roy J, Navas ML, Sonie L. Invasions by annual bromegrasses: a case study challenging the homoclime approach to invasions. In: Grove RN, di Castri F, eds. Biogeography of Mediterranean Invasions. Cambridge: Cambridge University Press, UK 1991: 205-221.

[74] Sales F. Taxonomy and nomenclature of *Bromus* sect. *Genea*. Edinburg J Bot 1993; 50: 1-31.

[75] Sales F. A reassessment of the *Bromus madritensis* complex (Poaceae): a multivariate approach. Isr J Plant Sci 1994; 42: 245-255.

[76] Sales F. Evolutionary tendencies in some annual species of *Bromus* (*Bromus* L. sect *Genea* Dum. (Poaceae). Bot J Linnean Society 1994; 115: 197-210.

[77] Sanchez Anta MA, Gallego MF, Navarro AF. Aportaciones al conocimiento cariologico de las Cistaceas del centro-occidente Espanol. Stvd. Bot. (Salamanca) 1986; 5: 195-202.

[78] Sanchez Anta MA, Gallego MF, Navarro AF. Aspectos anatomicos de la epidermis de algunas especies de *Bromus* L. y su cariologia. Acta Botanica Barcinon 1988; 37: 335-344.

[79] Scholz H. Zur Systematik der Gattung *Bromus* L., subgenus *Bromus* (Gramineae). Willdenowia 1970; 6: 139-160.

[80] Scholz H. Bemerkungen uber *Bromus madritensis* und *B. rubens* (Gramineae). Willdenowia 1981; 11: 249-258.

[81] Scholz H. Der *Bromus-pectinatus*-Komplex (Gramineae) im Nahen und Mittleren Osten. Bot Jahrb Syst 1981; 102: 471-495.

[82] Smith PM. The *Bromus mollis* aggregate in Britain. Watsonia 1968; 6: 327-344.

[83] Smith PM. Taxonomy and nomenclature of the brome-grasses (*Bromus* L. s.l.). Notes RBG Edinburgh 1970; 30: 361-375.

[84] Smith PM. Serology and species relationships in annual bromes (*Bromus* L. sect. *Bromus*). Ann Bot 1972; 36: 1-30.

[85] Smith PM. Observations on some critical Brome grasses. Watsonia 1973; 9: 319-332.

[86] Smith PM. *Bromus* L. In: Tutin TG, ed. Flora Europea 1980: Cambridge Univ. Press, Cambridge. UK 5: 182-189.

[87] Smith PM. Ecotypes and subspecies in annual bromes-grasses (*Bromus*, Gramineae). Bot Jahrb Syst 1981; 102: 497-509.

[88] Smith PM. Proteins mimicry and microevolution in grasses. In: Jensen U, Fairbrothers DE, eds. Proteins and Nucleic Acids in Plant Systematics. Berlin: Springer, 1983: 311-323.

[89] Smith PM. *Bromus* L. In: Davis PH, ed. Flora of Turkey and the East Aegean Islands. Edinburgh, UK: University Press, 1985: 272-301.

[90] Smith PM. Observations on Turkish brome-grasses. Some new taxa, new combinations and notes on typification. Notes RBG Edinburgh 1985; 491-501.

[91] Smith PM, Sales F. *Bromus* L. sect. *Bromus*: Taxonomy and relationships of some species with small spikelets. Edinburgh J Bot 1993; 50: 149-171.

[92] Soltis DE, Soltis PS. Molecular Data and the Dynamic Nature of Polyploidy. Crit Rev in Plant Sci 1993; 12: 243-273.

[93] Soreng RJ, Davis JI. Phylogenetics and character evolution in the grass family (Poaceae): simultaneous analysis of morphological and chloroplast DNA restriction site character sets. Bot Rev 1998; 64: 1-85.

[94] Spalton LM. A new subspecies of *Bromus hordeaceus* L. (Poaceae). Watsonia 2001; 23: 525-531.

[95] Spalton LM. An analysis of the characters of *Bromus racemosus* L., *B. commutatus* Schrad. and *B. secalinus* L. (Poaceae). Watsonia 2002; 24: 193-202.

[96] Spalton L. A key to Bromeae in the Mediterranean climatic zones of Southern Europe, South West Asia, and North Africa. BSBI News 2004; 95: 22-26.

[97] Stace CA. A new flora of the British Isles. Cambridge, UK: Cambridge University Press. 1997.

[98] Stebbins GL. The evolutionary significance of natural and artificial polyploids in the family Graminea. Hereditas, Suppl 1949; 461-485.

[99] Stebbins GL. Cytogenetics and evolution of the grass family. Am J Bot 1956; 43: 890-905.

[100] Stebbins GL. Chromosomes and evolution in the genus *Bromus* (*Gramineae*). Bot Jahrb Syst 1981; 102: 359-379.

[101] Stebbins GL, Tobgy HA. The cytogenetics of hybrids in *Bromus*. I. Hybrids within the section *Ceratochloa*. Am J Bot 1944; 31: 1-11.

[102] Tzvelev NN. Zlaki SSSR (Poaceae USSR). Leningrad Rus. Fed.: Nauka, 1976.

[103] Thunberg CP. *Bromus japonicus* in Flora japonica sistens plantas insularum japonicarum secumdum. Systemae sexuale emenfatum dedactas. Lipsiae 1784; 52-54.

[104] Vogel KP, Moore KJ, Moser LW. Bromegrasses. In: Moser LE, ed. Cool-season forage grasses. Agron Monogr 34, ASA, CSSA, SSSA, Madison, USA: WI, 1996.

[105] Wilson D. Cytogenetic studies in the genus *Bromus*. PhD. thesis, Univ. of Wales, Aberystwyth, 1956.

[106] Wu K, Jain S. Genetic and plastic responses in geographic differentiation of *Bromus rubens* populations. Can J Bot 1978; 56: 873-879.

CHAPTER

10

Phylogeny and Evolution of *Festuca* L. and Related Genera of Subtribe Loliinae (Poeae, Poaceae)

PILAR CATALÁN

High Polytechnic School of Huesca, University of Zaragoza, Department of Agriculture (Botany), Huesca, Spain

ABSTRACT

Festuca and the related grasses of subtribe Loliinae, are major temperate and cool-season forage and grassland species of the world. A review of the evolutionary history of the festucoids, based on recent phylogenetic evidence and on compiled data from different phenetic, genomic and molecular sources, is presented in this chapter. *Festuca* is resolved as a large paraphyletic assemblage with several genera (*Lolium*, *Vulpia*, and others) included within it. Morpho-anatomical, cytogenetic, hybridization and reproductive biological traits of *Festuca* and its close allies support an evolutionary trend from ancestral broad-leaved taxa with large genome sizes and little heterochromatin content, to more recently evolved fine-leaved taxa with small-sized genomes but heterochromatin-rich. Major heterochromatin losses seem to be correlated with the origin of some putatively derived annual lineages. Evolutionary rates differ significantly between the slow-evolving perennial lineages and the rapidly mutating annual lineages, indicating a release from stabilized selection. Hybridization, polyploidy and the acquisition of the annual habit have played a key role in the speciation processes at different evolutionary times. Allopolyploidy emerges as the most widespread mechanism for speciation across the festucoid lineages. Biogeographical analyses indicate a likely late

Address for correspondence: High Polytechnic School of Huesca, University of Zaragoza, Department of Agriculture (Botany), Ctra. Cuarte km 1, E-22071 Huesca (Spain). E-mail: pcatalan@unizar.es.

Tertiary origin of the diploid Loliinae in Eurasia, followed by successive colonizations and secondary polyploid radiations in the southern hemisphere accompanied by occasional transcontinental long-distance dispersal events. A more recent Quaternary origin is hypothesized for the high polyploid lineages that successfully colonized newly deglaciated areas. Phylogeographical profiles indicate a likely expansion of some Mediterranean and European fescues and ryegrasses in the Holocene-Neocene times.

Key Words: Cytogenetics and hybridization, evolution, *Festuca*, Loliinae, molecular and morphological phylogenies

Abbreviations: Ca. = Circa, p.p. = pro parte, s.l. = sensu lato, spp = species (plural)

INTRODUCTION

Festuca L. and its related genera form the subtribe Loliinae Dumort., one of the main lineages of the subfamily Pooideae. The most recent phylogenetic studies have revealed that the festucoids are monophyletic and that *Festuca*, as traditionally circumscribed, is not a natural genus, but a large paraphyletic assemblage of distinctly related lineages, with *Lolium*, *Vulpia* and several other genera included within it [28, 29, 77]. The evolutionary study of the festucoids is of great relevance as they represent one of the largest and most widely spread groups of temperate grasses of both ecological and economical importance.

Festuca is one of the largest genera of the grass tribe Poeae, accounting for more than 500 species distributed in all continents except Antarctica [92, 169]. The genus is widely distributed across the northern hemisphere and in grassland communities of the southern hemisphere, but restricted to higher altitudes in subtropical and tropical regions [39, 169]. *Festuca* species show the typical pooid-like spikelet with short glumes, several florets, and five-veined lemmas [39, 99, 166]. *Festuca* consists of herbaceous perennial plants, predominantly allogamous, with or without clonal reproduction [92]. The genus is characterized by having species with paniculate inflorescences, spikelets with subequal glumes and dorsally rounded lemmas, and caryopses with linear hilums [60]. These species are highly variable in both vegetative and reproductive traits that have traditionally been used to separate them into the 'broad-leaved' fescues, and the 'fine-leaved' fescues [60, 92, 126].

Festuca includes some of the most valuable forage grasses of cold and temperate climates of the northern hemisphere, like the meadow and tall fescues [*Festuca pratensis* Huds. and *F. arundinacea* Schreb. complex of *Festuca* subgen. *Schedonorus* (P. Beauv.) Peterm.], and the 'red' and 'ovina' fescues (*F. rubra* L. and *F. ovina* L. groups of *Festuca* subgen. *Festuca*). Montane species of *Festuca* have been used in the restoration of subalpine landscapes and ski slopes (*F. eskia* Raymond ex DC. and *F. gautieri* (Hack.) K. Richt. of *Festuca* subgen. *Festuca*), whereas other fescues are widely planted as ornamentals in gardens (*F. glauca* Vill. group of *Festuca* subgen. *Festuca*). The extensive worldwide ecological range covered by *Festuca* taxa and their abundance in some mountain ecosystems have been used to characterize several grassland phytocenological alliances dominated by fescues [118].

Taxonomic circumscription of *Festuca* and its close allies has changed over the last few centuries (Table 1). *Festuca* is a complex genus divided into several subgenera and sections [4, 5, 6, 7, 9, 60, 61, 62, 93, 94, 115, 165]. Conversely, several segregates of *Festuca*, which were included within this genus in the past (*Vulpia* C. C. Gmel., *Schedonorus* P. Beauv., *Drymochloa* J. Holub, and *Leucopoa* Griseb.), have been recognized as independent genera at different times [43, 58, 70, 71, 139, 140, 143, 167, 168]. One of the most comprehensive studies of *Festuca* was that by Hackel [60], who divided the European fescues into six sections based on characters associated with leaf vernation, leaf sheath, auricles, spikelets and floral bracts (lemma and palea), ovary hairiness and adherence of caryopsis to palea, among others. He also separated infrasectional groups (series) based on the type of shoot innovation and established the anatomical analysis of the leaf cross-sections as a useful approach to identify species and infraspecific taxa [60]. Later, this system was broadly accepted by festucologists [4, 5, 6, 7, 9, 60, 61, 62, 93, 94, 115, 165], who further divided the genus into several subgenera and sections. The most recent revisions of the world's fescues [4, 5, 6, 7, 8, 9] recognized 11 subgenera and several sections within each.

Nine of Alexeev's subgenera [*Festuca* subgenera. *Asperifolia* E.B. Alexeev, *Drymanthele* V. Krecz. & Bobrov, *Erosiflorae* E.B. Alexeev, *Leucopoa* (Griseb.) Hack., *Mallopetalon* (Döll) E.B. Alexeev, *Schedonorus* (P. Beauv.) Peterm., *Subulatae* (Tzvelev) E.B. Alexeev, *Subuliflorae* E.B. Alexeev, and *Xanthochloa* (Krivot.) Tzvelev] are assigned to the broad-leaved fescues, whereas two of them (*Festuca* subgen *Festuca* and *Helleria*

Table 1. Taxonomic treatments adopted by different authors for *Festuca* and other related genera of subtribe Loliinae

Hackel (1882-1906)	*Tzvelev* (1971-2000)	*Cotton & Stace* (1977) *Stace* (1981)	*Alexeev* (1977-1986)	*Kerguélen & Plonka* (1989)	*Clayton & Renvoize* (1986)	*Holub* (1998)	*Catalán et al.* (2004) *Müller & Catalán* (2005)
Festuca	*Festuca*	Stace (1981)	*Festuca*	*Festuca*	*Festuca*	*Festuca*	*Festuca*
subgenus	subgenus *Festuca*		subgenus *Festuca*	subgenus *Festuca*			subgenus *Festuca*[a]
EuFestuca	sect. *Festuca*						
sect. *Ovinae*							
-*Intravaginales*				sect. *Festuca*			sect. *Festuca*[a]
-*Extravaginales*				sect. *Aulaxyper*			sect. *Aulaxyper*[a]
vel *mixtae*							
sect. *Variae*	sect. *Variae*			sect. *Eskia*			sect. *Eskia*[a]
				sect. *Amphigenes*			sect. *Dimorphae*[a]
sect. *Subbulbosae*				sect. *Subbulbosae*			sect. *Subbulbosae*[b]
							sect. *Lojaconoa*[b]
			subgenus *Helleria*	subgenus *Helleria*			subgenus *Helleria*[a]
sect. *Bovinae*	subgenus *Schedonorus*		subgenus *Schedonorus*	subgenus *Schedonorus*		*Schedonorus*	subgenus *Schedonorus*[b]
	sect. *Subulatae*		subgenus *Subulatae*	subgenus *Subulatae*			subgenus *Subulatae*[ab]
sect. *Montanae*			subgenus *Drymanthele*	subgenus *Montanae*		*Drymochloa*	subgenus *Drymanthele*[b]
sect. *Scariosae*							sect. *Scariosae*[b]
							sect. *Pseudoscariosa*[b]
subgenus *Leucopoa*	subgenus *Leucopoa*		subgenus *Leucopoa*	subgenus *Hesperochloa*		*Leucopoa*	subgenus *Leucopoa*[ab]
	sect. *Leucopoa*		subgenus *Subuliflorae*	subgenus *Subuliflorae*			
	sect. *Amphigenes*		subgenus *Obtusae*	subgenus *Obtusae*			
	sect. *Breviaristatae*		subgenus *Erosiflorae*				
			subgenus *Mallopetalon*				
	subgenus *Xanthochloa*		subgenus *Xanthochloa*	subgenus *Xanthochloa*			

Table 1 contd.

Table 1 contd.

subgenus *Vulpia*	*Vulpia*	*Vulpia*	*Vulpia*	*Vulpia*	*Vulpia*	*Vulpia*	*Vulpia* [a]
		sect. *Vulpia*					
		sect. *Monachne*					
		sect. *Spirachne*					
		sect. *Loretia*					
		sect. *Apalochloa*					
Ctenopsis		*Ctenopsis*			*Ctenopsis*		*Ctenopsis* [a]
Micropyrum,		*Micropyrum*			*Micropyrum*		*Micropyrum* [a]
		Narduroides					*Narduroides* [a]
		Psilurus			*Psilurus*		*Psilurus* [a]
Wangenheimia, *Castellia*		*Wangenheimia*			*Wangenheimia* *Castellia*		*Wangenheimia* [a] *Castellia*
	Lolium		*Lolium*	*Lolium*	*Lolium*	*Lolium*	*Lolium* [b]
					Micropyropsis		*Micropyropsis* [b]

Abbreviations: [a]*Fine-leaved* Festuca; [b]*Broad-leaved* Festuca.

E.B. Alexeev) are circumscribed to the fine-leaved fescues [4, 5, 6, 7, 8, 9] (Table 1). *Festuca* subgenus *Drymanthele* has been considered to be of relict origin, reflected by their putatively primitive leaf-blade and panicle features [60, 70, 165]. It encompasses the tallest extant *Festuca* species, which bear remarkably wide leaf-blades, large culms and panicles, and live in forested habitats of the Holarctic region. By contrast, the remaining broad-leaved subgenera show more restricted geographical distribution patterns and mostly grow in more mesic ecosystems. The fine-leaved fescues have been traditionally interpreted as more recently evolved taxa [60, 70, 165] because of their slender habit and successful adaptation to a range of humid to xeric habitats. *Festuca* subgenus *Festuca* encompasses the broad *Festuca* sections *Festuca* (*F. ovina* group) and *Aulaxyper* Dumort. (*F. rubra* group), together accounting for the highest number of described species in the genus. These are widely distributed across the Holarctic region and in the highest altitudes and latitudes of the southern hemisphere. These large sections share several morpho-anatomical traits, including grains adnate to the paleas, short truncate ligules, and conduplicate innovation leaves [60, 162].

Lolium L. is separated from *Festuca* in most current floras, based on distinctive inflorescence traits, particularly spikelets sunk in the excavated rachis of the spike, each covered by a single glume, and a lemma with more than five veins [159]. *Lolium* encompasses ca. 10—12 species that are mostly native to the pan-Mediterranean region from the Middle East to Macaronesia. *Lolium* taxa show a trend from short perennials/biennials to annuals, and from allogamous to autogamous plants, though no infrageneric ranks have been recognized for them [159]. The genus includes species of high economic importance, like the perennial and Italian ryegrasses (*Lolium perenne* L. and *L. multiflorum* Lam.), extensively used for grazing, turf and amenity purposes, and the cereal weed *Lolium rigidum* Gaudin. The affinity between representatives of *Lolium* and *Festuca* subgenus *Schedonorus* has been demonstrated by spontaneous and artificial intergeneric crosses [80, 97], similarity in chromosome banding patterns [160], and a series of molecular studies on seed-proteins, plastid restriction sites, nuclear RFLP (restriction fragment length polymorphism) and RAPD (random amplified polymorphic DNA) markers, and nuclear and plastid DNA sequence analysis [2, 25, 28, 35, 55, 76, 96, 146]. *Lolium* also shares morphological traits with *Festuca* subgenus *Schedonorus*, such as the possession of falcate auricles. This character is also present in *Micropyropsis tuberosa* Romero-Zarco & Cabezudo, a monotypic, short lived perennial endemic to the western Mediterranean region [122], and in *Castellia tuberculosa* (Moris) Bor, a monotypic annual that occurs in the pan-Mediterranean–Asian area [39, 169].

Vulpia was erected as a new genus [85] and has been separated from *Festuca* since based on their annual habit, very unequal glumes, and long-awned lemma though none of these characters is absolute [43]. Up to five different sections have been recognized within *Vulpia* [*Apalochloa* (Dumort.) Stace, *Loretia* (Duval-Jouve) Boiss., *Monachne* Dumort., *Spirachne* (Hack.) Boiss., and *Vulpia*] (Table 1), based on breeding system, spikelet structure, and several floral traits [43, 141, 143]. *Vulpia* taxa show a trend towards reduction in size and increasing self-fertility through sections *Loretia*, *Monachne/Spirachne* and *Vulpia* [43]. *Vulpia* accounts for ca. 22 species native to the Mediterranean region and America; the chromosome number is fixed within each *Vulpia* species and shows a geographical distribution pattern [42]. Representatives of three sections of *Vulpia* (*Vulpia*, *Monachne* and *Loretia*) hybridize with representatives of *Festuca* section *Aulaxyper* (*Festuca rubra* complex) in the wild or in artificial crosses [3, 15] and show affinities in chromosome pairing [11].

Another 11 annual genera (*Castellia* Tin., *Catapodium* Link, *Ctenopsis* De Not., *Cutandia* Willk., *Desmazeria* Dumort., *Loliolum* Krecz. & Bobr., *Micropyrum* Link, *Narduroides* Rouy, *Sclerochloa* P. Beauv., *Vulpiella* (Batt. & Trab.) Burollet, and *Wangenheimia* Moench) mostly native to the Mediterranean region were grouped with *Vulpia* in the *Vulpia-Desmazeria* complex [143]. These genera were ranked as being allied to *Vulpia*, *Desmazeria*, or as an intermediate among them [143]. These genera plus *Psilurus*, a further monotypic isolated Mediterranean genus, are differentiated from one another by their leaf-blade characteristics, inflorescence type, floral and hilum length, and were considered to merit generic status [39, 141, 142] (Table 1). *Festuca* and its satellite genera constitute one of the three main Poeae lines. Clayton and Renvoize [39] suggested that Mediterranean annuals evolved from mountain-grassland perennials. Classical circumscriptions of the festucoids placed from nine to 15 genera in subtribe Festucinae C. Presl [39, 166]; nomenclatural priority favours Loliinae Dumort. over Festucinae [49, 137] as the correct subtribe name for the festucoids.

Festucoids have been subjected to repeated taxonomic splitting and lumping. The recent advent of molecular phylogenetics has affected traditional classifications in this group and that of close subtribes and genera in the tribe Poeae R. Br. [27, 45, 72, 137]. However, unexpected evolutionary relationships have been revealed and previous hypotheses have been reassessed from our molecular-based cladistic and Bayesian analyses [28, 29, 77, 162, 164]. Molecular phylogenetics have also provided a solid evolutionary framework to test evolutionary hypotheses related to speciation processes involving genome rearrangements, differential mutation rates, hybridization, and polyploidization [28, 30, 164]. In addition, molecular phylogenies allow speculation on life history traits connected with extinctions, migration and radiation events reflected in the biogeographical and phylogeographical patterns [14, 52, 76, 77]. A series of studies conducted at both macroevolutionary and microevolutionary scales have devised new scenarios for understanding the origin and divergence of the festucoid lineages and allies [30, 31, 77].

CLASSICAL AND MODERN CIRCUMSCRIPTIONS OF THE FESTUCOIDS

As with many studies of angiosperms [67], the phylogeny of subtribe Loliinae revealed by nuclear and plastid data sets does not totally agree with traditional classifications based on morphological traits [4, 5, 6, 7, 8, 9, 60, 166]. However, some of the resolved lineages presented in the successive phylogenetic studies of *Festuca* and allies [28, 29, 76, 77, 162, 164] are more or less concordant with previous taxonomic circumscriptions for these groups (Table 1; Fig. 1). Nonetheless, systematic classifications remain unclear because of the large paraphyly found within *Festuca*, the instances of incongruence detected between molecular and morphological data, and the uncertain ascription of several newly described species of *Festuca* from the American, Asian and Austro-New Zealand continents [36, 41, 147, 148, 149, 150, 151].

Recent proposals suggest different scenarios for the classification of the festucoids [29] ranging from a synthetic monophyly criterion that would favour the treatment of all Loliinae taxa as members of a large and morphologically diverse genus *Festuca*, to many smaller generic and infrageneric splits. One of these includes an evolutionary systematics-based criterion that is nomenclaturally conservative and that recognizes both monophyletic and paraphyletic groups (i.e. *Lolium*, *Schedonorus*) as

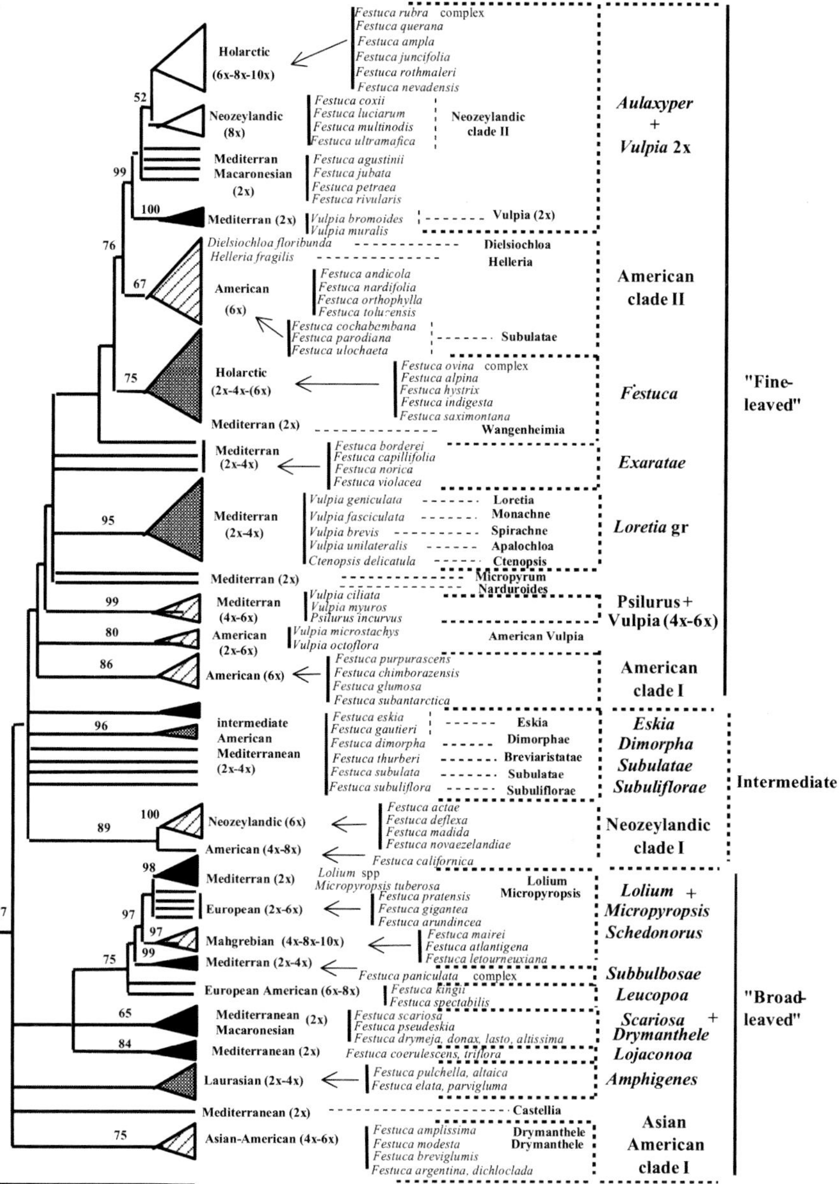

Fig. 1. *Summarized cladistic tree of the festucoids (subtribe Loliinae) based on combined analysis of ribosomal ITS and plastid trnTF sequences (taken from Catalán et al. [28], Inda et al. [77], and unpublished data). Numbers above branches correspond to bootstrap values. Geographic distribution and ploidy levels are indicated for the main lineages (ploidy increases from black (2x) to white (8x-10x) clades).*

independent taxonomic entities [29]. This criterion was applied in recent floras by those authors who recognize some of the broad-leaved lineages of Loliinae as unique genera outside of *Festuca* [140].

Plastid RFLP data [44] and ribosomal ITS sequences [35, 55, 162] first demonstrated that *Festuca* was paraphyletic and that *Lolium* and *Vulpia* were nested within *Festuca*. The closeness of *Lolium* to the broad-leaved *Festuca* of subgenus *Schedonorus* and that of *Vulpia* to the fine-leaved *Festuca* confirmed previous findings based on chromosome analyses and on artificial crosses [3, 15, 80, 100]. After combined analysis of chloroplast restriction sites and structural characters, *Catapodium*, *Cutandia* and *Desmazeria* were classified as belonging to a separate lineage, subtribe Parapholinae [137].

Further investigations on the relationship of *Festuca* and related genera based on independent and simultaneous analyses of nuclear and plastid DNA sequences have extended the range of paraphyly of both the broad and fine-leaved fescues [28, 29, 77, 164] (Fig. 1). The most exhaustive molecular studies have found a likely evolutionary trend from more ancestral broad-leaved *Festuca* lineages towards more recently derived fine-leaved ones [28, 77]. Polyphyletic *Vulpia* and other Mediterranean genera of ephemerals (*Ctenopsis*, *Micropyrum*, *Narduroides*, *Psilurus* and *Wangenheimia*), which are nested within the fine-leaved *Festuca* clade, American perennials *Helleria* and *Dielsiochloa*, a genus previously attributed to Aveneae [39], also fall within the fine-leaved group, whereas *Lolium* and *Micropyropsis* are included within the broad-leaved clade (Fig. 1). Previous results indicate that the sister clade Dactylidinae Stapf, which includes the orchard grass genus *Dactylis* L. plus *Lamarckia* Moench, and the Cynosurinae Fr./Parapholinae Caro (*Catapodium*, *Cutandia*, *Cynosurus* L., *Desmazeria*, *Hainardia* Greuter, *Parapholis* C. E. Hubbard and *Sphenopus* Trin.), are the closest relatives of Loliinae [28, 77]. Affinity of *Cutandia* to the Parapholiinae has been recently confirmed through further genome sequence analysis [77], solving previous misattributions of this genus to the *Vulpia* sect. *Loretia* 'assemblage' [28, 164].

A larger taxon sampling helped to illustrate the evolutionary history of the main *Festuca* lineages and relatives [29, 77, 111]. These studies concur with the previous hypotheses on the evolutionary trend from more ancestral broad-leaved *Festuca* lineages (subgenus *Drymanthele* + section *Scariosa* Hack.+ section *Pseudoscariosa* Krivot., section *Lojaconoa* Catalán & Müller) towards less ancestral broad-leaved groups (section

Subbulbosae Nyman, subgenus *Leucopoa*, subgenus *Schedonorus* + *Micropyropsis* + *Lolium*), and then to the more recently evolved crown group of fine-leaved festucoids (Fig. 1). However, the range of morphologically broad to intermediate size leaved lineages has been extended considerably within this last group, with several *Festuca* lineages (subgenus *Subulatae* p.p., subgenus *Subuliflorae*, section *Dimorphae* Müller & Catalán, plus narrow-leaved section *Eskia* Willk.) aligned basally to the fine-leaved clade. The successive divergences of lineages of the *Psilurus* + polyploid *Vulpia* section *Vulpia* group, the annual '*Loretia* assemblage' (*Vulpia* section *Loretia* + *Monachne* + *Spirachne* + *Apalochloa* plus *Ctenopsis*), and *Festuca* subsection *Exaratae* ultimately led to the most recently evolved *Festuca* section *Aulaxyper* + diploid *Vulpia* section *Vulpia* and *Festuca* section *Festuca* + *Wangenheimia* core clades (Fig. 1).

Geographic structure has been detected within the *Festuca* subgenus *Schedonorus* group [28, 76]. This group is separated into two subgroups: (1) a 'European' subclade that encompasses diploid *Festuca fontqueri* St.-Yves and *F. pratensis*, and hexaploid *F. arundinacea* and *F. gigantea* (L.) Vill., with *Micropyropsis* and diploid *Lolium* derived from within it, and (2) a 'Maghrebian' subclade, which includes tetraploid western Mediterranean *F. fenas* Lag. and *F. mairei* Hack. ex Hand.-Mazz. and their high polyploid North African derivatives (*F. arundinacea* var. *atlantigena* (St.-Yves) Auquier and *F. arundinacea* var. *letourneuxiana* Torrecilla & Catalán).

More exhaustive phylogenetic surveys using both nuclear and plastid genomes indicate a biogeographical pattern resulting in New Zealand and American clades of various taxonomic ranks that are distinctly resolved at intermediate placement between the broad and fine-leaved lineages (Neozeylandic clade I) and within the fine-leaved group (American clades I and II) (Fig. 1), paralleling the cases found in the large genus *Poa* L. (Poinae, Poeae) of bluegrasses [136]. The *Aulaxyper* clade is enlarged by several putatively relict, diploid Macaronesian elements (*Festuca agustinii* Linding., *F. jubata* Lowe, *F. petraea* Guthnick) that align basally with another Mediterranean diploid congener (*F. rivularis* Boiss.), and by another polyploid Neozeylandic (clade II) close to the Holarctic high polyploid clade (Fig. 1). Also an American *Helleria* + *Dielsiochloa* + *Festuca* subgenus *Subulatae* p.p. clade shows a close relationship to the red fescues [77] (Fig. 1).

The most updated studies indicate that *Lolium* and *Vulpia* have had different life histories and origins [76, 77]. The diploid and morphologically homogeneous ryegrasses probably diverged recently as they are resolved as monophyletic and derived from a European *Festuca* subgenus *Schenodorus* lineage. By contrast, the more variable and geographically widespread diploid to polyploid *Vulpia* species show a larger extended polyphyly and incompletely resolved relationships within one another, and to other fine-leaved *Festuca* lineages. Perhaps this is a result of a more ancestral and complex radiation pattern with several polytopic origins [77] (Fig. 1).

Other geographically isolated genera (*Austrofestuca* E. B. Alexeev, *Parafestuca* E. B. Alexeev), formerly classified within *Festuca* and more recently separated from it, have been found to be unrelated to *Festuca* and to the festucoids [39]. Molecular phylogenetic reconstructions have resolved Australian and New Zealand *Austrofestuca* as members of a core Poinae clade [56, 75], whereas the Macaronesian *Parafestuca* falls within a core Aveneae clade [116].

Current investigations on evolutionary relationships of Loliinae, within a broad supratribal Aveneae-Poeae framework, have shaped the monophyly of this lineage and its relative isolation with respect to the pooid and avenoid clades [31]. Recent studies have detected potential horizontal gene transfer of mitochondrial *Festuca* sequences into the less related genomes of *Secale* (Triticeae) and *Danthonia* (Danthonieae) (H. C. Ong, S. Chang, and J. D. Palmer, personal communication).

CONFLICTS AND CONCERTS BETWEEN MOLECULES AND MORPHOLOGY IN THE LOLIINAE

Conflicts between the Loliinae topologies recovered from different genomes have been interpreted as a consequence of interacting phenomena, such as lineage sorting or reticulation [28]. The existence of past hybridization events might obscure phylogenetic reconstruction within the temperate grasses [45, 91, 105, 137]. Chloroplast capture, allopolyploidy and lineage sorting are hypotheses invoked to explain the failure to reconcile topologies recovered from different genomes in Triticeae [91, 105] and these have been advocated to interpret the unexpected placements of some *Vulpia* and other ephemeral taxa within the fine-leaved clade and of *Festuca* subgenus *Leucopoa*, subgenus *Subulatae*, and section *Amphigenes* s.l. across the broad and fine-leaved clades of the festucoids [28, 77]. Paralogy might be another disturbing factor that affects phylogenetic reconstruction within a single data set, since pseudogene copies of ITS sequences have been detected in *Lolium* [55] and can be extended to other festucoid lineages [164].

Structural characters are believed to have arisen through different gene regulatory mechanisms or developmental pathways in grasses [84, 137] and, therefore, would be expected to be congruent with molecular phylogenies. However, discrepancies in phylogenetic reconstruction between molecular and structural evidence are frequent within Poaceae [59]. Incongruence between molecular-based and morphology-based approaches has been related to the innate plasticity of morphological features and their consequent higher homoplasy that makes them inappropriate to recover deep phylogenetic signals in macroevolutionary groups, such as the subfamily Pooideae [90]. A careful examination of a large set of morphological traits within Loliinae and close subtribes (primary synapomorphies, sensu de Pinna [47]) led to the discovery of several secondary synapomorphic characters that support more inclusive groups [29] (Fig. 2). These traits correspond to changes in qualitative characters that mark preferentially the deeper nodes of the tree and that could be associated with heterotopic changes, resulting from shifts in the expression of developmental pathways in different plant organs, as indicated by Kellogg [84].

Incongruences between molecules and morphology mostly affect the resolution of the broad-leaved *Festuca* lineages, some fine-leaved *Festuca*, and the *Vulpia* lineages [29]. Among them, the Holarctic subgenus *Drymanthele*, the Asian-North American subgenera *Leucopoa* and *Subulatae*, the European section *Amphigenes*, and the Western Mediterranean section *Subbulbosae* are resolved as polyphyletic taxa. The Eurasian and Mediterranean subgenus *Schedonorus* appears to be paraphyletic with *Lolium* and *Micropyropsis* included within it (Fig. 1). *Festuca* subgenera *Drymanthele*, *Leucopoa*, and *Subulatae*, as well as section *Amphigenes* share several foliar characters related to the 'broad-leaved' syndrome. *Leucopoa* (section *Leucopoa*) differs from the others in being dioecious. *Festuca* sections *Subbulbosae*, *Lojaconoa*, *Scariosae*, and *Pseudoscariosa* show intermediacy in their characters. Another striking finding is the wide polyphyly shown by species of the 'transclade' subgenera *Leucopoa* and *Subulatae*, with taxa exhibiting broad-leaved traits but nested both at the base of the fine-leaved clade and close to the *F. rubra* group clade (Fig. 1). The artificiality of various broad-leaved hierarchic ranks (i.e. *Festuca* subgenera *Subulatae* and *Drymanthele*) was already manifested in previous studies [28, 36, 44].

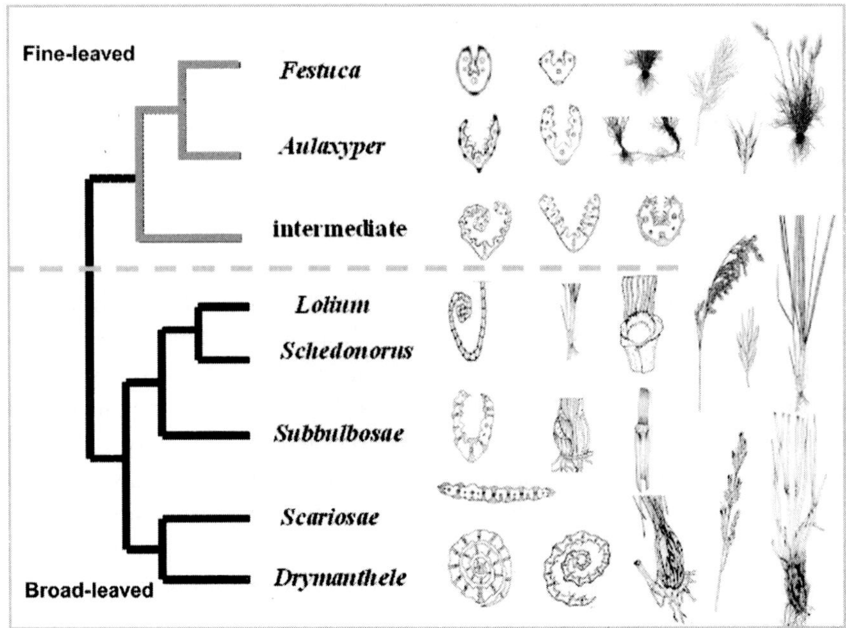

Fig. 2. *Evolution of morphological characters across several representatives of broad-leaved (subgenus Drymanthele, section Scariosae, subgenus Schedonorus + Lolium), intermediate (section Eskia, section Dimorphae, subgenus Leucopoa p.p.), and fine-leaved (sections Aulaxyper and Festuca) Loliinae lineages. Branch darkness indicates character-states associated with the 'broad-leaved syndrome' (flat leaves, sclerenchyma trabeculate, extravaginal shoots, cataphylls present, convolute to supervolute vernation), branch lightness indicates those associated with the 'fine-leaved syndrome' (thin leaves, sclerenchyma non-trabeculate, intravaginal (and mixed) innovation shoots, cataphylls mostly absent, plicate vernation).*

The most recently derived *Festuca* sections *Festuca* and *Aulaxyper* are resolved as monophyletic and with moderate support for groups that include the type species and closest relatives (Fig. 1). The most unexpected resolution is that obtained for *Vulpia*, which shows a high polytomy for members of typical section *Vulpia*, with species taxonomically similar to each other but with chromosome and geographical races distinctly related to the red-fescues (Mediterranean diploids), to the American subclade of fine-leaved fescues [American *V. octoflora* (Walter) Rydb. and *V. microstachys* (Nutt.) Munro], or in an unresolved but supported Mediterranean polyploid clade with *Psilurus* (tetraploids, hexaploids). A further resolved but differently supported lineage is that of representatives of *Vulpia* sections *Loretia*, *Monachne*, *Spirachne*, *Apalochloa* plus *F. plicata* Hack.; which also incorporates *Ctenopsis* (Fig. 1). Relationships of other annual genera, i.e. *Micropyrum*, *Wangenheimia*, *Narduroides* are not satisfactorily resolved as they show distinct and poorly supported relationships in nuclear vs. plastid trees [28, 77].

Simultaneous cladistic analysis of morphological and molecular data within Loliinae and close subtribes was conducted to evaluate the phylogenetic signal of selected morpho-anatomical traits [37, 43, 60, 90, 94, 102, 103, 125, 141, 143, 159, 169, 171]. This analysis resulted in an overall lack of resolution for the Loliinae and its allies [29]. However, better resolution was obtained when the analysis was restricted to Loliinae representatives and secondary synapomorphies were detected for several lineages. The presence of a long linear hilum was found to be synapomorphic for the festucoids (except for a reversal in *Wangenheimia*). The thickened base of the leaf sheath is shared by *Festuca* sections. *Subbulbosae* and *Lojaconoa*. The trend towards reduced spike-type inflorescences, an excavated rachis, and a single glume has evolved at least three times in the *Lolium*, *Psilurus*, and *Hainardia* lineages. The fine-leaved *Festuca* subgenera *Festuca* and *Helleria* share a perennial habit, conduplicate vernation, and an awned lemma. The basal *Festuca* sections *Eskia* and subgenus *Helleria* have common traits, such as a hairy ovary tip, and some of the intermediate *Festuca* subsections *Festuca* and *Exaratae* representatives have leaf blades that possess a continuous abaxial ring of sclerenchyma. The more recently diverged *Festuca* section *Aulaxyper* s. l. group is characterized by an admixture of intravaginal and extravaginal innovation shoots and reduced cataphylls. The *Festuca* subgenus *Schedonorus* + *Micropyropsis* + *Lolium* clade is characterized by falcate auricles and awned lemmas –unique features within the broad-leaved group– whereas representatives of *Festuca* section *Amphigenes* and *Festuca* subgenera *Drymanthele*, *Leucopoa*, and *Subulatae* lack falcate auricles and awned lemmas, but, like *Festuca* subgenus *Schedonorus* + *Micropyropsis* + *Lolium*, present innovation leaves with complete abaxial and adaxial sclerenchyma trabecules. The basalmost groups also bear prominent cataphylls, convolute to supervolute vernation, and creeping extravaginal rhizomes [29] (Fig. 2). These results partly agree with the hypothesis brought forward by Hackel (1882), Holub (1984), and Tzvelez (1982) on the general evolutionary trend from more primitive robust broad-leaved *Festuca* towards more advanced slender fine-leaved

Festuca [60, 70, 166]. Interpretation of the character state changes along the broad-leaved *Festuca* clade indicates that the 'broad-leaved' foliar syndrome could be plesiomorphic and could also have evolved several times towards the opposite trend along the clade. The lack of resolution for these, however, and the presence of old lineages of intermediate taxa in most subclades, raised doubts about their ancestry [28].

Concert between molecules and morphology has been evidenced more convincingly at microevolutionary scale in *Festuca* and *Lolium* [13, 14, 51, 163]. Quantitative morphological changes might be postulated to characterize the shallow nodes of the trees in Poeae, as indicated by Kellogg [84]. Highly variable molecular markers and quantitative morphological traits have been useful tools in characterizing the most recently derived lineages of Loliinae. Different allozymic combinations have been found for strains of *Lolium perenne* and *L. rigidum* across Europe [13], and multilocus RAPD phenotypes have detected taxonomic structure among the four polyploid microspecies of the *Festuca brachyphylla* complex in the arctic island of Svalbard [51]. Combined multivariate analysis of RAPD phenotypes and of quantitative morphological traits showed *F. baffinensis* Polunin, *F. brachyphylla* Schult. & Schult. f., *F. hyperborea* Holmen & Fredericksen, and *F. edlundiae* S. Aiken, Consul & Lefkov. to be different genetic entities [51]. The potential value of the highly polymorphic markers (RAPD) fails, however, when recovering genetic affinities of more removed Loliinae groups [35, 146]. Similarly, quantitative morphological traits are useless in phylogenetic reconstructions due to the inherent problems of homology presented by any proposed system of coding of continuous characters [117]. However, phenetic analyses of morphometric characters have provided a baseline to detect taxonomic structure in the Mediterranean species of *Festuca* section *Eskia* [163], and, combined with RAPD data, to clarify the hybrid origin of wild sterile *Festuca* taxa (*F.* x *picoeuropeana* Nava, *F.* x *souliei* St.-Yves) from their respective putative parents (*F. eskia* x *F. gautieri* and *F. eskia* x *F. quadriflora* Honck.) in hybrid zones of the northern Iberian and Pyrenean mountain ranges [108, 157].

SPECIATION AND EVOLUTIONARY TRENDS IN LOLIINAE: C-VALUES, KARYOTYPE EVOLUTION, HYBRIDIZATION AND POLYPLOIDY

Cytogenetic and hybridization studies within and among *Festuca*, *Lolium* and *Vulpia* provided the key for the analysis of speciation processes and evolutionary trends within subtribe Loliinae [3, 11, 12, 20, 46, 48, 63, 73, 78, 79, 80]. These results have been be re-interpreted in the light of our present phylogenetic knowledge.

Festuca and its close allies show uniformity in the chromosome base number x = 7, which is otherwise characteristic of Poeae and other tribes of the Pooideae core clade [69, 138]. The lack of aneuploid series on the number of chromosomes per set has been interpreted as evidence of a relatively recent origin of Poeae [27, 72, 138]. However, the large number of ploidy levels reported in *Festuca* and some relatives, i.e. *Vulpia*, was interpreted as a consequence of having an old evolutionary history among different lineages [46, 164]. Ploidy levels range from diploids to duodecaploids in *Festuca* [28, 48, 162] though diploids and polyploids are differentially distributed across the main subgenera and sections. Some of the purportedly oldest broad-leaved *Festuca* lineages consist exclusively (*Drymanthele*, *Lojaconoa*, *Scariosae* and *Pseudoscariosa*) or predominantly (*Subbulbosae*) of diploids. Within the fine-leaved *Festuca*, the basal lineages (*Eskia*, *Dimorpha*) also show higher percentages of diploid taxa. Polyploidy is extensive in *Festuca*, up to 70% [48, 100]. Most polyploids belong to the largest and more recently evolved fine-leaved taxa of *Festuca* subgenus *Festuca*; however, some intermediate and broad-leaved lineages also incorporate diploids and polyploids (*Schedonorus*) or are formed exclusively by distinct polyploid ranges (*Leucopoa*, *Subulatae*) [28].

The evolution of polyploidy has clearly emerged parallel to the main lineages of the broad and fine-leaved *Festuca*, though the mechanisms experienced by each lineage probably occurred at different times. Within the fine-leaved fescues, the two largest groups (*F. ovina* and *F. rubra*) show opposite patterns of ploidy. In the *F. ovina* group, most taxa are diploids or low polyploids (tetraploids), in the red fescues (*F. rubra*), diploids and tetraploids are rare compared to the more abundant hexaploids and higher ploidy level taxa [3]. High ploidy levels also dominate some of the broad-leaved lineages of *Festuca* subgenus *Schedonorus*, i.e. *Festuca arundinacea* complex, and *Leucopoa* [*F. kingii* (S. Watson) Cassidy].

The closest relatives of *Festuca* show differences in chromosome numbers. Within the fine-leaved lineage, most of the ephemeral genera are diploids (*Ctenopsis*, *Micropyrum*, *Narduroides*, *Wangenheimia*), *Psilurus* is tetraploid and *Vulpia* exhibits three ploidy levels (from diploid to hexaploid) [42, 143]. The relict distribution of these endemic

Mediterranean genera and their low ploidy levels have been interpreted as evidence of ancient origin [164]. The variable ploidy-levels in the geographically widespread species of *Vulpia* suggest a longer evolutionary history. By contrast, within the broad-leaved lineage, diploid *Lolium* species could be considered as diverging recently from a *Schedonorus* lineage that has maintained its diploid level up to the present day [28, 100].

Karyotype evolution in subtribe Loliinae is also concordant with hypotheses recovered from our molecular data (Fig. 3). *Festuca* shows striking differences in genome sizes between the broad and fine-leaved lineages [20, 131, 132]. Among diploid *Festuca* species and their closest allies (*Lolium*, *Vulpia*), there is considerable variation in chromosome size and nuclear DNA quantity [20, 131, 132], showing a trend from the large chromosomes and high C-values of the more ancestral broad-leaved *Drymanthele* and *Scariosae* taxa, through intermediate chromosomes and medium C-values of the more advanced broad-leaved *Schedonorus* and *Lolium* taxa, to the small chromosomes and low C-values of the more recent fine-leaved *Festuca*, *Aulaxyper* and *Vulpia* taxa [20] (Fig. 3). Karyotypic C-banding patterns are highly correlated with this variation, though surprisingly, heterochromatin content is inversely correlated with genome size in *Festuca* and its allies [11, 46, 160]. Cytogenetic analyses demonstrated a 2.5-fold range decrease in chromosome size coupled with a 7.5-range increase in absolute heterochromatin content from primitive broad-leaved *Festuca* lineages (*Drymanthele*, *Scariosae*, *Subbulbosae*) to the more advanced fine-leaved *Festuca* lineages (*Schedonorus*, *Eskia*, *Festuca*, *Aulaxyper*) [11, 46].

A full range of C-banding patterns have been found across the Loliinae lineages [11, 46, 160] (Fig. 3), paralleling other pooid groups such as Triticeae [133] and *Avenea* [121]. Karyotypes of ancestral broad-leaved *Scariosae* and *Subbulbosae* lineages are characterized by near metacentric chromosomes with very little heterochromatin distributed in a few bands (Fig. 3), resembling the *Triticum* D-genome [46]. The ancestral *Drymanthele* lineage shows larger and more unequal chromosomes with a slight increase in heterochromatin resulting in the addition of small centromeric and a few intercalary bands [46] (Fig. 3). The more advanced *Schedonorus* and *Lolium* lineages have smaller submetacentric chromosomes and intermediate levels of heterochromatin distributed among centromeric and intercalary bands [46, 160] (Fig. 3),

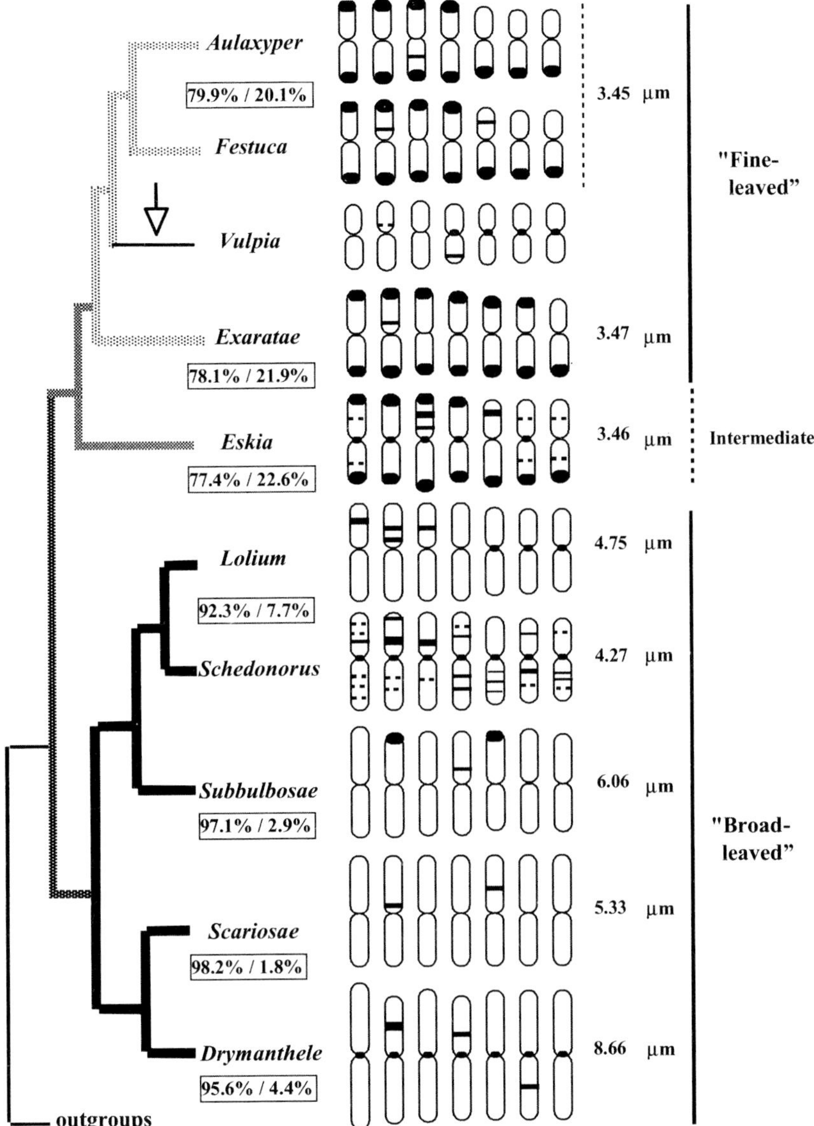

Fig. 3. *Analysis of karyotype evolution in Loliinae. Variation in chromosome C-banding patterns, mean chromosome length (µm) and percentages of total euchromatin/total heterochromatin per karyotype is mapped for the main broad-leaved (Festuca subgenus Drymanthele (F. drymeja), section Scariosae (F. scariosa), section Subbulbosae (F. paniculata), subgenus Schedonorus (F. pratensis), plus Lolium (Lolium spp.), intermediate (Festuca section Eskia (F. pumila), and fine-leaved (Festuca subsection Exaratae (F. norica), section Festuca (F. ovina), section Aulaxyper (F. rubra), plus Vulpia (V. fasciculata) Loliinae lineages. The arrow indicates a potential reversal to massive telomeric heterochromatin loss in Vulpia. Karyotype data was retrieved from Bailey & Stace [11], Dawe [46], and Thomas [160].*

similar to those of the *Triticum* A-genome [133]. The most recently evolved fine-leaved *Festuca* and *Aulaxyper* lineages have the smallest submetacentric chromosomes and the highest contents in heterochromatin that is primarily located in the telomeric bands [11, 46] (Fig. 3), as those of the *Secale*-genome [46]. The mapping of structural karyotype characters onto the molecular topologies is in agreement with a scenario that predicts a strong reduction in chromosome size and a major increase in heterochromatin has occurred during the evolutionary history from more ancestral broad-leaved lineages towards the more recently evolved fine-leaved lineages of Loliinae (Figs. 1 and 3). A more detailed evaluation of karyotype evolution shows an intermediate C-banding pattern between those of the *Schedonorus* and *Lolium* taxa and those of the core fine-leaved *Festuca* and *Aulaxyper* taxa, which is present in the basal *Eskia* lineages of the fine-leaved clade, consisting of a mixture of intercalary and telomeric bands [46] (Fig. 3). The karyotype of the more advanced *Exaratae* taxa is similar to those of the more recently evolved *Aulaxyper* and *Festuca* species [46] (Fig. 3).

In contrast, the *Vulpia* karyotype shows small metacentric chromosomes with few heterochromatic bands that are mostly centromeric [11] (Fig. 3). This is in agreement with the low genome size reported for these plants [20]; however, it does not show any affinities with the karyotypic profiles of its closest sister, *Festuca* section *Aulaxyper* (Fig. 1). The most parsimonious interpretation for the remarkable distinctiveness of the *Vulpia* karyotype is that the strong reduction observed in its genome size can be correlated with a massive loss of telomeric heterochromatin coupled with a slight gain of centromeric bands. Perhaps this occurred during the acquisition of the annual habitat from a perennial chromatin-rich *Aulaxyper* ancestor.

Loliinae chromosome banding patterns appear to be lineage-specific to species-specific (Fig. 3), meeting the criteria for a discriminating systematic and evolutionary tool [46, 121]. Karyotype profiles have been proposed to reflect the breeding affinities of the intercrossable groups within the subtribe [46]. This is certainly accomplished in the intergeneric x *Festulolium* crosses where karyotypic patterns are similar in both *Schedonorus* and *Lolium* parents [46, 160] that share a common ancestry (Fig. 1). The *Drymanthele* + *Scariosae* lineages have had successful interfertile artificial crosses between the karyotype of *Drymanthele* vicariants, *Festuca drymeja* Mert. & W.D.J. Koch, and *F. donax* Lowe [22, 46], these being resolved as sister groups [77]. In contrast, there are the less successful and sterile crosses between *F. drymeja*, *F. lasto* Boiss., and *F. altissima* All., which are more karyotypically diverse [22, 23, 46] and also more distant to one another on the phylogenetic trees [28, 77]. The genetic similarities of *Festuca scariosa* to species of *Drymanthele* as tested through artificial crosses [22, 23], is seen in their affinities in banding patterns [46], which in turn is supported by close phylogenetic placement [28, 77]. Despite these apparent coincidences, karyotype patterns are not always correlated with breedings affinities in Loliinae. Within the fine-leaved lineage, intergeneric x *Festulpia hubbardii* Stace & Cotton crosses are produced spontaneously though the *Vulpia* and *Aulaxyper* parents show strongly different C-banding profiles [11]. However, the *Vulpia* lineages are closely related to the red fescues [28, 164] (Fig. 1). Karyotype dissimilarities between *Aulaxyper* and *Vulpia* could be explained as a result of extensive heterochromatin loss in the annual *Vulpia* lineages, which probably has no effect on their ability to occasionally intercross [12].

Concordance between molecular phylogenetics and speciation processes along the festucoid lineages are also supported by cytogenetic data. Deeper insights into the genomic relatedness of several Loliinae lineages were obtained from FISH (fluorescence in situ hybridization) and GISH (genomic in situ hybridization) techniques [12, 63, 73]. The mapping of ribosomal DNA sites depicted decreasing degrees of relatedness within the *Drymanthele* lineage, from similar mapped *Festuca drymeja* and *F. donax*, to less similar *F. lasto*, and to the more distant mapping patterns of *F. altissima* [63]. This is in total agreement with the resolution of our molecular topology [77]. However, the speculations brought forward to favour either *Festuca scariosa* or *F. altissima* as the potential sources for the unknown *F. arundinacea* genome, and *F. scariosa* as the potential genome donor of *F. mairei* [63], first postulated from morphological and breeding affinities [22, 23], do not agree with the divergent phylogenetic placement found for these groups [28, 77] (Fig. 1). More exhaustive GISH procedures have confirmed the participation of the diploid *Festuca pratensis* and tetraploid *F. fenas* in the allohexaploid *F. arundinacea* and have demonstrated, despite initial indications [78], the non-relatedness of *Lolium* to this polyploid [73].

The impact of reticulation and polyploidy on the evolutionary processes of most angiosperms groups has been largely documented [135, 144, 152]. Recurrent introgression and polyploidization are common in many plant families and have been postulated as microevolutionary scenarios for cladogenesis [134]. Grasses account for some of the highest percentages of polyploid taxa of hybrid origin [153]. Hybridization and polyploidy are commonly found in subtribe Loliinae where intergeneric hybrids between *Festuca* and *Lolium* (x *Festulolium*) and between *Festuca* and *Vulpia* (x *Festulpia*) occur spontaneously in the wild [3, 79, 97]. The extent of hybridization events among different *Festuca* groups and their closest allies (Fig. 4) can be used to trace past evolutionary processes obscured by biased dilution of parental traits in newly formed hybrid lineages [16, 80]. Spontaneous crosses between different festucoid lineages indicate the limits of their reproductive barriers [79, 97], whereas artificial crosses are a measure to assess the degree of genomic similarity [3, 17, 78, 80, 81, 82].

Within the broad-leaved clade, hybridization is broadly extended across the *Schedonorus* + *Micropyropsis* + *Lolium* lineage [78, 81, 82] though spontaneous crosses have been detected among closely related taxa in other broad-leaved groups, such as those involved in the origin of the *Subbulbosae* microspecies [95]. Success in artificial crosses was obtained among diploid members of *Festuca* subgenus *Drymanthele* that show wide geographic vicariance (*F. drymeja* and *F. donax*) and between *F. scariosa* (*Scariosae*) and taxa of subgenus *Drymanthele* [21, 22], thus confirming their close phylogenetic relationship [28, 77] (Fig. 1). The high ploidy levels manifested by representatives of *Leucopoa* p.p. (*F. kingii*, *F. spectabilis* Jan.) might also indicate a hybrid polyploid origin.

The most recently evolved *Schedonorus* + *Micropyropsis* + *Lolium* group shows a high number of spontaneous hybrids, especially those derived from intergeneric crosses between 'European' *Schedonorus* and *Lolium*. Natural hybrids between perennial and biennial species of *Lolium* (*L. perenne* and *L. multiflorum*) and three species of 'European' *Schedonorus* (*F. pratensis, F. arundinacea, F. lenas* (*F. gigantea*) occur widely in sympatry in all six specie combinations, though those hybrids are not completely sterile [80, 97]. However, all *Lolium* species can intercross with each other resulting in fertile offsprings [79, 80, 159]; as can the three previously mentioned representatives of 'European' *Schedonorus* [80]. Diploid *F. pratensis* and tetraploid F. *glaucescens* have been recognized as the genome donors of hexaploid *F. arundinacea* within the European clade based on data on the meiotic behaviour of F1 hybrids and derivatives [21, 32, 33, 101]. This was corroborated by RFLP probing [172] and genomic in situ hybridization [73]. Based on morphological similarities and chromosome pairing analyses of artificial

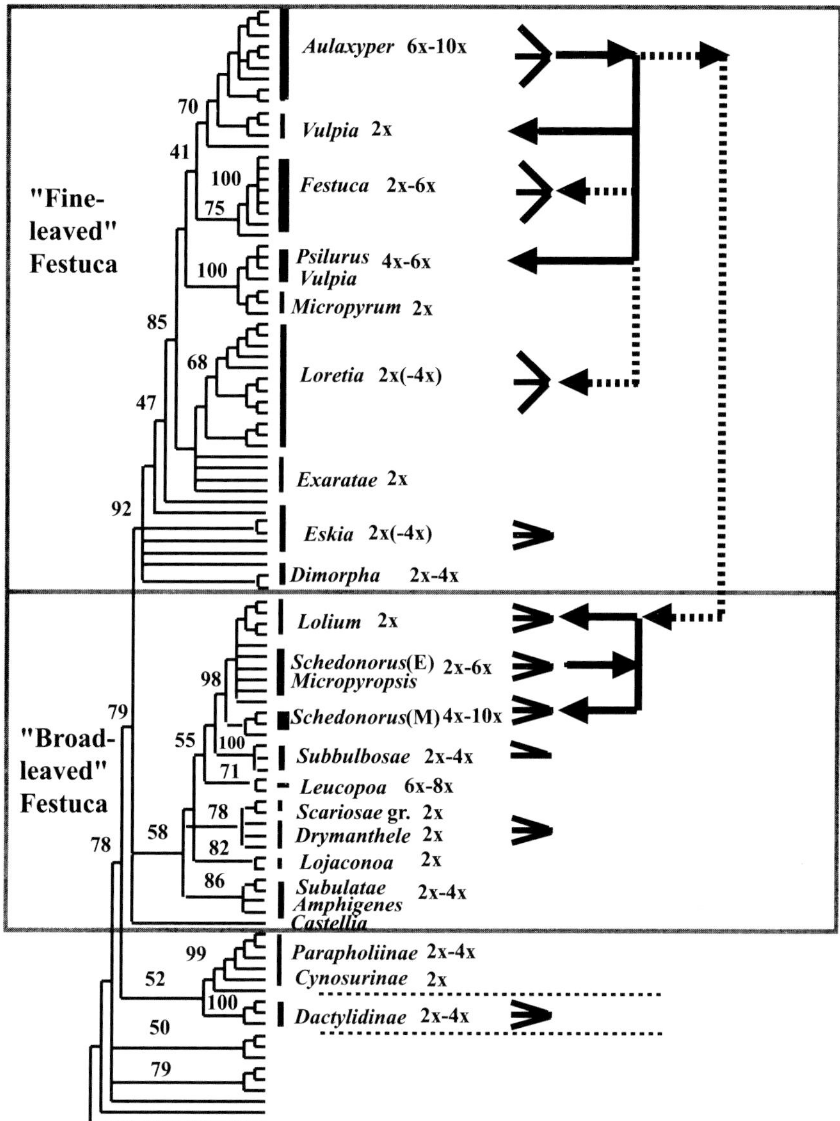

Fig. 4. *Mapping of hybridization events of Loliinae taxa onto the combined nuclear and plastid DNA-most parsimonious tree (modified from Catalán et al. [28]).*

hybrids, tetraploid *F. mairei* and *F. fenas* of the 'Maghrebian' clade have been postulated as putative parents of the highly polyploid taxa, *F. arundinacea var. atlantigena* and *F. arundinacea var. letourneuxiana* [34,

172]. However, the ultimate diploid ancestors of those lineages are either extinct or have not yet been identified. According to the present data and the phylogenetic resolution observed within the *Schedonorus* + *Micropyropsis* + *Lolium* group, it seems likely that a *F. pratensis*-type ancestor was involved in the origin of this diploid lineage and that *Lolium* stemmed out from it. Conversely, recurrent hybridization and amphipolyploidy probably gave rise to the separate polyploid-rich *Schedonorus* complexes of the 'European' and 'Maghrebian' subclades [76].

Extensive reticulation has also been detected within the fine-leaved clade where intergeneric crosses are common between representatives of *Festuca* section *Aulaxyper* and *Vulpia* [3]. Hybridization is relatively widespread within the perennial fine-leaved groups when the parental taxa are close enough, as manifested in the separate diploid to polyploid *Eskia*, *Festuca*, and *Aulaxyper* assemblages, though most of the resulting intra- and intersectional hybrids are usually highly sterile [15, 81, 82, 163, 164] (Fig. 4). By contrast, interspecific crosses are rare within the annual fine-leaved groups and artificial attempts among representatives of the annual *Vulpia* lineages were less successful resulting in sterile offsprings [15]. Higher rates of success were observed, however, in both spontaneous and artificial crosses between polyploid perennial and diploid to polyploid annual lineages, particularly those involving the red fescues. Representatives of the perennial hexaploid and octoploid *F. rubra* complex hybridize spontaneously with annual representatives of *Vulpia* sections *Vulpia* and *Monachne*, and artificially with those of *Vulpia* section *Loretia* in different ploidy combinations [3, 15, 145]. The *F. rubra* taxa could also hybridize with perennial species of *Festuca* section *Festuca* [15, 81]. Though most of these crosses produced sterile hybrids, backcross derivatives to the male *F. rubra* parent were fully fertile [3]. Increased genetic recruitment through recurrent introgression between representatives of *Festuca* section *Aulaxyper* and those of *Vulpia* and *Festuca* section *Festuca* has been interpreted as one of the major evolutionary trends within these recently evolving fine-leaved groups [17].

The crucial role played by the red fescues in the reticulation processes found within the Loliinae extends far beyond the fine-leaved group. Rare, intergeneric crosses between representatives of the *F. rubra* complex and *Lolium* have been documented either as spontaneous or as artificial derivatives [78, 80, 81, 113]. The *Aulaxyper* group is an important assemblage of species and might be one of the major contributors to the speciation process in festucoids. Their ability to repeatedly backcross with diploid or low polyploid taxa from other lineages, acting as pollen donors and resulting in offsprings that mostly resemble these paternal parents, could explain the highly heterogeneous substitution rates observed within this lineage [30] and close phylogenetic placement near some annual (*Vulpia*, *Micropyrum*) and perennial (*Helleria*) lineages [28, 164]. Some of the highly polyploid taxa of this *Aulaxyper* complex are probably products of recent speciation fostered by introgression and polyploidization (Fig. 4).

Extended genomic introgression might explain the close evolutionary relationships observed between *Festuca* section *Aulaxyper* and representatives of *Vulpia* within the fine-leaved clade (Fig. 4). GISH analysis of natural and synthetic F1 x *Festulpia* hybrids demonstrated that telomeric heterochromatic regions of the *Festuca rubra* chromosomes are not present in the *Vulpia* genome, whereas euchromatic regions of the *Vulpia* genome hybridize to some extent to non-heterochromatic regions of the *Festuca* chromosomes [12]. This evidence further supports the proposed hypothesis of a potential loss of fine-leaved *Festuca*-specific repeated telomeric DNA sequences in *Vulpia* during the evolutionary trend towards reduction in genome size and acquisition of an annual life-cycle (Fig. 3). The occurrence of some heterogenetic pairing of *Festuca* and *Vulpia* chromosomes in meiotic bivalents of x *Festulpia hubbardii* plants [11, 12] supports their genetic proximity. However, the resolution of relationships between *Festuca* section *Aulaxyper* and the different *Vupia* lineages (Fig. 1) is not satisfactorily explained, since representatives of several lineages of *Vulpia* sections *Vulpia* and *Monachne,* that spontaneously intercross with representatives of the *F. rubra* group, show different degrees of affiliation [28, 77, 164] (Fig. 1). These apparent incongruences might be caused by different recurrent intergeneric and interploidal introgressions of *Festuca* and *Vulpia* nuclear genomes, coupled with concurrent chloroplast captures that have obscured the derivation of clear phylogenetic hypotheses. Assortative matching and recurrent polyploid formation are common phenomena in several groups of angiosperms [134] and may be the likely hidden mechanisms behind the multiple origins of the diverging *Vulpia* lineages [164]. The similarities among the diploid representatives of *Vulpia* section *Vulpia* to *Festuca* section *Aulaxyper* indicate their common genomic ancestry, whereas the polyploid *Vulpia* lineages might have arisen at different evolutionary times. Although the identity of their potential genome donors has not been established yet, the close genomic affinities of the red fescues with the polyploid *Vulpia* lineages via spontaneous intergeneric crosses and their morphological similarity to cosectional diploid taxa points towards the contribution of the *Aulaxyper* + diploid *Vulpia* lineage in their origin [164].

The secondary origin of polyploidy in *Festuca* and close allies can also be deduced from our phylogenetic trees. For those lineages that include diploid and polyploid taxa, i.e. *Schedonorus* + *Micropyropsis* + *Lolium* clade within the broad-leaved lineage and the *Aulaxyper* lineage within the fine-leaved clade, diploid species always align basally (Fig 1). The nature of polyploidy across *Festuca* has long been debated [78]. The regularity in size, symmetry and banding patterns of individual karyotypes [46] and the high percentages of regular meiotic pairings [113] have been interpreted as evidence of autopolyploidy for some *Festuca* lineages, i.e. *Festuca* sections *Festuca* and *Aulaxyper*. However, in other cases, the slight asymmetry of karyotypic profiles and the ratio between homologous vs. homeologous and heterologous pairing in artificial interspecific hybrids have been postulated as evidence of segmental polyploidy (*Festuca* sections *Festuca* and *Subbulbosae*) [46, 100] or true allopolyploidy (*Festuca* subgenus *Schedonorus*) [32, 33, 101]. This last mechanism was corroborated by the explanation of the origin of hexaploid *Festuca arundinacea* from diploid *F. pratensis* and tetraploid *F. fenas* through different cytogenetic analyses [63, 73, 172]. Based on meiotic behaviour of artificial hybrids, this mechanism has also been proposed for the derivation of the polyploid taxa of the *Schedonorus* lineage [33] and the *Aulaxyper* lineage [78]. Ribosomal DNA mapping has also confirmed that taxa previously considered to be autopolyploids, such as the tetraploid *F. apennina* De Not., are of hybrid origin with one genome from diploid *F. pratensis* and another genome of unknown origin (J.A. Harper, personal communication).

The regulatory mechanism of chromosome pairing in *Festuca* seems to be under genetic control in the allopolyploids [78] and apparently evolved in parallel along the broad and fine-leaved festucoid lineages (Fig. 4). Based on evidence compiled from chromosome pairing of intergeneric *F. arundinacea* x *Lolium multiflorum* Lam. and *F. rubra* x *Lolium perenne* hybrids and derived amphidiploids, Jauhar [78] inferred that the diploid-like meiosis of the hexaploid *F. arundinacea* and *F. rubra* species is regulated by a gene system that represses homeologous pairing in disomic dosage, but is ineffective in hemizygous dosage, similar to the genetic-controls observed in hexaploid wheats and oats. The extent of this mechanism across the festucoids was indirectly confirmed by the heterologous pairing observed in the wild and artificial x *Festulpia hubbardii* hybrids [11, 12]. The analysis of intraspecific *F. arundinacea* crosses from different geographical ecotypes allowed Jauhar [78] to discern the presence of this regulatory system acting in the diallelic homozygous fertile crosses from its absence in the monoallelic hemizygous sterile ones. The extension of this regulatory system in subtribe Loliinae probably favoured the stabilization of allopolyploids originating from closely related parental species in different broad-leaved and fine-leaved lineages. Breakdown of this genetic control caused by occasional intergeneric or interspecific hybridizations could have given rise to semi-fertile hybrids which may have represented windows of opportunity for further speciation via introgression or amphidiploidization in the Loliinae.

EVOLUTIONARY RATES ACROSS THE FESTUCOIDS

Bayesian phylogenetic analyses allowed inferences on the differences of evolutionary rates across the Loliinae lineages based on differences in their respective branch lengths [28, 164]. Grasses have shown to be one of the most rapidly evolving lineages within monocotyledons and to support in most cases, though not always, to the Minimum Generation Time (MGT) hypothesis. This predicts that annual or short-lived perennials, which propagate rapidly, show higher mutational rates than their respective long-lived perennial congeners, indicating a release from stabilized selection [53, 54]. Different evolutionary trends in rate heterogeneity have also been associated with the Speciation Rate (SR) hypothesis, which assumes that higher cladogenetic events are associated with higher substitutional rates [18, 19]. A Reticulation-Polyploidization (RP) hypothesis to explain the higher fitness and adaptive success shown by polyploid festucoids with respect to their close diploid relatives in connection with their differential mutation rates was proposed by Catalán et al. [30].

Relative rate tests [119, 120] performed between the main lineages of Loliinae and its close allies for the ITS and plastid *trn*LF genome regions made it possible to test those evolutionary hypotheses [30, 164]. A strong correlation was found between substitution rates in the nuclear and plastid genome regions and the life-cycle strategies shown by these groups [30, 164]. Evolutionary rates within the Loliinae vary enormously showing a general trend from slowly-evolving perennial lineages towards rapidly-evolving annual ones. The broad-leaved groups evolve in general terms at a lower pace than the fine-leaved ones. The broad-leaved perennial assemblages of *Leucopoa*, *Drymanthele*, and *Subbulbosae*, which include some of the tallest and most robust representatives of *Festuca*, showed the lowest rates of substitutions, followed by those of *Scariosae*, *Pseudoscariosa*, and *Lojaconoa*, and then by those of the more rapidly evolving *Schedonorus* + *Micropyropsis* + *Lolium* lineage. Within the last group, the European *Schedonorus* evolves at a lower pace, followed by *Micropyropsis* and the Maghrebian *Schedonorus* group, and then by *Lolium*, with the annual *Lolium* taxa showing the most accelerated substitution rates of all the broad-leaved groups and as fast as (or faster than) some of the fine-leaved groups [30].

Groups of intermediate placement in our phylogenetic trees (*Dimorpha*, *Leucopoa* p.p.) also show intermediate rates of mutation between the slowly-evolving broad-leaved groups and the fast-evolving fine-leaved groups. The more advanced fine-leaved *Festuca* groups show a trend towards higher substitutional rates in both nuclear and plastid sequences. Rate heterogeneity ranges are lowest in the old relict perennial groups (*Eskia*, *Exaratae*), intermediate in the more recently evolved perennials (*Festuca*, *Helleria*, *Aulaxyper*), and fastest in the newly derived annual lineages. Within the last group, different assemblages of the genus *Vulpia* and its close allies show significant differences from most of the remaining fine-leaved groups [30, 164].

Annual lineages evolve significantly faster than the perennial ones within both Loliinae groups, thus supporting the MGT hypothesis [30]. MGT mechanisms are operating in the rapidly evolving ephemeral groups of these grass lineages even if the biological factors that regulate these processes have not yet been deciphered [54, 112]. The changes in the evolutionary rates between slow-evolving perennials and fast-evolving annuals are interpreted as the probable consequence of a release from stabilized selection followed by the ephemeral groups that allowed the acquisition of rapid adaptive changes to new environmental habitats [57]. Our preliminary analyses also favour the speciation rate hypothesis within Loliinae, as indicated by the higher diversifying rates shown by the higher accelerated fine-leaved Loliinae lineages compared to the lower-diversification rates of the slower mutational broad-leaved lineages [30]. Interpretation of the wide array of monotypic and small-sized genera described within the fine-leaved *Festuca* clade might be a direct consequence of the higher speciation rates developed by these ephemeral groups [164].

Increased evolutionary rates are correlated with increased levels of ploidy [30], supporting the reticulation/polyploidization hypothesis for some festucoid lineages. This evolutionary scenario has developed independently along the two main clades of Loliinae, as exemplified by the *Aulaxyper* and the Maghrebian *Schedonorus* groups within the fine and broad-leaved *Festuca* lineages, respectively. These two groups encompass highly polyploid taxa (8x-10x) that are presumably derived from their respective lower-ploidy-level relatives. Diploid perennial lineages display the lowest mutational rates in both broad and fine-leaved *Festuca* lineages and the broad-leaved ones evolve significantly more slowly. Conversely, diploid annual taxa tend to show higher mutational rates than the ephemeral polyploids [30]. Although hybridization may have equally affected diploid and polyploid lineages, it is more common within the latter groups; this circumvented the new sterility barriers via recurrent introgression and polyploidization [3, 79, 144, 153]. Polyploidization is expected to increase the rate of variability of the nuclear genome concordantly with the accumulation of more gene copies [135] but the plastid genome should have lower levels of variability. However, concurrent rates of nucleotide substitutions in the two genomes, detected for both diploid and polyploid lineages of the Loliinae, may indicate other concerted nuclear and cytoplasm replication mechanisms in these plant cells [54].

High mutation rates may have negative consequences on phylogenetic reconstructions due to the loss of deep phylogenetic signal to give undesirable results such as long-branching attraction and site saturation effects [68, 170]. The lack of resolution observed at some subbasal nodes of our festucoid tree (Fig. 1) and some unexpected relationships could be associated with disturbance caused by increased levels of homoplasy displayed by the most rapidly evolving groups [30]. In *Lolium* the annual taxa show high accelerated mutational rates significantly different from most of the slowly-evolving perennial lineages. The high levels of morphological variability detected in *L. rigidum* and *L. canariense* Steud., which moved some authors to describe different infraspecific and specific taxa out of those complexes [130, 159], are thus correlated with their higher susbtitution rates. Within *Vulpia*, homoplasy may be enhanced in these highly evolving lineages altering the phylogenetic inference. However, the concurrent reconstruction of a consistent *Loretia* + *Monachne* + *Spirachne* + *Apalochloa* + *F. plicata* + *Ctenopsis* clade supports a common ancestor for all these *Vulpia* lineages (the '*Loretia* assemblage') except for typical section *Vulpia* [30, 164].

The switch from perennial towards annual life-cycle probably represents a more ancestral evolutionary phenomenon, experienced independently in the two main festucoid lineages as manifested by the mainly diploid nature of the ephemeral lineages (Fig. 4). This scenario could be concurrent with strong genome rearrangements caused by several heterochromatin losses, deduced from cytogenetic analyses [11, 12]. Reticulate processes involving recurrent hybridization and polyploidy probably constitute secondary evolutionary events that have affected some of the most recently evolved perennial lineages of broad and fine-leaved *Festuca* and a few annual fine-leaved lineages. Both polyploids and their diploid parents show similar karyotypic and ribosomal DNA mapping profiles [46, 161], which indicate less evident genome rearrangements. Although acceleration rates are significantly different across the annual lineages, the polyploid complexes show relatively higher rates, indicating that reticulation perhaps fostered the substitutional rates of these groups through the addition of new gene-pools [30]. Both scenarios agree, partially, with the speciation-rate hypothesis, as annuals and highly polyploid taxa show a wider array of taxa within each lineage than their congeneric or cosectional relatives. However, the annual lineages show distinctive traits that were used to classify them as different genera [39, 169], and the polyploid complexes are formed by a series of microtaxa that can hardly be morphologically differentiated from each other [102]. These lines of evidence add support for an older divergence of the annual lineages from their respective common ancestors and for a secondary and more recent divergence of the polyploid lineages.

BIOGEOGRAPHICAL AND PHYLOGEOGRAPHICAL PATTERNS OF FESTUCOID LINEAGES

Biogeography

Molecular studies provide the basis for discerning the biogeographical patterns shown by different Loliinae lineages [13, 14, 28, 52], confirming previous hypotheses based on geographical distribution of both regional endemics [64] and chromosome races [48]. Most recent phylogenetic surveys [28, 77] allow us to distinguish primary and secondary radiation centres located in different areas of the Old and New World, and to postulate evolutionary scenarios for the expansion of the festucoid lineages from the northern hemisphere towards the southern hemisphere (Fig. 5).

The Mediterranean–Asian region was the likely primary centre of speciation for both broad and fine-leaved lineages, since most diploid Loliinae representatives are endemic to this region and align basally in their respective clades [28]. Diploid ephemeral taxa (*Lolium*, *Vulpia*, *Narduroides*, *Wangenheimia*, *Ctenopsis*, *Micropyrum*) are native to the pan-Mediterranean-Asian area, whereas old relict diploid perennial lineages (*Subulbosae*, *Drymanthele*, *Scariosae*, and *Pseudoscariosa* within the broad-leaved clade, and *Eskia* and *Exaratae* within the fine-leaved clade) are distributed in the Mediterranean region and, to some extent, in Eurasia [28]. Karyological reports indicate that almost all the American and southern hemisphere fescues studied are polyploid [41, 48]. Diploid taxa predominate in the circum-Mediterranean region and across the main Asian mountain ranges; tetraploids are prevalent in northern and Central America; hexaploids in Africa; and octoploids with hexaploids constitute the main contingents in the austral regions of Patagonia [48] and New Zealand [41] (Fig. 5). Dubcovsky & Martinez [48] highlighted the noticeable absence of diploid species of *Festuca* in the southern hemisphere.

A clear north-to-south cline of increasing ploidy levels can be envisaged from the pan-Mediterranean–Asian belt towards America and the African/Austropacific continents (Fig. 5). This cline is paralleled by a similar south-to-north increase in ploidy levels from the temperate pan-Mediterranean–Asian regions towards the arctic zone where tetraploids, hexaploids and higher polyploids predominate over diploids [48]. However, the different northern and southern hemisphere polyploid lineages probably had different evolutionary origins.

Combined molecular phylogenies of Loliinae recover the sequential divergences of polyploid American and New Zealand *Festuca* clades that are nested in an intermediate position between the broad and fine-leaved lineages or within the last clade [77] (Fig. 1). All the Holarctic polyploid lineages are derived from clades where the basal diploid taxa are of pan-

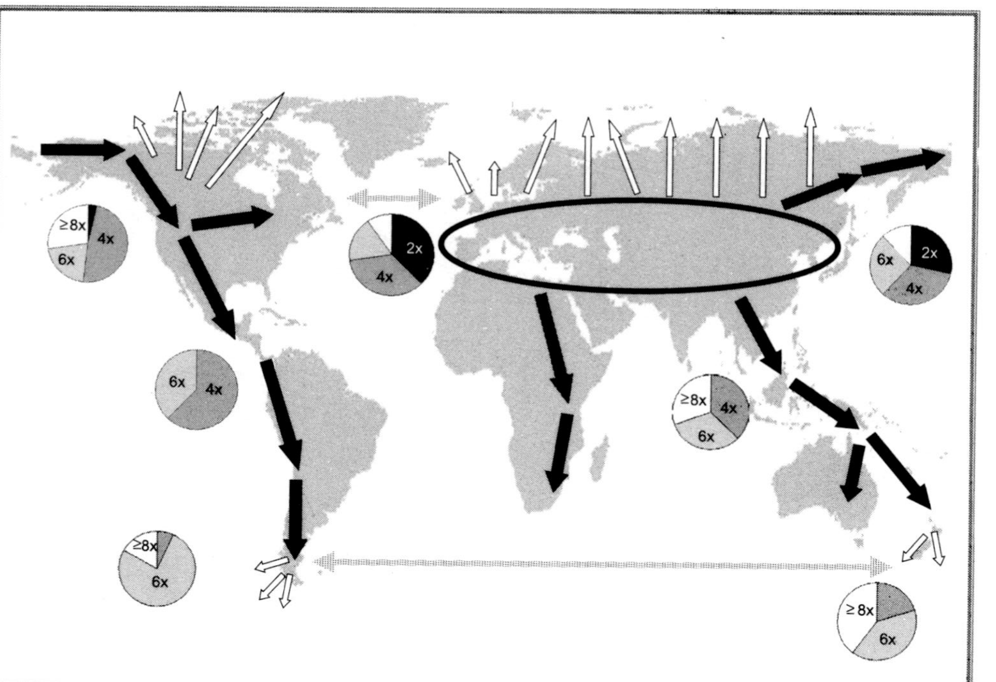

Fig. 5. *Biogeographical hypothesis of primary and secondary radiation centres of Festuca and related genera, based on phylogenetic inference and on chromosomal geographical distribution. Eurasia is postulated as the likely primary centre of divergence of the diploid festucoids in the late Tertiary. Potential colonization pathways towards America, Africa, and the Austropacific region are indicated by black arrows. Secondary speciation centres of polyploid lineages related to the Quaternary climatic changes are indicated by white arrows. Occasional long distance dispersal events in the periantarctic and the Holarctc areas are indicated by grey arrows. Circles indicate the relative percentages of ploidy levels in each continental area (modified from Dubcovsky & Martínez [48]).*

Mediterranean–Asian origin (Fig. 1). Thus, it could be speculated that the Mediterranean and European-Asian region (Eurasia) was the centre of origin for the old relict *Festuca* elements and that different migration routes allowed the more aggressive polyploid lineages to colonize other continents where successive radiations increased the spectrum of the extant taxa (Fig. 5). This scenario is likely to have predated the Pliocene and Pleistocene glaciations and could be associated with late Tertiary transcontinental land bridges that connected the Asian and American continents through the Beringian passage and the North American and South American masses after the closure of the Caribbean pass [77]. According to the present phylogenetic evidence (Fig. 1), colonization of the American continent probably took place on at least six separate occasions and that of the Neozeylandic region on at least two occasions, each followed by consecutive secondary radiations [77]. This hypothesis is concordant with molecular clock estimates based on nucleotide substitution rates of the festucoid plastid genome that estimate nine mya (million years ago) for the divergence of the broad and fine-leaved lineages [35], indicating a late Miocene origin for the basal lineages of *Festuca*, and two mya for the divergence of the *Schedonorus* + *Lolium* lineage [35, 52], indicating a Quaternary origin for this more recently evolved group.

Similar ancestral biogeographical north-to-south colonization and radiation scenarios have been postulated for the widely distributed grass genera *Poa* [136] and *Brachypodium* [129]. This hypothesis could be extended to other pooid groups where diploids predominate in the northern hemisphere, but are absent south of the equator [48, 64, 98]. The nesting of polyploid South American and New Zealand lineages within both the broad and fine-leaved lineages (Fig. 1) is better interpreted as a result of distinct colonization and secondary divergence episodes rather than of recent repeated long distance dispersal events. However, one of these lineages supports a floristic link between the New World and New Zealand [77], as evidenced by the sister-group relationships between the Neozeylandic clade I and the western North American *Festuca californica* Vasey (Fig. 1). Although the present taxon sampling of *Festuca* s. l. does not allow further speculation on this hypothesis, a major radiation of New Zealand taxa from colonizers that dispersed from North America to South America, and then to the Neozeylandic region rather than from Asia, has been proposed for the bluegrasses (*Poa*), based on a biogeographical analysis of plastid restriction site data [136]. Several Asian-western American and amphi-neotropical disjunctions observed in some Loliinae clades, such as those of the broad-leaved Asian-American clade I (Fig. 1), suggest that the number of dispersal events between the Old World to North America and then to South America might be more numerous than what is detected with the present data [77].

The more recent secondary radiation associated with the oscillatory climatic changes of the Quaternary probably fostered the speciation of the Holarctic polyploid lineages and some of the southern hemisphere groups (Fig. 5). Clear examples of the former case can be found within the Mediterranean and Eurasian broad-leaved *Schedonorus* and fine-leaved *Aulaxyper* lineages, which show a broad array of ploidy levels ranging from diploids to decaploids (Fig. 1). Several documented cases support the hypothesis that polyploid plants were more successful than diploids in colonizing deglaciated areas [24, 154, 155], and that recurrent polyploid formation spanned over different scales in time and space during the Quaternary from low-level to high-level polyploid taxa [1, 24]. Successive cycles of contraction and expansion of populations concomitant with the advances and retreats of the ice cover ultimately led to repeated contacts between previously isolated divergent lineages resulting in increasingly intricate polyploid complexes through hybridization and polyploidy [1, 24, 65, 66]. In *Schedonorus* and *Aulaxyper* lineages, putatively old and scarce diploid species are outnumbered by different series of allopolyploid taxa, now widespread in the previously glaciated areas of central and northern Europe and the highest Mediterranean mountain peaks. The highly heterozygous nature of allopolyploids can help buffer against inbreeding and genetic drift. Polyploidization has been hypothesized as the main evolutionary force that favoured the survival of the newly arisen taxa during the coldest periods of the glaciations, as well as their subsequent success in colonizing the deserted areas across a wide range of ecological habitats [24, 152, 153].

The high polyploid complexes found in the subantarctic peripheral area that was exposed to dramatic changes during Quaternary glaciations might mirror the pattern observed in the Holarcticone. Secondary cladogenic events possibly resulted in the diversification of closely related lineages of *Festuca*, such as the hexaploid South American clade II and the hexaploid Neozeylandic clade II [77] (Fig. 1). Some circum-antarctic taxa share with other circum-arctic congeners their ability to reproduce vegetatively, a common character in polar plants, and of inbreeding, an uncommon derived trait in *Festuca* [41, 51]. These traits probably facilitated the colonization and survival of festucoids in the severe climates.

Phylogeography

Using ITS phylogenetic inference [76], plastid DNA molecular clock estimates [35, 52], and pollen to seed flow ratios deduced from plastid RFLP and allozyme data [13, 14], more detailed phylogeographic scenarios have been postulated for the origin and dispersal of the *Lolium* lineages and *Festuca pratensis* (Fig. 6). Charmet et al. [35] and Fjellheim et al. [52] concur that early Pliocene divergence of the *Festuca pratensis* and *Lolium* lineages from their common ancestor occurred approximately two mya. However, they differ in the proposed postglacial colonization patterns followed by each of these lineages that were probably connected with the expansion of agriculture during the Holocene in the European-Mediterranean area [14, 52].

The plastid data revealed that both the native races of meadow fescues (*F. pratensis*) and eight species of *Lolium* probably diversified at the beginning of the Pleistocene (one mya) but experienced strong bottlenecks during the most recent glaciations (Riss and Würm, < 150,000 ya), resulting in the extermination of most European populations, except those sheltered in the circum-Mediterranean glacial refugia [35, 52]. The phylogeographic patterns deduced for *F. pratensis* suggest that the glacial survivors were isolated in three southern European refugia located in the Iberian Peninsula, in southeastern Europe and in western Caucasus, and that the contemporary variation detected in *F. pratensis* could be explained by northwards expansion of the Iberian haplotype along western Atlantic Europe and the southeastern haplotype migration towards central and northern Europe, coupled with a lack of expansion of the Caucasian haplotype (Fig. 6). This migration scenario is concordant with that postulated for other higher plants that were sheltered in the three main southern European Mediterranean refugia [40, 158].

Molecular evidence was used to postulate a distinct origin and migration scenario for the *Lolium* species implying human intervention during the spread of agriculture from the Fertile Crescent towards Europe and the Mediterranean basin after the last glaciation [14]. However, such

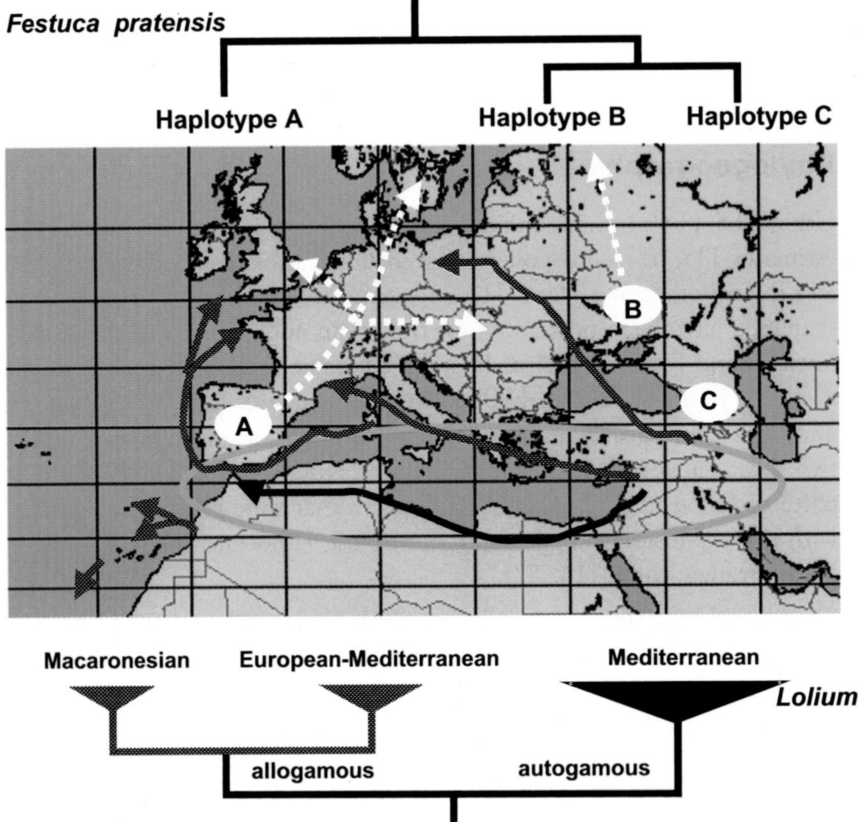

Fig. 6. *Phylogeographic patterns of Festuca pratensis and Lolium in their native pan-Mediterranean and European areas. F. pratensis: Postglacial expansion of western and easten European plastid haplotypes from their respective Iberian and eastern European glacial refugia, and non-expansion of the Caucasian haplotype (white circles and arrows) (taken from Fjellheim et al. [52]). Lolium: Potential ribotype distribution area of an ancestral Mediterranean Lolium lineage (grey circle) that diverged into present Mediterranean autogamous and allogamous taxa, and further split and colonization pattern of a Macaronesian allogamous lineage (thin dotted arrows) (taken from Inda et al. [77] and unpublished data). Hypotheses on recent Holocene-Neocene expansions of ryegrass plastid haplotypes and allozymic profiles associated with the spread of the agriculture from the Middle East to the western Mediterranean area and to Europe (L. rigidum: North African migratory route (black arrows); L. perenne: South European and Central European migratory routes (thick dotted arrows) (taken from Balfourier et al. [14]).*

a recent Middle East origin and Neocene dispersal of the *Lolium* taxa are in disagreement with nuclear ITS phylogenies and with geographical data that indicate *Lolium* lineages probably diverged and expanded before the Holocene [76] (Fig. 6). The origin of the *Lolium* taxa has been controversial as different hypotheses have been proposed for the alternative earlier splits of perennial vs. annual species and of outbreeding vs. inbreeding species [14, 35, 50, 100]. Classical hypotheses regarded the perennial/biennial outbreeders (i.e. *Lolium perenne*, *L. multiflorum*) as the most primitive species of the genus, that through progressive reduction gave rise to the annual outbreeders (*L. rigidum*) and then to the annual inbreeders (i.e. *L. temulentum* L., *L. remotum* Schrank, *L. persicum* Boiss. & Hohen.) [50]. However, karyotype analysis indicated that the inbreeding species had 40% more DNA than the outbreeding species, and that *Lolium perenne* features a derived lineage characterized by the shortest chromosomes in *Lolium* with the lowest amounts of heterochromatin [100, 160]. The apparent earlier divergence of the self-fertilizing *Lolium* species, with respect to the more recent origin of the outbreeding species, has also been supported by RAPD [146] and plastid [35] data, though this evolutionary trend does not agree with the one proposed for the fine-leaved *Festuca* lineages and *Vulpia* [43, 46, 164].

Despite previous proposals on the evolution of *Lolium*, exhaustive molecular phylogenetic studies of the *Schedonorus* + *Micropyropsis* + *Lolium* group recovered a clear ribotypic reproductive and geographical structure for the ryegrasses indicating the earlier split of the two autogamous vs. allogamous Mediterranean lineages followed by a further branching off of an allogamous Macaronesian clade (*Lolium canariense*, *L. edwardii* H. Scholz, Stierst. & Gaisb.) [76] (Fig. 6). It has been also demonstrated that the annual lineages (i.e. *L. canariense*, *L. rigidum*) mutate faster than the perennial lineages (*L. perenne*), arguing for a more rapid adaptive trend of the ephemeral lineages irrespective of their divergence times [30].

Based on the distribution of *Lolium perenne* and *L. rigidum* plastid haplotypes in the European and pan-Mediterranean areas and on calculations of pollen/seed flow ratio, Balfourier et al. [13, 14] concluded that the low values for these wind-pollinated ryegrass species apparently favoured a seed dispersal scenario, which was estimated to be concomitant with the expansion of the agriculture in the Neocene.

Three different eastern to western man-mediated migratory clines were hypothesized for the dispersal of the *Lolium rigidum* (north-African pathway) and the *L. perenne* (south-European and central-European pathways) plastid haplotypes [14] (Fig. 6). This Holocene-Neocene scenario of a historical single origin of all *Lolium* species from an inbreeding ancestor in the Fertile Crescent and subsequent expansion towards the west in conjunction with agriculture, disagrees with the present geographic distribution of native *Lolium* species [159] and with topologies inferred from nuclear data [76] (Fig. 6). Despite the monophyletic origin recovered for the *Lolium* lineage (Fig. 1), the divergence of the allogamous Macaronesian endemic clade (*L. canariense, L. edwardii*) (Fig. 6) probably predated the expansion of agriculture in the Old World. The Macaronesian *Lolium canariense* and the El Hierro island endemic *L. edwardii* probably represent old splits of an annual outbreeding *Lolium* lineage that colonized the Canary, Madeira, and Cap Verde isles from the near western European-African continent, rather than from the more distant eastern Mediterranean or Middle Eastern areas, as documented for most of the Macaronesian endemic angiosperms [26]. The presence of native *Lolium canariense* in the Macaronesian isles that were not populated before the 15th century (Madeira) clearly refutes the hypothesis of man-mediated transport of all *Lolium* species in Holocene-Neocene times and supports an earlier preglacial or interglacial expansion scenario for the *Lolium* lineages along the Mediterranean basin [76]. However, very recent eastern to western dispersal routes fostered by early Neocene farmers might have occurred for some native continental species that could be spread as crop weeds (*L. rigidum*) or as animal fodder (*L. perenne*) [13, 14].

CONCLUSION

Our most recent and exhaustive phylogenetic surveys of the festucoids have covered a broad range of subgeneric and sectional ranks of *Festuca* and its close allies, making it possible to build a comprehensive framework to interpret a classification of the evolutionary history of subtribe Loliinae [29, 77]. The extended phylogenetic analyses have developed concomitantly with other crucial sources of data on genome structure, hybridization and polyploidization, which together have brought forward new hypotheses on macroevolutionary and microevolutionary speciation events for different lineages and taxa of this important group of grasses. Ongoing research, however, focuses on the analysis of other genome sources, such as nuclear single copy genes, mitochondrial and plastid genes, which, similar to other grass tribes [85, 86, 87, 106, 107, 114], could facilitate a better understanding of the evolutionary mechanisms implied in the divergence of both deep and terminal branches of the festucoids.

Detailed analyses of genomic interaction between different gene copies, genome complements and cytoplasmic genes [104] could be decisive for the unravelling of the origins of the polyploids in this festucoid group where allopolyploidy seems to have been the driving evolutionary force [79]. Further research on gene duplication, gene expression, epistasis and pleiotropy could shed light onto the phenotypic and adaptive differences observed in some Loliinae species, as indicated for various Poaceae taxa [85, 87, 88]. Developmental genetics, investigated through genome colinearity and functional genome analysis, is a promising field of research explored in several grass groups of economic importance [86, 87], including Loliinae [10, 83]. It has provided new insights into the qualitative and quantitative morphological differences observed among taxa as results of different models of gene mutations and gene expression-repression mechanisms in different organs and life-stages [88, 89]. Comparative genome mapping detected highly orthologous and colinear *Lolium perenne* and *Festuca pratensis* linkage groups that also showed a syntenic relationship to the homeologous Triticeae, rice and oat genomes [10, 83]. Similarities in karyotype C-banding patterns, regulatory chromosome pairing genes and major nuclear genome arrangement support the hypothesis of the grasses (or at least the most recently evolved Pooideae) as a single functional genetic unit. The genetic basis for the heterochronic to heterotopic phenetic changes detected in a few model grass plants analyzed so far [84, 86] extends beyond those cases and is present in the Loliinae [74, 110].

Another interesting, though yet unexplored, field of evolutionary research is that of coevolution between some Loliinae lineages and their endosymbiotic fungi [109, 123, 124, 128, 156]. Endophytes of genus *Epichloë* range in a continuum from vertically transmitted seedborne asexual mutualists (called *Neotyphodium*) to horizontally transmitted sexual obligated pathogenic antagonists [128]. The horizontally transmitted species have haploid genomes and speciate cladistically (bifurcate branching), whereas the strictly seedborne mutualists are interspecific hybrids (reticulate) with selective advantage for the symbiotic habit [109]. Despite the cross-specific and cross-tribal jumps of sexual *Epichloë* species, they only express their sexual state, when symbiotic with a particular host genus or tribe [127]. Different authors have suggested that the strong mutualistic stamp of endophytes and their hosts could be largely historical and system-based [38, 123]. Some evidences presented by Moon et al. [109] indicated that there might be some history of association between asexual *Neotyphodium* taxa and *Schedonorus* + *Lolium* taxa (*F. arundinacea*, *Lolium rigidum*, *Lolium* spp.), as four hybrid endophytes showed *tub2* and *tef1* alleles that joined in a well-supported '*Lolium*-associated' clade.

Acknowledgement

I am indebted to Clive Stace, John Bailey, Jochen Müller and Paul Peterson for their valuable insightful comments on this manuscript, and for fruitful discussions and collaborative work on the study of the evolution of the festucoids carried out for many years. I wish to convey my sincere gratitude to Pedro Torrecilla, Jose Angel López-Rodriguez, Luis Inda and Jose Gabriel Segarra-Moragues for stimulating shared research, to Jeff Palmer, Siri Fjellheim, Christian Brochmann, John Harper, Kari Saikkonen and Christopher Schardl for valuable information, and to the Spanish Ministry of Science and Education (ANEP Project Grant Nos. BOS2000-0996 and REN2003-02818/GLO) and the Aragon Government (Project Grant Bioflora B39) for scientific support.

[1] Abbott RJ, Brochmann C. History and evolution of the arctic flora: in the footsteps of Eric Hultén. Mol Ecol 2003; 12: 299-313.

[2] Aiken SG, Gardiner SE, Bassett HC, Wilson BL, Consaul LL. Implications from SDS-PAGE analyses of seed proteins in the classification of taxa of *Festuca* and *Lolium* (Poaceae). Biochem Syst Ecol 1998; 26: 511-533.

[3] Ainscough MM, Barker CM, Stace CA. Natural hybrids between *Festuca* and species of *Vulpia* section *Vulpia*. Watsonia 1986; 16: 143-151.

[4] Alexeev EB. K sistematike asiatskich ovsjaniz (*Festuca*). I. Podrod *Drymanthele*, *Subulatae*, *Schedonorus*, *Leucopoa* (To the systematics of Asian fescues (*Festuca*). I. Subgenera *Drymanthele*, *Subulatae*, *Schedonorus*, *Leucopoa*). Bjull Moskovsk Obshch Isp Prir Otd Biol 1977; 82: 95-102.

[5] Alexeev EB. K sistematike asiatskich ovsjaniz (*Festuca*). 2. Podrod *Festuca* (To the systematics of Asian fescues (*Festuca*). 2. Subgenus *Festuca*). Bjull Moskovsk Obshch Isp Prir Otd Biol 1978; 83: 109-122.

[6] Alexeev EB. Novye podrody i sekzii ovsjaniz (*Festuca* L.) Severnoj Ameriki i Meksiki (*Festuca* L. Subgenera et sectiones novae ex America boreali et Mexica). Novosti Sist Vyssh Rast 1980; 17: 42-53.

[7] Alexeev EB. Novye taksoni roda *Festuca* (Poaceae) is Meksiki i Zentral'noj Ameriki (The new taxa of the genus *Festuca* (Poaceae) from Mexico and Central America). Bot Zhurn (Moscow & Leningrad) 1981; 66: 1492-1501.

[8] Alexeev EB. Novye rody slakov (New genera of grasses). Bjull Moskovsk Obshch Isp Prir Otd Biol 1985; 90: 102-109.

[9] Alexeev EB. Ovsjanizy (*Festuca* L., Poaceae) Venezuely, Kolumbii i Ekvadora (*Festuca* L. (Poaceae) in Venezuela, Colombia et Ecuador). Novosti Sist Vyssh Rast 1986; 23: 5-23.

[10] Alm V, Fang C, Busso CS, et al. A linkage map of meadow fescue (*Festuca pratensis* Huds.) and comparative mapping with other Poaceae species. Theor Appl Genet 2003; 108: 25-40.

[11] Bailey JP, Stace CA. Chromosome banding and pairing behaviour in *Festuca* and *Vulpia* (Poaceae, Pooideae). Pl Syst Evol 1992; 182: 21-28.

[12] Bailey JP, Bennett ST, Bennett MD, Stace CA. Genomic *in situ* hybridisation identifies parental chromosomes in the wild grass hybrid x *Festulpia hubbardii*. Heredity 1993; 71: 413-420.

[13] Balfourier F, Charmet G, Ravel C. Genetic differentiation within and between natural populations of perennial and annual ryegrass (*Lolium perenne* and *L. rigidum*). Heredity 1998; 81: 100-110.

[14] Balfourier F, Imbert C, Charmet G. Evidence for phylogeographic structure in Lolium species related to the spread of agriculture in Europe. A cpDNA study. Theor Appl Genet 2000; 101: 131-138.

[15] Barker CM, Stace CA. Hybridization in the genera *Vulpia* and *Festuca*: the production of artificial F1 plants. Nordic J Bot 1982; 2: 435-444.

[16] Barker CM, Stace CA. Hybridization in the genera *Vulpia* and *Festuca* (Poaceae): The characteristics of artificial hybrids. Nordic J Bot 1984; 4: 289-302.

[17] Barker CM, Stace CA. Hybridization in the genera *Vulpia* and *Festuca* (Poaceae): meiotic behaviour of artificial hybrids. Nordic J Bot 1986; 6: 1-10.

[18] Barraclough TG, Savolainen V. Evolutionary rates and species diversity in flowering plants. Evolution 2001; 55: 677-683.

[19] Barraclough TG, Harvey P, Nee S. Rate of *rbc*L gene sequence evolution and species diversification in flowering plants (angiosperms). Proc Roy Soc London, Ser B 1996; 263: 589-591.

[20] Bennett MD, Leitch IJ. Nuclear DNA C-values database (release 1.0, Sept 2001) 2001; htpp://www.rbgkew.org/uk/cval/homepage.html.

[21] Borrill M. Studies in *Festuca*. 3. The contribution of *F. scariosa* to the evolution of polyploids in section *Bovinae* and *Scariosae*. New Phytol 1972; 71: 523-532.

[22] Borrill M, Kirby M, Morgan WG. Studies in *Festuca*. 11. Interrelationships of some diploid ancestors of the polyploid broad-leaved fescues. New Phytol 1977; 78: 661-674.

[23] Borrill M, Kirby M, Morgan WG. Studies in *Festuca*. 12. Morphology, distribution, and cytogenetics of *F. donax*, *F. scariosa* and their hybrids, and the evolutionary significance of the fertile amphidiploid derivative. New Phytol 1980; 86: 423-439.

[24] Brochmann C, Brysting AK, Alsos IG, et al. Polyploidy in arctic plants. Biol J Linn Soc 2004; 82: 521-536.

[25] Bulinska-Radomska Z, Lester RN. Intergeneric relationships of *Lolium*, *Festuca* and *Vulpia* (Poaceae) and their phylogeny. Pl Syst Evol 1988; 159: 217-227.

[26] Carine MA, Russell SJ, Santos-Guerra A, Francisco-Ortega J. Relationships of the Macaronesian and Mediterranean floras: molecular evidence for multiple colonizations into Macaronesia and back-colonizations of the continent in *Convolvulus* L. (Convolvulaceae). Amer J Bot 2004; 91: 1070-1085.

[27] Catalán P, Kellogg EA, Olmstead RG. Phylogeny of Poaceae subfamily Pooideae based on chloroplast *ndh*F gene sequences. Mol Phylogen Evol 1997; 8: 150-166.

[28] Catalán P, Torrecilla P, López-Rodríguez JA, Olsmtead RG. Phylogeny of the festucoid grasses of subtribe Loliinae and allies (Poeae, Pooideae) inferred from ITS and trnL-F sequences. Mol Phylogen Evol 2004; 31: 517-541.

[29] Catalán P, Torrecilla P, López-Rodríguez JA, et al. A systematic approach to subtribe Loliinae (Poaceae: Pooideae) based on phylogenetic evidence. In: Columbus JT, Friar EA, Hamilton CW, et al. eds. Monocots: Comparative biology and evolution. 2 vols. Claremont: Rancho Santa Ana Botanic Garden 2006a; (in press).

[30] Catalán P, Torrecilla P, López-Rodríguez JA, Müller J. Molecular evolutionary studies shed new lights on the relationships of *Festuca*, *Lolium*, *Vulpia* and related grasses (Loliinae, Pooideae, Poaceae). Current Taxonomic Research on the British & European Flora. BSBI. (2006b).

[31] Catalán P, Quintanar A, Gillespie L, et al. Evolutionary analysis of the Poaceae subfamily Pooideae tribal complex Aveneae-Poeae: Systematic and biogeographic implications. XVII International Botanical Conference. Vienna 2005; http://www.ibc2005.ac.at/ (and unpublished data).

[32] Chandrasekharan P, Thomas H. Studies in *Festuca*, 5. Cytogenetic relationships between species of *Bovinae* and *Scariosae*. Z Pflanzenzücht 1971a; 65: 345-354.

[33] Chandrasekharan P, Thomas H. Studies in *Festuca*, 6. Chromosome relationships between *Bovinae* and *Scariosae*. Z Pflanzenzücht 1971b; 66: 76-86.

[34] Chandrasekharan P, Lewis EJ, Borrill M. Studies in *Festuca*. 2. Fertility, relationships between species of sections *Bovinae* and *Scariosae*, and their affinities with *Lolium*. Genetica 1972; 43: 375-386.

[35] Charmet G, Ravel C, Balfourier F. Phylogenetic analysis in the *Festuca-Lolium* complex using molecular markers and ITS rDNA. Theor Appl Genet 1997; 94: 1038-1046.

[36] Chen X, Aiken SG, Dallwitz MJ, Boucherd P. Systematic studies of *Festuca* (Poaceae) occurring in China compared with taxa in North America. Can J Bot 2003; 81: 1008-1028.

[37] Chicouene D. Realité et perspectives dans la description morphologique des festuques. In: Portal R, ed. *Festuca* de France. 1999; France: Vals près Le Puy: pp. 22-32.

[38] Clay K, Scharld C. Evolutionary origins and ecological consequences of endophyte symbiosis with grasses. Amer Nat 2002; 160: 99-127.

[39] Clayton WD, Renvoize SA. Genera graminum: Grasses of the world, Kew Bull, Add Ser 1986; 13: 1-389.

[40] Comes HP, Kadereit JW. The effect of Quaternary climatic changes on plant distribution and evolution. Trends Pl Sci 1998; 3: 432-438.

[41] Connor HE. *Festuca* (Poeae: Gramineae) in New Zealand. I. Indigenous taxa. New Zealand J Bot 1998; 42: 253-262.

[42] Cotton R, Stace CA. Taxonomy of the genus *Vulpia* (Gramineae). I. Chromosomes numbers and geographical distribution of the world species. Genetica 1976; 46: 235-255.

[43] Cotton R, Stace CA. Morphological and anatomical variation of *Vulpia* (Gramineae). Bot Notiser 1977; 130: 173-187.

[44] Darbyshire SJ, Warwick, SI. Phylogeny of North American *Festuca* (Poaceae) and related genera using chloroplast DNA restriction site variation. Canad J Bot 1992; 70: 2415-2429.

[45] Davis JI, Soreng RJ. Phylogenetic structure in the grass family (Poaceae) as inferred from chloroplast DNA restriction site variation. Amer J Bot 1993; 80: 1444-1454.

[46] Dawe JC. Sectional survey of Giemsa C-banded karyotypes in *Festuca* L. (Poaceae: Festuceae). Systematic and evolutionary implications. Dissertation. Austria: University of Vienna, 1989; 1-198.

[47] De Pinna MCC. Concepts and tests of homology in the cladistic paradigm. Cladistics 1991; 7: 367-394.

[48] Dubcovsky J, Martinez A. Distribución geográfica de los niveles de ploidía en *Festuca*. Parodiana 1992; 7: 91-99.

[49] Dumortier BCJ. Observations sur les graminées de la flore Belgique 1824; Belgium: Tournay: J. Casterman: pp. 1-153.

[50] Essad S. Étude génétique et cytogénétique des espèces *Lolium perenne* L., *Festuca pratensis* Huds. et leurs hybrids 1962; France: Institute National de la Recherche Agronomique, Ser. A No. Orsay No. d'Ord. 8.

[51] Fjellheim S, Elven R, Brochmann C. Molecules and morphology in concert. II. The *Festuca brachyphylla* complex (Poaceae) in Svalbard. Amer J Bot 2001; 88: 869-882.

[52] Fjellheim S, Rognli OA, Fosnes K, Brochmann C. Phylogeographic history of the widespread meadow fescue (*Festuca pratensis* Huds.) inferred from chloroplast DNA sequences. J Biogeogr 2006; (in press).

[53] Gaut BS, Muse SV, Clark WD, Clegg MT. Relative rates of nucleotide substitutions at the *rbc*L locus of monocotyledonous plants. J Mol Evol 1992; 35: 292-303.

[54] Gaut BS, Clark LG, Wendel JF, Muse SV. Comparisons of molecular evolutionary process at *rbc*L and *ndh*F in the grass family (Poaceae). Mol Biol Evol 1997; 14: 769-777.

[55] Gaut BS, Tredway LP, Kubik C, et al. Phylogenetic relationships and genetic diversity among members of the *Festuca-Lolium* complex (Poaceae) based on ITS sequence data. Pl Syst Evol 2000; 224: 33-53.

[56] Gillespie LJ, Soreng RJ. A phylogenetic análisis of the bluegrass genus *Poa* based on cpDNA restriction site data. Syst Bot 2005; 30: 84-105.

[57] Givnish TJ. Adaptative radiation and molecular systematics: Issues and approaches. In: Givnish TJ, Systma KJ, eds. Molecular evolution and adaptive radiation. Cambridge, UK: Cambridge University Press; 1997: 1-54.

[58] Gmelin CC. Flora badensis alsatica. Vol. 1. Germany: Karlsruhe: Müller, 1805; xxxii + 1-768.

[59] Grass Phylogeny Working Group (GPWG). Phylogeny and subfamilial classification of the grasses (Poaceae). Ann Missouri Bot Gard 2001; 88: 373-457.

[60] Hackel E. Monographia Festucarum Europearum 1882. Kassel and Berlin: T Fischer, 1882; 1-216, 4 lam.

[61] Hackel E. Gramineae. In: Engler HGA, Prantl KAE, eds. Die Natürlichen Pflanzenfamilien, Teil 2, Abteilung 2. Leipzig: Engelmann, 1887; 1-97.

[62] Hackel E. Gramineae novae. Repert Nov Spec Regni Veget 1906; 2: 69-72.

[63] Harper JA, Thomas ID, Lovatt JA, Thomas HM. Physical mapping of rDNA sites in possible diploid progenitors of polyploid *Festuca* species. Pl Syst Evol 2004; 245: 163-168.

[64] Hartley W. Studies on the origin, evolution, and distribution of the Gramineae 4: the subfamily festucoideae. Austral J Bot 1973; 21: 201-234.

[65] Hewitt GM. Some genetic consequences of ice ages, and their role in divergence and speciation. Biol J Linn Soc 1996; 58: 247-276.

[66] Hewitt GM. The genetic legacy of the Quaternary ice ages. Nature 2000; 405: 907-913.

[67] Hillis DM, Wiens JJ. Molecules versus Morphology in Systematics. Conflicts, Artifacts, and Misconceptions. In: Wiens JJ, ed. Phylogenetic Analysis of Morphological Data. Washington: Smithsonian Institution Press, Smithsonian Series in Comparative Evolutionary Biology, 2000; 1-19.

[68] Hillis D, Mable BK, Mortiz C. Applications of molecular systematics: the state of the field and a look to the future. In: Hillis DM, Moritz C, Mable BK, eds. Molecular Systematics, second edition. Sunderland: Sinauer Associates, 1996; 515-543.

[69] Hilu KW. Phylogenetics and chromosomal evolution in the Poaceae (grasses). Austral J Bot 2004; 52: 13-22.

[70] Holub J. New Genera in Phanerogamae [1-3]. Fol Geobot Phytotax 1984; 19: 95-99.

[71] Holub J. Reclassifications and new names in vascular plants 1. Preslia 1998; 70: 97-122.

[72] Hsiao C, Chatterton NJ, Asay KH, Jensen KB. Molecular phylogeny of the Pooideae (Poaceae) based on nuclear rDNA (ITS) sequences. Theor Appl Genet 1995; 90: 389-398.

[73] Humphreys MW, Thomas HM, Morgan WG, et al. Discriminating the ancestral progenitors of hexaploid *Festuca arundinacea* using genomic in situ hybridisation. Heredity 1995; 75: 171-174.

[74] Humphreys J, Harper JA, Armstead IP, Humphreys MW. Introgression-mapping of genes for drought resistance transferred from *Festuca arundinacea* var. *fenas* into *Lolium multiflorum*. Theor Appl Genet 2005; 110: 579-587.

[75] Hunter AM, Orlovich DA, Lloyd KM, et al. The generic position of *Austrofestuca littoralis* and the reinstatement of *Hookerochloa* and *Festucella* (Poaceae) based on evidence from nuclear (ITS) and chloroplast (*trnL-trnF*) DNA sequences. New Zealand J Bot 2004; 42: 253-262.

[76] Inda LA, Santos A, Sequeira M, Catalán P. Phylogeography of native Mediterranean and Macaronesian taxa of the *Lolium/Micropyropsis/Schedonorus* complex (Loliinae, Poeae, Poaceae) based on analyses of DNA sequences. Plant Evolution in

Mediterranean Climate Zones, IX IOPB Meeting 2004; http://www.jardibotanic.org/marc.html (and unpublished data).

[77] Inda LA, Segarra-Moragues JG, Peterson PM, et al. Phylogenetic studies and the radiation of the New and Old World festucoids (Loliinae, Pooideae, Poaceae). XVII International Botanical Conference. Vienna 2005; http://www.ibc2005.ac.at/ (and unpublished data).

[78] Jauhar PP. Genetic regulation of diploid-like chromosome pairing in the hexaploid species *Festuca arundinacea* Schreb. and *F. rubra* L. (Gramineae). Chromosoma 1975; 52: 363-382.

[79] Jauhar PP. Cytogenetics of the *Festuca-Lolium* complex; Berlin Heidelberg: Springer, 1993.

[80] Jenkins TJ. Interspecific and intergeneric hybrids in herbage grasses. Initial crosses. J Genet 1933; 28: 205-264.

[81] Jenkins TJ. Interspecific and intergeneric hybrids in herbage grasses, XV. The breeding affinities of *Festuca rubra*. J Genet 1955a; 53: 125-130.

[82] Jenkins TJ. Interspecific and intergeneric hybrids in herbage grasses, XIV. The breeding affinities of *Festuca ovina*. J Genet 1955b; 53: 118-124.

[83] Jones ES, Mahoney NL, Hayward MD, et al. An enhanced molecular marker based genetic map of perennial ryegrass (*Lolium perenne*) reveals comparative relationships with other Poaceae genomes. Genome 2002; 45: 282-295.

[84] Kellogg EA. Are macroevolution and microevolution qualitatively different? Evidence from Poaceae and other families. In: Cronk QCB, Bateman RM, Hawkins JA, eds. Developmental Genetics and Plant Evolution 2002a; 70-84.

[85] Kellogg EA. Inflorescence diversification in the panicoid "bristle grass" clade (Paniceae, Poaceae): Evidence from molecular phylogenies and developmental morphology. Amer J Bot 2002b; 89: 1203-1222.

[86] Kellogg EA. Evolution and developmental traits. Current Opinions Pl Biol 2004a; 7: 92-98.

[87] Kellogg EA. The evolution of nuclear genome structure in seed plants. Amer J Bot 2004b; 91: 1709-1725.

[88] Kellogg EA. Genetic control of branching in foxtail millet. PNAS 2004c; 101: 9045-9050.

[89] Kellogg EA. Heterogeneous expression patterns and separate roles of the SEPALLATA gene LEAFY HULL STERILE1 in grasses. Pl Cell 2004d; 16: 1692-1706.

[90] Kellogg EA, Watson L. Phylogenetic studies of a large data set. I. Bambusoideae, Andropogonodae, and Pooideae (Gramineae). Bot Rev 1993; 59: 273-343.

[91] Kellogg EA, Appels R, Mason-Gamer RJ. When gene trees tell different stories, incongruent gene trees for diploid genera of the Triticeae (Gramineae). Syst Bot 1996; 21: 321-347.

[92] Kerguélen M, Plonka F. Les *Festuca* de la Flore de France (Corse comprise). Bull Soc Bot Centre-Ouest N S Num Spéc 1989; 10: 1-368.

[93] Krechetovich VI, Bobrov EG. *Festuca* L. s. str. In: KomarovVL, Rozhevits RY, Shishkin BK, eds. Flora SSSR. Vol. 2. Leningrad: Akademija Nauk SSSR, 1934; 497-535.

[94] Krivotulenko U. Novye sekzii roda *Festuca* L. (Generis *Festuca* L. sectiones novae). Bot Mat (Leningrad) 1960; 20: 48-67.

[95] Küpfer P. Recherches sur les liens de paréntė entre la flore orophile des Alpes et celle des Pyrénées. Boissiera 1974; 23: 1-322.

[96] Lehväslaiho H, Saura A, Lokki J. Chloroplast DNA variation in the grass tribe Festuceae. Theor Appl Genet 1987; 74: 298-302.

[97] Lewis EJ. *Festuca* L. x *Lolium* L. = x *Festulolium* Aschers. & Graebner. In: Stace CA, ed. Hybridization and the flora of the British Isles. London: Academic Press, 1975; 547-552.

[98] Macfarlane TD. Poaceae subfamily Pooideae. In: Soderstrom TR, Hilu KW, Campbell CS, Barkworth ME, eds. Grass Systematics and Evolution. Washington: Smithsonian Institution Press; 1988; 265-276.

[99] Macfarlane TD, Watson L. The classification of Poaceae subfamily Pooideae. Taxon 1982; 31: 178–203.

[100] Malik CP, Thomas PT. Karyotypic studies in some *Lolium* and *Festuca* species. Caryologia 1966; 19: 167-196.

[101] Malik CP, Thomas PT. Cytological relationships and genome structure of some *Festuca* species. Caryologia 1967; 20: 1-39.

[102] Markgraf-Dannenberg, I. *Festuca* L. In: Tutin TG, Heywood VH, Burges NA, et al. eds. Flora Europaea vol. 5. Cambridge, UK: Cambridge University Press: 1980; 125-153.

[103] Markgraf-Dannenberg, I. *Festuca* L. In: Davis PH, ed. Flora of Turkey and the East Aegean Islands vol. 9. Edinburgh, UK: University Press: 1985; 400-442.

[104] Mason-Gamer RJ. Reticulate evolution, introgression, and intertribal gene capture in an allohexaploid grass. Syst Biol 2004; 53: 25-37.

[105] Mason-Gamer RJ, Kellogg EA. Phylogenetic analysis of the Triticeae using the starch synthase gene, and a preliminary analysis of some North American Elymus species. In: Jacobs SWL, Everett J, eds. Grasses. Systematics and Evolution. Melbourne: CSIRO, 2000; 102-109.

[106] Mason-Gamer RJ, Weil CF, Kellogg EA. Granule-bound starch synthase: structure, function, and phylogenetic utility. Mol Biol Evol 1998; 15: 1658-1673.

[107] Mason-Gamer RJ, Orme NL, Anderson CM. Phylogenetic analysis of North American *Elymus* and the monogenomic Triticeae (Poaceae) using three chloroplast DNA data sets. Genome 2002; 45: 991-1002.

[108] Mirones V. Estudios morfológicos y análisis polínico en *Festuca* sect. *Eskia* Willk. (Gramineae). Dissertation; University of Zaragoza, Spain. 2000; 1-97.

[109] Moon CD, Craven KD, Leuchtmann A, et al. Prevalence of interspecific hybrids among asexual fungal endophytes of grasses. Mol Ecol 2004; 13: 1455-1467.

[110] Moore BJ, Donnison IS, Harper JA, et al. Molecular tagging of a senescence gene by introgression mapping of a stay-green mutation from *Festuca pratensis*. New Phytol 2005; 165: 801-806.

[111] Müller J, Catalán P. Notes on infrageneric classification of *Festuca* L. (Gramineae). Taxon 2006.

[112] Muse SV. Examining rates and patterns of nucleotide substitutions in plants. Pl Mol Biol 2000; 42: 25-43.

[113] Nilsson F. Ein spontaner Bastard zwischen *Festuca rubra* und *Lolium perenne*. Hereditas 1933; 18: 1-15.

[114] Petersen G, Seberg O. Molecular evolution and phylogenetic application of *DMC1*. Mol Phyl Evol 2002; 22: 43-50.

[115] Piper CV. North American species of *Festuca*. Contr US Natl Herb 1906; 10: 1-48.

[116] Quintanar A, Castroviejo S, Catalán P. Phylogeny of tribe Aveneae Dumort. (Pooideae, Poaceae) inferred from nuclear and chloroplast sequence analysis. XVII International Botanical Conference. Vienna 2005; http://www.ibc2005.ac.at/ (and unpublished data).

[117] Reid G, Sidwell K. Overlapping variables in botanical systematics. In: Macleod N, Forey PL, eds. Morphology, Shape and Phylogeny. London: Taylor & Francis; Systematics Association Special Volume Series 64: 2002; 53-66.

[118] Rivas-Martinez S, Díaz-González TE, Fernández-González F, et al. Vascular plant communities of Spain and Portugal. Itin Gebot 2002; 15: 5-922.

[119] Robinson M, Gouy M, Gautier C, Mouchiroud D. Sensitivity of the relative-rate test to taxonomic sampling. Mol Biol Evol 1998; 15: 1091-1098.

[120] Robinson-Rechavi M, Huchon D. RRTree: Relative-Rate Test between group of sequences on a phylogenetic tree. Bioinformatics 2000; 16: 296-297.

[121] Rodionov AV, Tyupa1 NB, Kim EC, et al. Genome composition of the autotetraploid oat species *Avena macrostachya* determined by comparative analysis of the ITS1 and ITS2 sequences: on the oat karyotype evolution on early events of oats species divergence. Russ J Genet 2005; 41: 1-11.

[122] Romero-Zarco C, Cabezudo B. *Micropyropsis*, género nuevo de Gramineae. Lagascalia 1983; 11: 94-99.

[123] Saikkonen K, Faeth SH, Helander M, Sullivan TJ. Fungal endophytes: a continuum of interactions with host plants. Ann Rev Ecol Syst 1998; 29: 319-343.

[124] Saikkonen K, Wäli P, Helander M, Faeth SH. Evolution of endophyte-plant symbioses. Trends Pl Sci 2004; 9: 275-280.

[125] Saint-Yves A. Les *Festuca* (subgen. *Eu-Festuca*) de l'Afrique du Nord et des Isles Atlantiques. Candollea 1922; 1: 1-63.

[126] Saint-Yves A. Tentamen. Claves analyticae Festucarum veteris orbis (subgen. Eu-Festucarum) ad subspecies, multas varietates et nonullas subvarietates usque ducentes. Rev Bretonne Bot Pure Appl 1927; 2: 1-124.

[127] Schardl CL, Leuchtmann A, Chung KR, et al. Coevolution by common descent of fungal symbionts (*Epichloë* spp.) and grass hosts. Mol Biol Evol 1997; 14: 133-143.

[128] Schardl CL, Leuchtmann A, Spiering MJ. Symbioses of grasses with seedborne fungal endophytes. Ann Rev Plant Biol 2004; 55: 315-340.

[129] Schippmann U. Revision der europäischen Arten der Gattung *Brachypodium* Palisot de Beauvois (Poaceae). Boissiera 1991; 45: 1-249.

[130] Scholz H, Stierstorfer C, Gaisberg MV. *Lolium edwardii* sp. nova (Gramineae) and its relationship with *Schedonorus* sect. *Plantynia* Dumort. Feddes Repert 2000; 111: 561-565.

[131] Seal AG. DNA variation in *Festuca*. Heredity 1983; 50: 225-236.

[132] Seal AG, Rees H. The distribution of quantitative DNA changes associated with the evolution of diploid Festuceae. Heredity 1982; 49: 179-190.

[133] Sears ER. Wheat cytogenetics. Ann Tev Genet 1969; 3: 451-468.

[134] Soltis DE, Soltis PS. Molecular data and the dynamic nature of polyploidy. Crit Rev Pl Sci 1993; 12: 243-273.

[135] Soltis DE, Soltis PS. Polyploidy: recurrent formation and genome evolution. Trends Ecol Evol 1999; 14: 348-352.

[136] Soreng RJ. Chloroplast-DNA phylogenetics and biogeography in a reticulating group: study in *Poa* (Poaceae). Amer J Bot 1990; 77: 1383-1400.

[137] Soreng RJ, Davis JI. Phylogenetic structure in Poaceae subfamily Pooideae as inferred from molecular and morphological characters: misclassification versus reticulation. In: Jacobs SWL, Everett J, eds. Grasses: Systematics and evolution. Melbourne: CSIRO, 2000; 61-74.

[138] Soreng RJ, Davis JI, Doyle JJ. A phylogenetic analysis of chloroplast DNA restriction site variation in Poaceae subfam. Pooideae. Pl Syst Evol 1990; 172: 83-97.

[139] Soreng RJ, Terrell EE. Taxonomic notes on *Schedonorus*, a segregate genus from *Festuca* or *Lolium*, with a new nothogenus x*Schedololium* and new combinations. Phytologia 1998; 83: 85-88.

[140] Soreng RJ, Terrell EE. *Schedonorus* P. Beauv. In: Soreng RJ, Peterson PM, Davidse Judziewicz GEJ, eds. Catalogue of the New World Grasses (Poaceae): IV. Subfamily Pooideae. Contr US Natl Herb 2003; 48: 1-730.

[141] Stace CA. Changing concepts of the genus *Nardurus* Reichenb.(Gramineae). Bot J Linn Soc 1978a; 76: 344-350.

[142] Stace CA. Notes on *Cutandia* and related genera. Bot J Linn Soc 1978b; 76: 350-352.

[143] Stace CA. Generic and infrageneric nomenclature of annual Poaceae: Poeae related to *Vulpia* and *Desmazeria*. Nordic J Bot 1981; 1: 17-26.

[144] Stace CA. Hybridization and the plant species. In: Urbanska KM, ed. Differentiation patterns in higher plants. London: Academic Press: 1987; 115-127.

[145] Stace CA, Al-Bermani AKKA. Earliest records for two x *Festulpia* combinations. Watsonia 1989; 17: 363-364.

[146] Stammers M, Harris J, Evans GM, et al. Use of random PCR (RAPD) technology to analyse phylogenetic relationships in the *Lolium/Festuca* complex. Heredity 1995; 74: 19-27.

[147] Stancik D. New endemic taxa of *Festuca* from the Colombian Sierra Nevada de Santa Marta. Preslia 2003a; 75: 339-347.

[148] Stancik D. Las especies del género *Festuca* (Poaceae) en Colombia. Darwiniana 2003b; 41: 93-153.

[149] Stancik D. New taxa of *Festuca* (Poaceae) from Ecuador. Fol Geobot 2004; 39: 97-110.

[150] Stancik D, Peterson PM. Two new species of *Festuca* from South America (Poaceae: Loliinae: sect. *Subulatae*). Sida 2002; 20: 21-29.

[151] Stancik D, Peterson PM. *Festuca dentiflora* (Poaceae: Loliinae: sect. *Glabricarpae*), a new species from Peru and taxonomic status of *F. presliana*. Sida 2003; 20: 1015-1022.

[152] Stebbins GL. Variation and Evolution in Plants. New York: Columbia University Press: 1950; 1-643.

[153] Stebbins GL. Cytogenetics and evolution of the grass family. Amer J Bot 1956; 43: 890-905.

[154] Stebbins GL. Polyploidy and the distribution of the arctic-alpine flora: new evidence and new approach. Bot Helv 1984; 94: 1-13.

[155] Stebbins GL. Polyploidy, hybridization, and the invasion of new habitats. Ann Missouri Bot Gard 1985; 72: 824-832.

[156] Sullivan TJ, Faeth SH. Gene flow in the endophyte *Neotyphodium* and implications for coevolution with *Festuca arizonica*. Mol Ecol 2004; 13: 649-656.

[157] Sus AM. Análisis de los fenómenos de hibridación entre representantes de *Festuca* sect. *Eskia* Willk. (Gramineae) utilizando marcadores hipervariables (RAPD, RFLP) y secuencias de DNA (ITS, *trn*LF). Dissertation; University of Zaragoza, Spain. 2000; 1-57.

[158] Taberlet P, Fumagalli L, Wust-Saucy AG, Cosson JF. Comparative phylogeography and postglacial colonization routes in Europe. Mol Ecol 1998; 7: 453-464.

[159] Terrell EE. A taxonomic revision of the genus *Lolium*. US Dept Agric Tech Bull 1968; 1-1392.

[160] Thomas HM. The Giemsa C-band karyotypes of six *Lolium* species. Heredity 1981; 46: 263-267.

[161] Thomas HM, Harper JA, Meredith MR, et al. Physical mapping of ribosomal DNA sites in *Festuca arundinacea* and related species by in situ hybridisation. Genome 1997; 40: 406-410.

[162] Torrecilla P, Catalán P. Phylogeny of broad-leaved and fine-leaved *Festuca* lineages (Poaceae) based on nuclear ITS sequences. Syst Bot 2002; 27: 241-251.

[163] Torrecilla P, López-Rodríguez JA, Stancik D, Catalán P. Systematics of *Festuca* sects. *Eskia* Willk., *Pseudatropis* Kriv., *Amphigenes* (Janka) Tzvel., *Pseudoscariosa* Kriv., and *Scariosae* Hack. based on analysis of morphological characters and DNA sequences. Pl Syst Evol 2003; 239: 113-139.

[164] Torrecilla P, López-Rodríguez JA, Catalán P. Phylogenetic relationships of *Vulpia* and related genera (Poeae, Poaceae) based on analysis of ITS and *trnL–F* sequences. Ann Missouri Bot Gard 2004; 91: 124-158.

[165] Tzvelev NN. K sistematike i filogenii ovsjaniz (*Festuca* L.) flory SSSR. I. Sistema roda i oshoriye naprav'enija evoljuzii (On the taxonomy and phylogeny of genus *Festuca* L. of the U.S.S.R. flora. I. The system of the genus and main trends of evolution). Bot Zhurn (Moscow & Leningrad) 1971; 56: 1252-1262.

[166] Tzvelev NN. Poryadok zlaki (Poales). *Zhizn Rast* 1982; 6: 341-378.

[167] Tzvelev NN. Ob obeme i nomenklature nekotorykh rodov sosudistykh rastenii Evropeiskoi Rossii. (On the size and nomenclature of some genera of the vascular plants of European Russia). Bot. Zhurn. (Moscow & Leningrad) 1999; 84: 109-118.

[168] Tzvelev NN. Combinationes novae taxorum plantarum vascularium. Novosti Sist Vyssh Rast 2000; 32: 181-183.

[169] Watson L, Dallwitz MJ. The Grass Genera of the World. Wallingford, UK: C. A. B. International: 1992; 1-1038.

[170] Wendel JF, Doyle JJ. Phylogenetic incongruence: window into genome history and molecular evolution. In: Soltis DE, Soltis PS, Doyle JJ, eds. Molecular Systematics of Plants II. DNA Sequencing; Boston: Kluwer Academic, 1998; 265-296.

[171] Willkomm M. *Festuca*. In: Willkomm M, Lange L, eds. Prodromus Florae hispanicae Vol. I. Stuttgart: E. Schweizerbarth. 1861; 95-96.

[172] Xu WW, Sleper DA. Phylogeny of tall fescue and related species using RFLPs. Theor Appl Gen 1994; 88: 685-690.

CHAPTER

11

Molecular Genetic and Cytogenetic Evidences Supporting the Genome Relationships of the Genus *Avena*

ARACELI FOMINAYA, YOLANDA LOARCE, M LUISA IRIGOYEN and ESTHER FERRER

Department of Cell Biology and Genetics, University of Alcalá, Campus Universitario, ES-28871-Alcalá de Henares, Madrid, Spain

E-mail: araceli.fominaya@uah.es.

ABSTRACT

The genus *Avena* is an ideal system for investigating genomic organization and co-evolution of different genomes in nuclei to which they are common. This genus includes species with different degrees of ploidy and diverse genome composition. Cytological and cytogenetic studies based on chromosome number, morphology and pairing behaviour in hybrids, have been used to explain the evolutionary pathway of its polyploid species. Recently, molecular genetic and cytogenetic approaches, including the analysis of DNA sequences, have been used to disclose genome relationships within the genus. In the present study, fluorescence in situ hybridization using ribosomal sequences and two satellite DNA sequences, As120a (specific to the A-genome chromosomes) and Am1 (specific to the C-genome chromosomes), were used to identify the genome composition of a new tetraploid species. The results shed new light on the putative ancestors of the polyploid species. This work also reviews the present understanding of the evolution of the genus *Avena*, emphasizing the differences among its species and their genomes, as shown by molecular genetic and cytogenetic techniques.

Key Words: *Avena*, molecular cytogenetic, molecular analysis, *A. insularis*

INTRODUCTION

Understanding the genome relationships between and within plant species is very useful to cytogeneticists, plant breeders, evolutionists and molecular biologists. Since the publication of the monograph *Cytogenetics of Oats* [65], continued efforts have been devoted to understanding the genome composition of oat species (diploid, tetraploid and hexaploid species). The common cultivated oat (*Avena sativa* L.) is the world's sixth most important cereal crop, grown in many areas where climatic conditions are unfavorable to the production of major cereals [55, 69]. Oats is a major crop for three main reasons: (1) it shows broad adaptation, (2) it can provide both human food and animal fodder, and (3) it has a higher concentration of well-balanced proteins than other grains [59].

Attempts have been made to establish the genome relationships between the diploid and polyploid species. Genome differentiation was initially based on descriptions of karyotypes and the results of cytological studies on interspecific hybrids [65]. Modern technology has, however, provided a more comprehensive image of these genome relationships. For example, close and distant relationships between specific genomes have been revealed by genomic in situ hybridization (GISH) [7, 25,45] and fluorescence in situ hybridization (FISH) with molecular probes [13, 21, 31, 51]. In addition, the relationships among *Avena* species with different ploidy levels have been identified through the use of molecular markers, such as restriction fragment length polymorphisms (RFLPs) [1, 51, 54, 57, 68], amplified fragment length polymorphisms (AFLPs) [10], random amplified polymorphic DNA (RAPDs) [10], and microsatellites [47, 48]. This chapter reviews the present status of the genome relationships of *Avena* species based on molecular genetic and cytogenetic studies.

OAT GENOMES: AN OVERVIEW

The genus *Avena* contains four different genomes, A, B, C and D, each with a basic chromosome number of seven. The A and C genomes are represented by diploid species, but no diploid species with B or D genomes are known. The B and A genomes occur only in the combination AABB in wild oat species, while D occurs with A and C as AACCDD, a hexaploid form seen in wild and cultivated oats. The C genome also occurs in combination with A as AACC, a tetraploid form of wild oats. On the basis of karyotype analyses [11, 65], chromosome pairing affinities [71] and molecular markers [10], five A-genome diploids and two C-genome diploids have been described. Table 1 shows

Table 1. Oat genomes identified to date

Ploidy	*Species*	*Genomes*
2x	A. *strigosa* Shreb	AsAs
	A. *hirtula* Lag.	AsAs
	A. *wiestii* Steudel	AsAs
	A. *hispanica*	AsAs
	A. *brevis* Roth	AsAs
	A. *matritensis*	AsAs
	A. *lusitanica*	AsAs
	A. *atlantica* Baum et Fedak	AsAs
	A. *damascena* Rajhathy et Baum	AdAd
	A. *longiglumis* Durieu	AlAl
	A. *prostrata* Ladizinsky	ApAp
	A. *canariensis* Baum Rajhathy et Sampson	AcAc
	A. *eriantha* Dur. (A. *pilosa* M.Bieb)	CpCp
	A. *clauda* Dur.	CpCp
	A. *ventricosa* Bal. ex Cross.	CvCv
4x	A. *barbata* Pott ex Link	AABB
	A. *vaviloviana* (Malz.) Mordí	AABB
	A. *abyssinica* Hochst	AABB
	A. *agadiriana* Baum et Fedak	AABB
	A. *maroccana* Gdgr. (A. *magna* Murphy et Terrell)	AACC
	A. *murphyi* Ladizinsky	AACC
	A. *insularis*	AACC
	A. *macrostachya* Bal ex Coss. et Dur.	CCCC
6x	A. *sativa* L.	AACCDD
	A. *byzantina* C. Koch	AACCDD
	A. *sterilis* L.	AACCDD
	A. *fatua* L.	AACCDD

the oat genomes confirmed to date. All the representative species are annual inbreeders with the exception of *A. macrostachya*, which is an outbreeding perennial autotetraploid [43].

The genome size of oats has been determined by Feulgen microdensitometry (Table 2). In diploid species, it ranges from 4.0 to 5.6 picograms (pg), the A genome being the smallest and the C genome the largest [6]. In tetraploid species it ranges from 8.5 to 9.7 pg, the AABB genome being the smallest and AACC the largest [6]. The genome of the cultivated *A. sativa* (13.2-13.7 pg) is the smallest of the hexaploid cereal species; for example, it is about 21% smaller than that of the cultivated wheat *Triticum aestivum* (15.7-17.3 pg) [6]. One half of the total oat genome occurs as repeat sequences [3]. It is estimated that the flow cytometry [4] reduces the 1C genome of *A. sativa* to only 11.725 pg (equivalent to 11,315 Mb), which is about 33% smaller than that of hexaploid wheat (*c.*15,966 Megabase) [4]. Thus, the average

Table 2. Nuclear DNA content of 16 *Avena* species determined by Feulgen microdensitometry [6]

Species	*Genome*	*DNA Content*	
		pg/1C	*pg/2C*
A. strigosa Shreb	AsAs	4.0 – 5.0	8.0 – 10.0
A. hirtula Lag.	AsAs	4.4 – 4.9	8.8 – 9.8
A. wiestii Steudel	AsAs	4.9 – 5.1	9.8 – 10.3
A. brevis Roth	AsAs	4.5 – 4.7	8.9 – 9.5
A. longiglumis Dur.	AlAl	4.9 – 5.3	9.8 – 10.6
A. eriantha Dur.	CpCp	4.7 – 5.5	9.5 – 11.0
A. clauda Dur.	CpCp	5.3	10.6
A. ventricosa Bal ex Coss.	CvCv	5.6	10.9
A. barbata Pott ex Link	AABB	8.9 – 9.3	17.8 – 18.5
A. vaviloviana Malz.	AABB	8.5 – 9.2	17.0 – 18.4
A. abyssinica Hochst	AABB	8.9 – 9.0	17.9 – 18.0
A. maroccana Gdgr	AACC	9.3 – 9.7	18.6 – 19.4
A. sativa L.	AACCDD	13.2 – 13.7	26.5 – 27.5
A. byzantina C. Koch	AACCDD	13.5 – 13.7	27.1 – 27.4
A. sterilis L.	AACCDD	13.7 – 14.3	27.3 – 28.6
A. fatua L.	AACCDD	12.9 – 14.2	25.7 – 28.3

chromosome size of this wheat is 760 Mb, while the average size of the *A. sativa* chromosomes is 539 Mb. This is comparable to the average size of a *Pisum sativum* (563 Mb) [4] chromosome.

Although the large size of the cultivated oat genome hinders the development of genetic maps with well established linkage groups, a number of maps based on several types of molecular marker have been produced [9, 18, 30, 61, 74]. The most comprehensive hexaploid map was produced using the cross *A. byzantina* cv. Kanota x *A. sativa* cv. Ogle [58, 72]. In its current version it shows a total of 1,166 markers including RFLPs, AFLPs, RAPDs, SSRs, isozymes and seed proteins grouped into 29 linkage groups. Efforts are being made by using approaches such as monosomic analysis, to reduce this number to the 21 expected [14].

GENOME DIFFERENTIATION AMONG DIPLOID SPECIES

Genome differentiation studies were initially based on descriptions of karyotypes and cytological observations made on interspecific hybrids. Early karyotype studies recognized two basic genomes for the diploid species: the A genome with isobrachial chromosomes, and the C genome with heterobrachial chromosomes [63]. This variation in chromosome morphology has also been demonstrated by C-banding. The chromosomes of A-genome species mainly show telomeric bands, whereas those from the C-genome species are characterized by higher chromatin condensation and the presence of several intercalary bands [11]. Chromosome pairing behaviour at meiosis in the interspecific AC hybrid (represented by *A. strigosa* x *A. eriantha*) involves a high number of unpaired chromosomes and shows partial homology/homeology between the A and C genome chromosomes [46, 56].

This lack of homology has also been borne out by several biochemical and molecular markers. Three systematic approaches have established a very clear separation of the A and C genomes. The first was based on the analysis of biochemical traits such as seed proteins and leaf isozymes. In an extensive gel electrophoresis study of seed proteins, a lack of correlation (r=0.04) was found among electrophoretic bands from the A- and C-genome diploid species [33]. In leaf isozyme studies, a group of electrophoretic bands characteristic of the C-genome diploids was absent from the A-genome diploids [8, 67]. The second approach was based on the examination of the number and chromosomal location of ribosomal

DNA loci (18S-5.8S-26S and 5S rDNA). As a rule, 18S-5.8S-26S rDNA loci are found in the nucleolus-organizing regions (NORs). They are cytologically visible by conventional C-banding and Ag-NOR (silver staining of NORs) techniques as secondary constrictions on satellited chromosomes. Using these techniques, two pairs of satellited chromosomes were identified in both A- and C-diploid species, except for *A. ventricosa*, which has only one satellited chromosome pair [11, 65]. FISH and Southern hybridization analyses using heterologous probes have obtained direct information on the differences between ribosomal loci. FISH employing wheat pTa71 (to detect NOR loci) and pTa794 (to detect 5S loci), rDNA probes showed the A-genome diploid species as having two pairs of 5S loci on both arms of one pair of satellited chromosomes, while the C-genome diploid species showed one pair of 5S loci on the long arm of one pair of subtelocentric chromosomes carrying an extra rDNA loci on the same long arm [50]. In Southern analysis using a maize rDNA probe for *Eco*RI-digested DNAs from diploid species, the C-genome diploids were found to have a 10.5-kb *Eco*RI fragment not present in the intergenic spacer (IGS) of the A-genome diploid species [26]. A third approach was based on the isolation of several genome-specific repeat sequences from A-genome diploid species and the use of RAPD and AFLP molecular markers. Four different repeat DNA sequences were isolated from an *A. strigosa* genomic library. As120a satellite DNA was found to be present in A-genome diploid species (with the exception of *A. longiglumis* and *A. damascena*), and absent in C-genome diploid species in experiments involving the hybridization of the pAs120a clone in Southern blots containing *Mun*I-digested genomic DNA from diploid species [51]. Similar information has been reported using two dispersed repeat sequences, As14 and As121 [53] and an LTR (long terminal repeat) fragment of a Ty1-*copia*-retrotransposon [52]. More recently, major clustering differences between the A and C genomes have been detected from phenograms produced with AFLP and RAPD markers. All A-genome diploid taxa were clustered together, while the C-genome diploids formed an outer branch [10].

Structural differences between the A genomes of diploid species have been revealed by the study of genomic relationships through chromosome pairing and fertility. Crosses involving *A. longiglumis* and *A. hirtula*, *A. strigosa* or *A. wiestii* resulted in the failure of chromosome pairing and high levels of hybrid sterility [20, 64, 70]. The cross between

A. prostrata and *A. longiglumis* is partially fertile [34]. *A. canariensis* x *A. damascena* hybrids show complete and regular pairing, but they are nonetheless self-sterile [39]. Almost complete chromosome pairing is observed in *A. damascena* x *A. longiglumis* hybrids [39]. All other crosses between A-genome species produce hybrids with irregular chromosome pairing at meiosis [39, 42]. Based on these observations, the genomic symbols As, Ad, Al, Ap and Ac (Table 1) have been suggested to distinguish between the five different karyotypes of the A-genome diploid species [71].

Molecular studies have been one of the major sources of information concerning the divergence of the A-genome diploid species. In autoradiograms of Southern blots of genomic DNAs digested with a single restriction endonuclease and hybridized with either a satellite repeat sequence (As120a) [51], or cDNA clones encoding seed storage proteins and α-amylase [1], divergence was observed between three endemic species (*A. longiglumis*, *A. damascena* and *A. canariensis*) and the other A-genome diploids. Information on the clustering of RFLP, RAPD and AFLP molecular markers has led to the genomes of *A. nuda* [57], *A. lusitania* and *A. matritensis* [10] being designated as As, although no karyotypic classification is currently available for these species. Based on genetic similarities estimated from RAPDs and RFLPs, Nocelli et al. [57] proposed that the A-genome species be divided into two main groups, the first with the As- and Ap-, and the second with Al-, Ad- and Ac-genome species. This conclusion is supported by studies on the similarity and genome organization of resistance gene analogue sequences (RGAs). The RFLP patterns of *A. strigosa* RGAs in *A. longiglumis*, *A. damascene* and *A. canariensis* show clear differences, both in the number and size of hybridizing bands, with patterns of the *A. strigosa* group of species [23]. This classification, however, contrasts with the genetic similarities estimated by RAPD and AFLP markers reported by Drossou et al. [10]. These authors indicate the Ad-genome species to be closely related to the As-genome species, with the Ap-genome species being the most distant. They also suggest the evolutionary sequence Ap → Al → Ad/Ac → As for the speciation of the A-genome diploids.

The C-genome diploids are represented by two karyotypes, Cp and Cv, corresponding to *A. eriantha*/*A. clauda* and *A. ventricosa*, respectively [11, 65]. Chromosome pairing and molecular markers have revealed structural differences between the Cp and Cv genomes. *A. eriantha* x *A.*

clauda hybrids undergo regular meiosis in which seven bivalents are formed. In contrast, hybrids from crosses of A. *ventricosa* with the former show irregular meiosis and are completely sterile [65], reflecting the rearrangements of chromosome structure that differentiate the Cp and Cv genomes. This differentiation is also supported by RAPD and AFLP marker patterns, which show A. *eriantha* and A. *clauda* to be more closely related to each other than to A. *ventricosa* [10].

GENOME DIFFERENTIATION AMONG TETRAPLOID SPECIES

The tetraploid species can be classified into four groups based on their karyotypes and patterns of chromosome pairing. The first group includes A. *barbata*, A. *vaviloviana* and A. *abyssinica*. These are genetically uniform and possess the AABB genome [65]. Cytological studies of meiotic chromosome in hybrids between AABB and AsAs autotetraploids show the B genome chromosomes to have large segments homologous to those of the As genome [66]. The C-banding patterns of the B genome chromosomes are similar to those of the A genome chromosomes [12]. In addition, GISH experiments employing total A. *strigosa* genomic DNA as a probe have uniformly labelled all 28 chromosomes of A. *barbata* [45] and A. *vaviloviana* [31], reflecting the close relationship between the A and B genomes. In dendrograms produced from AFLP [10] and RGA [23] data, the AABB tetraploids cluster closely with the As-genome diploids. Accordingly, the genomic designation AAA'A' has been suggested for these species. However, FISH experiments employing the As120a A genome-specific satellite sequence as a probe labelled only 14 of the 28 chromosomes of A. *barbata* and A. *vaviloviana* [21]. This shows the presence of two different chromosome sets and supports the AABB genomic designation of these species. This has been confirmed by assigning the satellited chromosomes to individual genomes, using the satellite itself and two ribosomal probes in simultaneous and sequential FISH analyses. Differences between A. *barbata* and A. *vaviloviana* genomes have also been revealed by FISH and Southern blotting techniques [21].

The second group includes A. *agaridiana* from Morocco [5]. Cytological studies of chromosome pairing in triploid hybrids between A. *agaridiana* and A genome diploid species have revealed the chromosomes of A. *agaridiana* to show residual homology with those of the A genome

[41]. Chromosome pairing studies in hybrids with A. *barbata* [41] revealed some structural similarities among chromosomes of both species. However, the A. *agaridiana* chromosomes do not closely match those of any of the described diploid or tetraploid species in terms of their arm ratios and C-banding patterns, although their overall C-band appearance resembles that of the A, B and D groups of chromosomes [27]. Consequently, the identity of the genomes remains undefined and requires further study.

The third group of tetraploids includes A. *maroccana* and A. *murphyi*, which have the AACC genome [65]. Cytological studies of chromosome pairing in hybrids of AACC tetraploids and A. *strigosa* indicate the As genome of the diploids to be partially homologous with one of the genomes of the tetraploid species [65]. Isozyme analyses have related A. *eriantha* to A. *maroccana* and A. *murphyi* [8]. The C-banding patterns identified two sets of chromosomes with one or the other of the two heterochromatin distribution patterns described for the chromosomes of A- and C-genome diploids [12]. GISH experiments employing either total A. *strigosa* or A. *eriantha* genomic DNA as a probe [19, 25, 45], and FISH using Am1 (a satellite sequence specific to the C-genome chromosomes isolated from A. *murphyi*) [13], confirmed the AACC genomic constitution of these species. These studies detected that at least seven or eight chromosome pairs in A. *maroccana* and A. *murphyi*, respectively, were involved in intergenomic interchanges between the A and C genomes.

A new tetraploid species, A. *insularis* Ladiz. [37] shares the gross morphological characteristics of the AACC tetraploid species. However, chromosome pairing in hybrids of A. *insularis* and A. *maroccana* showed little affinity between their chromosomes [37]. Similar results were obtained with A. *insularis* x A. *murphyi* hybrids [38]. The C-banding patterns show some structural similarities between the karyotype of A. *insularis* and those of the described AACC tetraploids, but are insufficient to support the unequivocal presence of the A and C genomes in this new tetraploid species [28]. In an effort to clarify this ambiguity, the present report describes the genomic constitution of A. *insularis* as revealed by FISH using probes pAm1 [13] and pAs120a [51]. Also, the genomic designation of each satellited chromosome was achieved by simultaneous FISH and reprobing the same metaphase plates with ribosomal pTa71 [15] and pTa794 [16].

When the rhodamine-labelled pAm1 probe was hybridized, seven pairs of the C-genome chromosomes were successfully labelled (red in Fig. 1a). Surprisingly, when the rhodamine-labelled pAs120a probe was hybridized with the same *A. insularis* metaphases, the 14 chromosome pairs appeared hybridized and showed a dispersed distribution of sequences (red in Fig. 1d). This is the first species studied to show a set of C-genome chromosomes that hybridize with the A-genome-specific As120a sequence in non-translocated areas. However, these results confirm the presence of two genomes in this tetraploid species: a C genome composed of the seven chromosome pairs that hybridized with pAm1, and seven other chromosome pairs initially assigned to the A genome based on hybridization with the pAs120a probe. When FISH was performed on the same metaphase plates with pTa 71 (green in Fig. 1c, to detect 18S-5.8S-26S rDNA loci) or pTa 794 (green in Fig. 1d, to detect 5S rDNA loci), one chromosome pair carrying both NOR and 5S loci, plus another pair carrying only a NOR locus, were detected. In both cases, they belonged to the A genome. Moreover, two chromosome pairs carrying 5S loci were detected (green in Fig. 1d). These were assigned to the C genome.

Both the satellited chromosome pairs of the A genome had terminally located Am1 homologous sequences, indicating the existence of two intergenomic translocations in the formation of *A. insularis*. Taking into account the description of the chromosomes of *A. sativa* after hybridization with the same probes [53], these two pairs of chromosomes should be identified as chromosomes 3D and 13D. Similarly, the 5S rDNA loci detected on the two chromosome pairs of the C genome closely resembled those of *A. sativa* [53] and should be identified as chromosomes 2C and 4C. These are involved in another two intergenomic translocations. Taken together, the FISH results of the present study show a relationship between the C genomes of *A. insularis* and the hexaploid species. However, the initially characterized A genome of *A. insularis* more closely resembles the D genome of the hexaploids than the A genome of these species. Karyotypic evidence [28] and the existence of high chromosomal pairing in hybrids between *A. insularis* and *A. sativa* [38], suggest these two species probably share two genomes. However, since pAs120a failed to hybridize with the C- and D-genome chromosomes of the C genome diploids, the AACC tetraploids and the AACCDD hexaploids [51], the C and D genomes of *A. insularis* should be considered modified and perhaps designated C'C'D'D'.

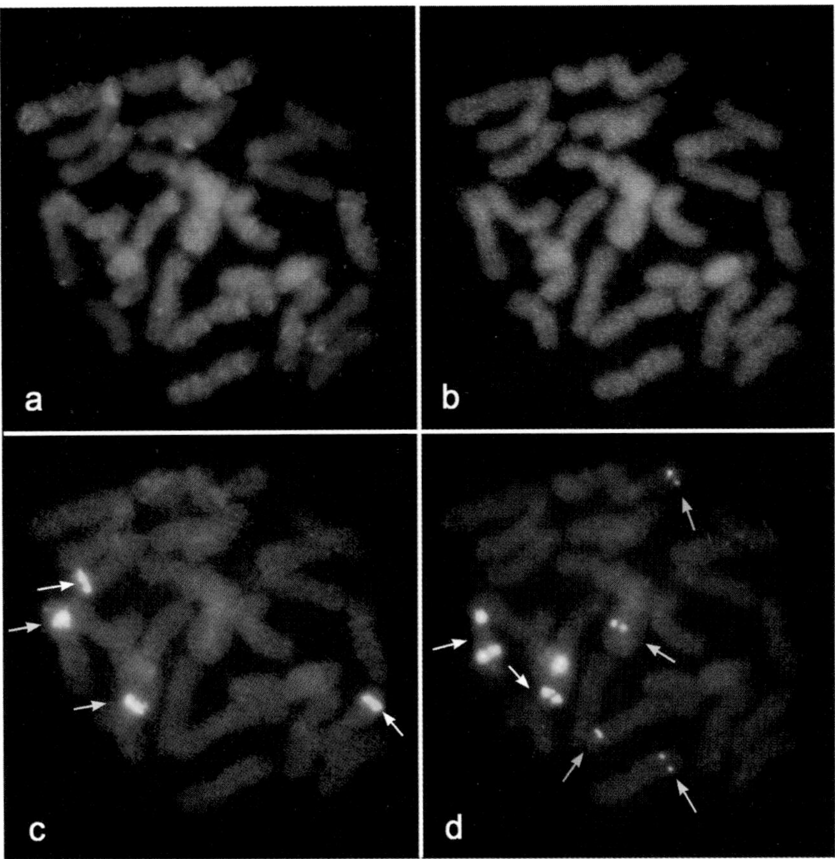

Fig. 1. *Fluorescent in situ hybridization (FISH) of mitotic metaphase plates of* Avena insularis *(a-d). (a) FISH with the rhodamine-labelled (red) pAm1 probe. (b) The same cell as in (a) shown after FISH with the rhodamine-labelled pAs120a probe (red). (c) The same cell as in (a) and (b) shown after FISH with the digoxigenin-labelled pTa71 (green) probe. Arrows indicate the chromosomes carried NOR loci. (d) The same cell as in (a), (b) and (c) shown after FISH with the digoxigenin-labelled pTa794 (green) probe. Arrows indicate C'-D' and D'-C' intergenomic translocations.*

The fourth group of tetraploid species includes A. *macrostachya*, an outbreeding perennial autotetraploid. Chromosome pairing in the hybrids of this species with either A. *sativa* or A. *murphyi* indicates that although there is some homology between the chromosomes of these species and those of A. *macrostachya*, it should be considered as little more than residual [40]. Chromosome pairing in triploid hybrids between this species and either A-genome or C-genome diploid species indicates that

A. *macrostachya* is more related to the C-genome diploids than to the A-genome diploids [40, 44, 60]. The latter's C-banding patterns are similar to those described for the C-genome diploid species [11, 62]. However, with the evidence currently available, the genomic constitution of this species remains doubtful.

GENOME DIFFERENTIATION AMONG HEXAPLOID SPECIES

The commonly recognized hexaploid species include four that are interfertile: two cultivated species, A. *sativa* and A. *byzantina*, and two wild species, A. *fatua* and A. *sterilis* [36]. Based on chromosome numbers, centromere position and chromosome pairing relationships, it has been proposed that all hexaploids share the same genomic composition of AACCDD [65]. In recent years, conventional and molecular cytogenetic techniques have been used to identify the three constituent genomes of hexaploid species. C-banding techniques have indicated that both the A and D genomes consist largely of euchromatic chromosomes, whereas the C-genome chromosomes contain largely heterochromatic regions [24, 49]. Fluorescent in situ hybridization with either genomic DNA [7, 19, 25, 45, 73], repeat DNA sequences specific to the C-genome chromosomes [13], or a combination of rDNA genes and cloned repeat DNA sequences [50, 53], indicate that the chromosomes of the D genome bear substantial similarities to those of the A genome. Discrimination among chromosomes of these genomes is possible using an A-genome-specific probe [51]. Together, these studies have confirmed the common genomic composition of the hexaploid species.

Molecular cytogenetic studies have also demonstrated that intergenomic translocations are present in the hexaploid species. Within species, some of these translocations seem to be common to all the cultivars or accessions studied; other translocations would have led to new genotypes. GISH has detected nine intergenomic translocations between the chromosomes of the A/D and C genomes in different cultivars of A. *sativa* [7, 25, 45], eight in accessions of A. *sterilis* [25] and A. *fatua* [73], and five in accessions of A. *byzantina* [25]. The existence of intergenomic translocations specific to a single or reduced number of hexaploid cultivars confirms the importance of these kinds of rearrangement in the evolution of the genus. As an example, FISH employing multiple cloned probes including pAs120a (which hybridizes exclusively with A-genome chromosomes), pAm1 (which hybridizes exclusively with C-genome chromosomes), pAs121 (which hybridizes exclusively with A- and D-genome chromosomes), pTa71 and pTa794, identified ten intergenomic translocations in the cultivar SunII of *A. sativa* [22]. These were: (i) between the A and C genomes (chromosome pair 5A), (ii) between the C and D genomes (pairs 1C, 2C, 4C, 10C and 16C), and (iii) between the D and C genomes (pairs 9D, 11D, 13D and 14D). The translocation involving chromosomes 10C and 14D differentiate this cultivar from other *A. sativa* cultivars [25, 53].

PUTATIVE PROGENITORS OF THE HEXAPLOID GENOMES

The evolution of *Avena* genomes has been a complex process involving divergence from a common diploid ancestor, followed by convergence, and subsequently by divergence at the polyploidy level [71].

It has been suggested that the evolution of the AACCDD genome involved two distinct steps. The first was the establishment of the tetraploid (AACC) by the hybridization of two diploid species (AA and CC), followed by a doubling of the chromosome number. This was followed by the hybridization of this tetraploid with a third diploid species, and subsequently by the doubling of chromosomes causing the triploid hybrid to become a hexaploid [43]. The tetraploid species *A. murphyi* was formerly favored as the donor of the AC genome to the hexaploid species. This conclusion was based on chromosome pairing in hybrids between hexaploid and tetraploid oats [32, 35], isozyme analysis [67], and the polymorphism detected using minisatellite and microsatellite sequences [47, 48]. However, the data provided by AFLP and RAPD markers more strongly point to *A. maroccana* [10].

The C-genome diploids have been subdivided into two groups, CpCp and CvCv. Presently there is substantial evidence that *A. eriantha* (CpCp genome) is closely related to the tetraploids and hexaploids. This conclusion was initially based on karyotypes [65] and C-banding [12, 24] analyses. The use of *A. eriantha* DNA in GISH showed the presence of this chromatin in the C-genome chromosomes of tetraploid and hexaploid species [7, 19, 25, 45, 73]. FISH employing the Am1 sequence isolated from *A. murphyi* showed this sequence to be present in *A. eriantha* chromosomes and the C-genome chromosomes of the tetraploid and hexaploid species [13]. Moreover, studies involving molecular markers such as minisatellites and microsatellites have shown that *A. clauda* (CpCp genome) is closely related to the AC-genome tetraploids and the hexaploids [48]. Nevertheless, two C-genome-specific sequences, AvsC-88 and AvsC-137 cloned from *A. sativa*, have been found in *A. ventricosa* (CvCv genome) and the AC-genome tetraploids and hexaploids [2], indicating that a relationship also exists between these two genomes.

The origin of the A genome in tetraploid and hexaploid species has been a point of controversy. In hexaploid species, initial comparisons of the conventional [65] and C-banding [24, 49] karyotypes of *A. sativa* and diploid *A. strigosa* showed the genome of the latter to very closely match the putative A genome of the hexaploid species. Chromosome pairing in tetraploid hybrids between *A. strigosa* x *A. sativa* was slightly higher than that observed in tetraploid hybrids involving other A-genome diploids [71]. This might indicate greater homology between *A. strigosa* and *A. sativa* than between this species and any other A-genome species. GISH employing *A. strigosa* DNA showed the presence of this chromatin in the A- and D-genome chromosomes [7, 19, 25, 45, 73]. Similarly, FISH employing several repeat sequences isolated from *A. strigosa*, such as dispersed [53] or Ty1-*copia*-retrotransposon [52, 54] elements, showed hybridization over 14 pairs of chromosomes corresponding to the A and D genomes. These studies, together with the physical position of the 5S ribosomal loci [50], show the A and D genomes of hexaploid species to be highly homologous to the *A. strigosa* genome. However, a satellite sequence, As120a, isolated from *A. strigosa* is able to differentiate between the A and D genomes of hexaploids, and can also distinguish between several A-genome diploid species. The As120a sequence is understood to be a sequence specific to the A genome [51].

A. canariensis has also been proposed the most likely progenitor and A-genome donor of the hexaploid species on the basis of genetic similarity data from the analysis of mini- and microsatellite polymorphisms [48]. However, this proposal disagrees with data obtained in Southern experiments which show the As120a sequence comes from the *A. canariensis* genome [51]. Therefore, *A. canariensis* (or some related species) could only be involved in the evolution of the hexaploids as a donor of the D genome.

In tetraploid species, disagreement exists over the implication of *A. strigosa* as the A genome donor. GISH experiments have shown that the chromosomes of AACC tetraploids contain chromatin of A. *strigosa* [25, 45] as well as various repeat sequences isolated from A. *strigosa* [52, 53, 54]. It has also been demonstrated by Southern experiments that A. *murphyi* contains DNA sequences isolated from A. *sativa* [2, 17]. However, the C-banding patterns of the A-genome chromosomes of the A. *murphyi* and A. *maroccana* species bear little resemblance to those of the A. *strigosa* species [12, 25]. Moreover, only three pairs of A/D-genome chromosomes of hexaploid species bear a close resemblance to the A-genome chromosomes of the two AACC tetraploid species [25]. This is corroborated by the results of a chromosome pairing study involving A. *strigosa* and the AACC-genome tetraploids; these found little evidence to support the presence of the A. *strigosa* genome in the tetraploid species [71]. When the distribution of chromosome markers, such as ribosomal loci and translocated segments, is studied in tetraploid and hexaploid species, a close resemblance is seen between chromosomes 4A and 8A of A. *murphyi* and the corresponding 3D and 13D chromosomes of A. *sativa* [50, 51, 52]. Together, these observations, plus the lack of any in situ hybridization of As120a with the tetraploid species [51], seem to indicate that the so-designated A genome in A. *murphyi* was donated by a diploid species bearing a genome more similar to the present day D genome of the hexaploid species. Therefore, a genomic designation of CCDD for A. *murphyi* is proposed.

Given the evidence currently available, the two above-mentioned steps in the evolution of the hexaploid species should probably be modified. Thus, the first step probably established a tetraploid CCDD (such as A. *murphyi* or a similar species) by the hybridization of a C-genome diploid species (such as A. *eriantha*) and a D-genome diploid species (presumably a diploid species with a modified A genome), followed by a doubling of the chromosome number. The second step involved the hybridization of this tetraploid with an A-genome species such as A. *strigosa*, forming a hexaploid by the doubling of the chromosomes of the resulting triploid hybrid. This new proposal concerning the formation of the hexaploid species is supported by the intergenomic translocations detected in hexaploid species, which appear to have occurred frequently during the evolution of *Avena* [7, 13, 19, 25, 45, 50, 51, 52, 53, 54, 73]. The discrimination of the three genomes of the hexaploids by in situ hybridization with probes specific for either the A-genome or D-genome chromosomes has allowed the number of intergenomic translocations in hexaploids to be determined. Thus, three

C-D and D-C translocations are found in cultivars of A. *sativa*, but so far only one C-A translocation has been detected [21, 53, 54]. Moreover, the two translocations that the cultivar SunII presents in excess also involve the D and C chromosomes. If translocations occurred throughout evolution at a steady rate, the larger number of rearrangements among C and D chromosomes could be interpreted as reflecting a longer "life in common" for these two genomes, rather than for either of these in combination with the A genome. This suggests that a tetraploid CCDD hybridized with A. *strigosa* to give rise to the AACCDD hexaploids.

It is interesting to note that none of the molecular reports mentioned in this review include mention of A. *insularis*. Although chromosome affinities between A. *insularis* and A. *sativa* have been described by both meiotic pairing of their hybrids [37] and chromosome marker analysis (this report), the hybridization patterns of the As120a sequence in the C'- and D'-genome chromosomes obtained in the present study indicate that A. *insularis* is more recent than A. *murphyi*. Consequently, A. *insularis* is probably not a candidate progenitor of the hexaploid species.

Acknowledgement

We thank the *Ministerio de Ciencia y Tecnología* of Spain (Grant No. PB95-0329 and AGL2003-02043) and *Universidad de Alcalá* (UAH GC2005/004) for supporting this study.

[1] Alicchio R, Aranci L, Conte L. Restriction fragment length polymorphism based phylogenetic analysis of *Avena* L. Genome 1995; 38: 1279-1284.

[2] Ananiev EV, Vales MI, Phillips RL, Rines HW. Isolation of A/D and C genome specific dispersed and clustered repetitive DNA sequences from *Avena sativa*. Genome 2002; 45: 431-441.

[3] Appels R, McIntyre CL. Cereal genome organization as revealed by molecular probes. Plant Mol Cell Biol 1985; 2: 235-252.

[4] Arumuganathan K, Earle ED. Nuclear DNA content of some important plant species. Plant Mol Biol Rep 1991; 9: 208-218.

[5] Baum RB, Fedak G. A new tetraploid species of *Avena* discovered in Morocco. Can J Bot 1985; 63: 1379-1385.

[6] Bennett MD, Smith JB. Nuclear DNA amounts in angiosperms. Proc R Soc Lond B Biol Sci 1976; 274: 227-274.

[7] Chen Q, Armstrong K. Genomic in situ hybridization in *Avena sativa*. Genome 1994; 37: 607-612.

[8] Craig IL, Murray BE, Rajhathy T. Leaf esterase isozymes in *Avena* and their relationship to the genomes. Can J Genet Cytol 1972; 14: 581-589.

[9] De Koeyer DL, Tinker NA, Wight CP, et al. A molecular linkage map with associated QTLs from a hul-less x covered spring oat population. Theor Appl Genet 2004; 108: 1285-1298.

[10] Drossou A, Katsiotis A, Leggett JM. et al. Genome and species relationships in genus *Avena* based on RAPD and AFLP molecular markers. Theor Appl Genet 2004; 109: 48-54.

[11] Fominaya A, Vega C, Ferrer E. Giemsa C-banded karyotype of *Avena* species. Genome 1988; 30: 627-632.

[12] Fominaya A, Vega C, Ferrer E. C-banding and nucleolar activity of tetraploid *Avena* species. 1988; 30: 633-638.

[13] Fominaya A, Hueros G, Loarce Y, Ferrer E. Chromosomal distribution of a repeated DNA from C-genome heterochromatin and the identification of a new ribosomal DNA locus in the *Avena* genus. Genome 1995; 38: 548-557.

[14] Fox SL, Jellen EN, Kianian SF, et al. Assignment of RFLP linkage groups to chromosomes using monosomic F_1 analysis in hexaploid oat. Theor Appl Genet 2001; 102: 320-326.

[15] Gerlach WL, Bedbrook JR. Cloning and characterization of ribosomal RNA genes from wheat and barley. Nucl Acids Res 1979; 7: 1869-1885.

[16] Gerlach WL, Dyer TA. Sequence organization of the repeated units in the nucleus of the wheat which contains 5S-rRNA genes. Nucl Acids Res 1980; 8: 4851-4865.

[17] Gupta PK, Giband M, Altosaar I. Two molecular probes characterizing the A and C genomes in the genus *Avena* (oats). Genome 1992; 35: 916-920.

[18] Groh S, Zacharias A, Kianian SF, et al. Comparative AFLP mapping in two hexaploid oat populations. Theor Appl Genet 2001; 102: 876-884.

[19] Hayasaki M, Morikawa T, Tarumoto I. Intergenomic translocations of polyploid oats (genus *Avena*) revealed by genomic in situ hybridization. Genes Genet Syst 2000; 75: 157-171.

[20] Holden JHW. Species relationships in *Avenae*. Chromosoma 1966; 20: 75-124.

[21] Irigoyen ML, Loarce Y, Linares C, et al. A. Discrimination of the closely related A and B genomes in AABB tetraploid species of *Avena*. Theor Appl Genet 2001; 103: 1160-1166.

[22] Irigoyen ML, Linares C, Ferrer E, Fominaya A. Fluorescence in situ hybridization mapping of *Avena sativa* L. cv. SunII and its monosomic lines using cloned repetitive DNA sequences. Genome 2002; 45: 1230-1237.

[23] Irigoyen ML, Ferrer E, Loarce Y. Cloning and characterization of resistance gene analogs from *Avena* species. Genome 2006; 49 (in press).

[24] Jellen EN, Phillips RL, Rines HW. C-banded karyotypes and polymorphisms in hexaploid oat accessions (*Avena* spp.) using Wright's stain. Genome 1993; 36: 1129-1137.

[25] Jellen EN, Gill BS, Cox TS. Genomic in situ hybridization differentiates between A/D- and C-genome chromatin and detects intergenomic translocations in polyploid oat species (genus *Avena*). Genome 1994; 37: 613-618.

[26] Jellen EN, Phillips RL, Rines HW. Chromosomal localization and polymorphisms of ribosomal DNA in oat (*Avena* spp.). Genome 1994; 37: 23-32.

[27] Jellen EN, Gill BS. C-banding variation in the Moroccan oat species *Avena agaridiana* (2n=4x=28). Theor Appl Genet 1996; 92: 726-732.

[28] Jellen EN, Ladizinsky G. Giemsa C-banding in *Avena insularis* Ladizinsky. Genet Res Crop Evol 2000; 47: 227-230.

[29] Jellen AN, Beard J. Geographical distribution of a chromosome 7C and 17 intergenomic translocation in cultivated oat. Crop Sci 2000; 400: 256-263.

[30] Jin H, Domier LL, Shen X, Kolb F. Combined AFLP and RFLP mapping in two hexaploid oat recombinant inbred populations. Genome 2000; 43: 94-101.

[31] Katsiotis A, Hagidimitriou M, Heslop-Harrison JS. The close relationship between the A and B genomes in *Avena* L. (Poaceae) determined by molecular cytogenetic analysis of total genomic, tandemly and dispersed repetitive DNA sequences. Ann Bot 1997; 79: 103-109.

[32] Ladizinsky G, Zohary D. Notes on species delimitation, species relationships and polyploidy in *Avena* L. Euphytica 1971; 20: 380-395.

[33] Ladizinsky G, Johnson BL. Seed protein homologies and the evolution of polyploidy in *Avena*. Can J Genet Cytol 1972; 14: 875-888.

[34] Ladizinsky G. The cytogenetic position of *Avena prostrata* among the diploid oats. Can J Genet Cytol 1973; 15: 443-450.

[35] Ladizinsky G, Fainstein R. Introgression between the cultivated hexaploid oat A. *sativa* and the tetraploid wild A. *magna* and A. *murphyi*. Can J Genet Cytol 1977; 19: 59-66.

[36] Ladizinsky G. Biological species and wild genetic resources in *Avena*. In: Mattson B, Layhagen R, eds. Proc 3rd Int Oat Conf. Lund, Sweden, Svalov AB, Sweden. 1988: 76-86.

[37] Ladizinsky G. A new species of *Avena* from Sicily, possibly the tetraploid progenitor of hexaploid oats. Genet Res Crop Evol 1998; 45: 263-269.

[38] Ladizinsky G. Cytogenetic relationships between *Avena insularis* (2n=28) and both A. *strigosa* (2n=14) and A. *murphyi* (2n=28). Genet Res Crop Evol 1999; 46: 501-504.

[39] Leggett JM. Morphology and metaphase chromosome pairing in three *Avena* hybrids. Can J Genet Cytol 1984; 26: 641-645.

[40] Leggett JM. Interspecific hybrids involving the perennial oat species *Avena macrostachya*. Can J Genet Cytol 1985; 27: 29-32.

[41] Leggett JM. Inter- and intra-specific hybrids involving the tetraploid species *Avena agaridiana* Baum et Fedak sp. nov. (2n=4x=28). In: Mattson B, Layhagen R eds. Proc 3rd Int Oat Conf. Lund, Sweden, Svalov AB, Sweden. 1988: 62-67.

[42] Leggett JM. Interspecific diploid hybrids in *Avena*. Genome 1989; 32: 346-348.

[43] Leggett JM. Classification and speciation in *Avena*. In: Marshall HG, Sorrells ME, eds. Oat science and technology. Madison, WI USA: Agrn Monogr 33 ASA, CSSA, 1992: 29-52.

[44] Leggett JM. Further hybrids involving the perennial autotetraploid oat *Avena macrostachya*. Genome 1992; 35: 273-275.

[45] Leggett JM, Markhand SM. The genomic structure of *Avena* revealed by GISH. In: Brandham PE, Bennett MD, eds. Proc Kew Chromosome Conf IV. Kew UK: 1995: 133-139.

[46] Leggett JM. Chromosome and genomic relationships between the diploid species *Avena strigosa*, *A. eriantha* and the tetraploid *A. maroccana*. Heredity 1998; 80: 361-363.

[47] Li CD, Rossnagel BG, Scoles GJ. The development of oat microsatellite markers and their use in identifying relationships among *Avena* species and oat cultivars. Theor Appl Genet 2000; 101: 1259-1268.

[48] Li CD, Rossnagel BG, Scoles GJ. Tracing the phylogeny of the hexaploid oat *Avena sativa* with satellite DNAs. Crop Sci 2000; 40: 1755-1763.

[49] Linares C, Vega C, Ferrer E, Fominaya A. Identification of C-banded chromosomes in meiosis and the analysis of nucleolar activity in *Avena byzantina* C. Koch cv 'Kanota'. Theor Appl Genet 1992; 83: 650-654.

[50] Linares C, González J, Ferrer E, Fominaya A. The use of double fluorescence in situ hybridization to physically map the positions of 5S rDNA genes in relation to the chromosomal location of 18S-5.8S-26S rDNA and a C genome specific DNA sequence in the genus *Avena*. Genome 1996; 39: 535-542.

[51] Linares C, Ferrer E, Fominaya A. Discrimination of the closely related A and D genomes of the hexaploid oat *Avena sativa* L. Proc Natl Acad Sci USA 1998; 95: 12450-12455.

[52] Linares C, Serna A, Fominaya A. Chromosomal organization of a sequence related to LTR-like elements of Ty1-*copia*- retrotransposons in *Avena* species. Genome 1999; 42: 706-713.

[53] Linares C, Irigoyen ML, Fominaya A. Identification of C-genome chromosomes involved in intergenomic translocations in *Avena sativa* L., using cloned repetitive DNA sequences. Theor Appl Genet 2000; 100: 353-360.

[54] Linares C, Loarce Y, Serna A, Fominaya A. Isolation and characterization of two novel retrotransposon of the Ty1-*copia* group in oat genomes. Chromosoma 2001; 110: 115-123.

[55] Murphy JP, Hoffman LA. The origin, history, and production of oat. In: Marshall HG, Sorrells ME, eds. Oat science and technology. Madison, WI USA: Agrn Monogr 33 ASA, CSSA, 1992: 1-28.

[56] Nishiyama I, Yabuno T. Meiotic chromosome pairing in two interspecific hybrids and a criticism of the evolutionary relationship of diploid *Avena*. Jpn J Genet 1975; 50: 443-451.

[57] Nocelli E, Giovannini T, Bioni M, Alicchio R. RFLP- and RAPD- based genetic relationships of seven diploid species of *Avena* with the A genome. Genome 1999; 42: 950-959.

[58] O´Donoughue LS, Kianian SF, Rayapati PJ, et al. A molecular linkage map of cultivated oat. Genome 1995; 38: 368-380

[59] Peterson DM. Composition and nutritional characteristics of oat grain and products. In: Marshall HG, Sorrells ME, eds. Oat science and technology. Madison, WI USA: Agrn Monogr 33 ASA, CSSA, 1992: 265-292.

[60] Pohler W, Hoppe HD. Homoeology between the chromosomes of *Avena macrostachya* and the *Avena* C genome. Plant Breed 1991; 106: 250-253.

[61] Portyanko VA, Hoffman DL, Lee M, Holland JB. A linkage map of hexaploid oat based on grass anchor DNA clones and its relationship to other oat maps. Genome 2001; 44: 249-265.

[62] Postoyko J, Hutchinson J. The identification of *Avena* chromosomes by means of C-banding. In: Lawes DA, Thomas H, eds. Proc 2nd Intl Oat Conf. University College of Wales, Welsh Plant Breeding Station, Aberystwyth. Amsterdam: Martinus Nijhoff Publishers: 1986: 50-51.

[63] Rajhathy T, Morrison JW. Chromosome morphology in the genus *Avena*. Can J Bot 1959; 37: 331-337.

[64] Rajhathy T. Chromosomal differentiation and speciation in diploid *Avena*. Can J Genet Cytol 1961; 3: 372-377.

[65] Rajhathy T, Thomas H. Cytogenetics of oats (*Avena* L.). Misc Publ Genet Soc Can 1974; 2: 1-90.

[66] Sadasivaiah RS, Rajhathy T. Genome relationships in tetraploid *Avena*. Can J Genet Cytol 1968; 10: 655-669.

[67] Sánchez de la Hoz P, Fominaya A. Studies of isozymes in oat species. Theor Appl Genet 1989; 77: 735-741.

[68] Solano R, Hueros G, Fominaya A, Ferrer E. Organization of repeated sequences in species of the genus *Avena*. Theor Appl Genet 1992; 83: 602-607.

[69] Sorrells ME, Simmons SR. Influence of environment on the development and adaptation of oat. In: Marshall HG, Sorrells ME, eds. Oat science and technology. Madison, WI USA: Agrn Monogr 33 ASA, CSSA, 1992: 115-164.

[70] Thomas H, Jones ML. Chromosomal differentiation in diploid species of *Avena*. Can J Genet Cytol 1965; 7: 108-111.

[71] Thomas H. Cytogenetics of *Avena*. In: Marshall HG, Sorrells ME, eds. Oat science and Technology. Madison, WI USA: Agrn Monogr 33 ASA, CSSA, 1992: 473-507.

[72] Wight CP, Tinker NA, Kianian SF, et al. A molecular marker map in 'Kanota'x 'Ogle' hexaploid oat (*Avena* spp) enhanced by additional markers and a robust framework. Genome 2003; 46: 28-47.

[73] Yang Q, Hanson L, Bennett MD, Leitch IJ. Genome structure and evolution in the allohexaploid weed *Avena fatua* L. (Poaceae). Genome 1999; 42: 512-518.

[74] Zhu S, Kaeppler HF. A genetic linkage map for hexaploid cultivated oat (*Avena sativa* L.) based on an intraspecific cross 'Ogle/MAM17-5'. Theor Appl Genet 2003; 107: 26-35.

Authors Index

Detailed Contents